Nonlinear Dynamic Phenomena in Mechanics

SOLID MECHANICS AND ITS APPLICATIONS
Volume 181

Aims and Scope of the Series

The fundamental questions arising in mechanics are: Why?, How?, and How much? The aim of this series is to provide lucid accounts written by authoritative researchers giving vision and insight in answering these questions on the subject of mechanics as it relates to solids.

The scope of the series covers the entire spectrum of solid mechanics. Thus it includes the foundation of mechanics; variational formulations; computational mechanics; statics, kinematics and dynamics of rigid and elastic bodies: vibrations of solids and structures; dynamical systems and chaos; the theories of elasticity, plasticity and viscoelasticity; composite materials; rods, beams, shells and membranes; structural control and stability; soils, rocks and geomechanics; fracture; tribology; experimental mechanics; biomechanics and machine design.

The median level of presentation is the first year graduate student. Some texts are monographs defining the current state of the field; others are accessible to final year undergraduates; but essentially the emphasis is on readability and clarity.

For other titles published in this series, go to
www.springer.com/series/6557

Jerzy Warminski, Stefano Lenci,
Matthew P. Cartmell, Giuseppe Rega,
and Marian Wiercigroch (Eds.)

Nonlinear Dynamic Phenomena in Mechanics

Springer

Editors

Jerzy Warminski
Lublin University of Technology
Nadbystrzycka 36
20-618 Lublin
Poland
E-mail: j.warminski@pollub.pl

Stefano Lenci
Polytechnic University of Marche
Department of Architecture,
Buildings and Structures
Via Brecce Bianche
60131 Ancona
Italy
E-mail: lenci@univpm.it

Matthew P. Cartmell
University of Glasgow
School of Engineering
James Watt South Building
G12 8QQ Glasgow
United Kingdom
E-mail: matthew.cartmell@glasgow.ac.uk

Giuseppe Rega
Sapienza University of Rome
Department of Structural and
Geotechnical Engineering
Via Antonio Gramsci 53
00197 Roma
Italy
E-mail: giuseppe.rega@uniroma1.it

Marian Wiercigroch
University of Aberdeen
School of Engineering
Centre for Applied Dynamics
Research (CADR)
King's College
AB24 3UE Aberdeen
United Kingdom
E-mail: m.wiercigroch@abdn.ac.uk

ISBN 978-94-007-2472-3 e-ISBN 978-94-007-2473-0

DOI 10.1007/978-94-007-2473-0

Library of Congress Control Number: 2011936650

Typeset & Cover Design: Scientific Publishing Services Pvt. Ltd., Chennai, India.

Printed on acid-free paper

9 8 7 6 5 4 3 2 1

springer.com

Preface

Nonlinear phenomena should play a crucial role in the design and control of engineering systems and structures as they can drastically change the prevailing dynamical responses. For example, bifurcations or transitions to irregularity (chaos) may completely alter intuitively expected behaviour. Dealing with nonlinear dynamic phenomena requires special analytical treatment and dedicated control techniques to harness the effects of unpredictable behaviour. In many cases, formulating an appropriate *nonlinear mathematical model* of a real structure or system would be essential to obtain a broad knowledge of the relevant response and the influence of its parameters.

Nonlinear mechanics is classical in origin, however its applications are modern and they are vast in science and engineering. In spite of the fact that linear models are in common use, in many practical problems certain key effects can be explained only by exploration of nonlinear models.

The origin of this book is a series of lectures given in the frame of the Transfer of Knowledge Project led by Prof. Tomasz Sadowski of Lublin University of Technology on 'Modern Composite Materials Applied in Aerospace, Civil and Sanitary Engineering: Theoretical Modelling and Experimental Verification' (contract MTKD-CT-2004-014058) and the FP7 Project 'Centre of Excellence for Modern Composites Applied in Aerospace and Surface Transport Infrastructure' (CEMCAST, FP7-REGPOT-2009 1, grant agreement No: 245479). The mentioned projects accommodated two groups of researchers: one, working in the field of modern materials mechanics, and the other, working in nonlinear dynamics, bifurcation, chaos theory and control. Some of the results obtained by the latter are included in this book, whose scope covers theoretical and applications-based problems of *nonlinear dynamics*. In the presented chapters the newest methods of nonlinear mechanics are applied to elucidate a rich variety of features of system response and the latest control techniques are used to enhance the dynamics or to reduce undesired responses. Besides composite structures and systems with controllable and adaptive properties, flexible structures and non-smooth problems are addressed, paying proper attention to real applications.

So called *small nonlinearities* very often result in large qualitative and quantitative changes in structural dynamics. As a classical example we may mention *pendulum-like systems* addressed in Chap. 1. In practice, a pendulum can be used as a dynamic absorber mounted in high buildings, bridges or chimneys. Swings of the pendulum can suppress oscillations of the primary structure which then oscillates with a very small amplitude or not at all. However, geometrical nonlinearities introduced by pendulum motion may change the system dynamics, and instead of the expected response we can observe a rapid increase of the oscillations of both the pendulum and the structure, leading to full pendulum rotation or chaotic dynamics. The reason for such behaviour is related to autoparametric coupling resulting in the occurrence of instability zones. To avoid such dangerous situations the proper selection of parameters or the introduction of semi-active *magnetorheological damping* is proposed. On the basis of the analytical solutions of a nonlinear two DOF model the resonance and instability regions are detected, and then chaotic oscillations, bifurcation points and transition paths from regular to chaotic vibrations are determined by numerical techniques. Theoretical results are validated by real experimental tests.

Nonlinear mechanics also has to be used to explain undesired response in *slender footbridges*. Motivation for research in this topic was the famous example of the *London Millenium Bridge* event. Strong horizontal vibrations, caused by synchronisation of pedestrian motions, were induced on its opening day. The problem of pedestrian induced lateral vibrations may occur in bridges of various structural types and materials. The parametric study presented in Chap. 2 allows a better understanding of the structural mechanics and also the detection of regions of increasing vibration. The observed phenomena can be explained by an analytical *nonlinear discrete-time model* based on the stroboscopic Poincaré map which then enables the location of instability regions and the prediction of the number of pedestrians required to trigger synchronisation of the structure. The analytical formula gives reliable values which are in good qualitative and quantitative agreement with real examples and observations.

Smart active or semi-active elements, like for example: magnetorheological dampers, piezoelectric patches or shape memory alloys actuators embedded inside the structure, together with robust control algorithms, may eliminate regions of dangerous behaviour. Also we may take advantage of the nonlinear phenomena to design an active structure to work more effectively. *Shape memory alloys* (SMAs) exhibits very interesting nonlinear thermo- mechanical properties such as the *shape memory effect* and *superelasticity*. Methodologies for integrating shape memory alloy elements are based on *active property tuning* (APT) and *active strain energy tuning* (ASET). Chapter 3 presents details of the modelling of the SMA effect and applications for SMA wires embedded in mechanical structures to control their dynamics. SMA elements integrated within composite beams or plates can be used for active modification of structure properties e.g. by affecting their natural frequencies. It is shown that the resonant characteristics of such hosts can be significantly altered by activation of the embedded elements. This concept is extended to shell-like structures, specifically tubular bearing housings used to locate flexible rotors, and also to more complex plate geometries in which the SMA is arranged in a periodic and repeating structure in order to control *multiple modes of vibration*.

Recent advancements in the theoretical and experimental research on the finite amplitude, resonant, forced dynamics of *sagged, horizontal or inclined, elastic cables* are presented in Chap. 4, by considering modelling, analysis, response, and, in particular, nonlinear/nonregular phenomena. Asymptotic solutions and a rich variety of features of *nonlinear multimodal interaction* occurring in various resonance conditions are comparatively discussed. Dynamical and mechanical characteristics of some of the principal, experimentally observed, responses are summarized, along with the relevant robustness, spatio-temporal features, and dimensionality. Challenging issues arising in the characterization of involved *bifurcation scenarios* resulting in transition to *complex dynamics* are addressed, and hints for proper *reduced-order modelling* in cable nonlinear dynamics are obtained based on both asymptotic solutions and experimental investigations, in the perspective of a profitable cross-validation of the observed nonlinear phenomena.

The importance of *non-smooth dynamical systems*, which are very common in engineering practice is discussed in Chap. 5. Mathematically, such systems can be considered as locally smooth and therefore the global solution is obtained by *stitching* local solutions, which can be determined by standard methods. If the dynamical system is piecewise linear then an implicit global analytical solution can be given, however the occurrences of non-smoothness have first to be determined. This leads to the necessity of solving a set of nonlinear algebraic equations. To illustrate non-smooth problems and the methodology for solving them, three mechanical engineering examples are studied: (i) a *vibro-impact system* in the form of a moling device, (ii) the influence of the opening and closing of a *fatigue crack* on the host system dynamics, and (iii) nonlinear interactions between a *rotor* and snubber ring system. The theoretical results have been obtained from the developed mathematical models and confirmed by experimental tests, with a good degree of correlation.

This book is aimed at a wide audience of engineers and researchers working in the field of nonlinear structural vibrations and dynamics, and undergraduate and postgraduate students reading mechanical, aerospace and civil engineering.

Lublin, Ancona, Glasgow, Roma, Aberdeen
December 2010

Jerzy Warminski
Stefano Lenci
Matthew Cartmell
Giuseppe Rega
Marian Wiercigroch

Acknowledgements

"The research leading to these results has received funding from the European Union Seventh Framework Programme (FP7/2007-2013), FP7 - REGPOT - 2009 - 1, under grant agreement No:245479".

Contents

List of Contributors

Jerzy Warminski, Krzysztof Kecik
Department of Applied Mechanics, Lublin University of Technology, Nadbystrzycka 36, 20-618 Lublin, Poland
E-mail: j.warminski@pollub.pl

Stefano Lenci, Laura Marcheggiani
Department of Architecture, Buildings and Structures, Polytechnic University of Marche, via Brecce Bianche, Ancona, 60131, Italy
E-mail: lenci@univpm.it

Matthew P. Cartmell[a], Arkadiusz J. Żak[b], Olga A. Ganilova[a]
[a]School of Engineering, University of Glasgow, James Watt South Building, Glasgow G12 8QQ, Scotland, United Kingdom
E-mail: matthew.cartmell@glasgow.ac.uk
[b]Szewalski Institute of Fluid Flow Machinery, Polish Academy of Sciences, Fiszera 14, 80-952 Gdańsk, Poland

Giuseppe Rega
Department of Structural and Geotechnical Engineering, Sapienza University of Rome, Via Antonio Gramsci 53, 00197 Roma, Italy
E-mail: giuseppe.rega@uniroma1.it

Marian Wiercigroch, Ekaterina Pavlovskaia
Centre for Applied Dynamics Research, School of Engineering, Aberdeen University, King's College, AB24 3UE, Aberdeen, Scotland, United Kingdom
E-mail: m.wiercigroch@abdn.ac.uk

Autoparametric Vibrations of a Nonlinear System with a Pendulum and Magnetorheological Damping

Jerzy Warminski and Krzysztof Kecik

Abstract. The chapter deals with autoparametric vibrations of a system composed of a nonlinear oscillator with an attached pendulum. Dynamics of the mechanical structure is studied analytically around the principal parametric resonance region, numerically and experimentally for a wide range of parameters. The influence of damping, nonlinear stiffness (hard and soft), amplitude and frequency of excitation on the system's behaviour is analysed in details. The obtained results show that the pendulum can be applied as a dynamical absorber. However, for selected parameters, near the main parametric resonance, instability, which transits the pendulum to chaotic oscillations or to a full rotation, occurs. Therefore, the application of a magnetorheological (MR) damper and a nonlinear spring is proposed to improve the dynamics and to control the response online. Periodic vibrations, chaotic motions or a full rotation of the pendulum obtained numerically are confirmed by the experiment. The chaotic nature of motion is determined from real signals by the attractor reconstruction and the recurrence plot calculation. The results show that the semi-active suspension may reduce dangerous motion and it also allows to maintain the pendulum at a given attractor or to jump to another one.

Keywords: Autoparametric vibrations, control, stability, chaos, parametric resonance, magnetorheological damping, nonlinearity, basins of attraction.

Jerzy Warminski
Department of Applied Mechanics, Lublin University of Technology, Nadbystrzycka 36, 20-618 Lublin, Poland
e-mail: `j.warminski@pollub.pl`

Krzysztof Kecik
Department of Applied Mechanics, Lublin University of Technology, Nadbystrzycka 36, 20-618 Lublin, Poland
e-mail: `k.kecik@pollub.pl`

J. Warminski et al. (Eds.): Nonlinear Dynamic Phenomena in Mechanics, SMIA 181, pp. 1–61.
springerlink.com

1 Introduction to Autoparametric Vibrations

Systems with time-varying parameters belong to a very important class in the field of structural dynamics. Many mechanical engineering problems are described by differential equations with periodically changing parameters. Therefore, vibrations generated by such systems are called parametric vibrations. Among many, as a classical example we can mention a pendulum with moving periodically point of suspension. Its mathematical model leads to the Mathieu equation.

Autoparametric systems belong to a special subclass of nonlinear dynamical structures of the broad group of parametrically induced vibrations. Their specific feature is that vibrations are caused by internal coupling of at least two subsystems. A pendulum attached to an oscillator subjected to vertical harmonic force is an example of such a system. The two substructures: an oscillator and a pendulum are coupled by inertia terms which play a role of parametric excitation. This specific coupling can lead to energy transfer between different vibration modes [25], as well as to resonances possible only in this specific problem [21]. Moreover, additional types of resonances, internal or combination ones, are possible [8] and, under some conditions, the system can transit to chaotic motion [5]. When a mechanical system consists of two or more coupled vibrating subsystems, vibrations of one of the components may destabilise the motion of the others.

In autoparametric vibrations, a small excitation can produce a large response basically when the frequency of excitation is close to one of the natural frequencies of the system. In actual engineering problems, the loss of stability depends on frequency tuning of the various components of the system, and on the interaction (the coupling) between them. The oscillator with an attached pendulum represents an interesting dynamical physical structure which is used in many mechanical and civil engineering applications. In a large number of problems it is used to diminish the vibration amplitudes. As an example of this we can mention a situation which occurs in aeroplanes where the aeroplane's engines are mounted under the wings by elastic suspenders. Vertical vibrations of the wing can, under certain conditions, initiate the swinging motion of the suspended engines. Autoparametric excitation, with the motion of the wings supplying the energy, can lead to violent vibrations of the engines, resulting in a fatal failure of the suspenders [34]. Nowadays, pendulum absorber is applied also to helicopters as one of the vibration suppression devices of helicopter's blades [23]. Moreover, special dampers working against earthquake are mounted in high buildings [29], mounted on bridges against river vortex [3] or on high chimneys where they are designed to reduce vibration induced by the wind [4].

Autoparametric vibrations which may occur in many mechanical structures are studied by many researchers. In paper [35] Vyas and Bajaj analyse dynamics of a resonantly excited single-degree-of-freedom linear system coupled to an array of non-linear autoparametric vibration absorbers (pendulums). They obtained, by method of averaging, the first order approximation of the non-linear response of the system. They showed that the frequency interval of the unstable single-mode response or the absorber bandwidth can be enlarged substantially, comparing it to that of a single pendulum absorber, by adjusting individually the internal mistuning of the pendulums. Their results show a significant enhancement in the performance of autoparametric vibration absorbers. A rigid pendulum attached by a hinge joint to a block-mass is studied by Hatwal et al. [11]. In this case, the restoring force of the secondary system occurred due to gravity or a torsional spring. Both cubic and quadratic nonlinearities are considered in the model. The method of the harmonic balance is applied to find the solution of equations and to show suppression of vibrations of the main system for taken configurations. However, in a later article Hatwal et al. [12] showed interesting time responses for a pendulum with chaotic behaviour in both numerical and experimental results. A stabilisation method for the $1/3$ order subharmonic resonance with an autoparametric vibration absorber is presented by Yabuno et al. in [42]. The subharmonic resonance was created by a nonlinear spring (cubic characteristic) and harmonic excitation. Damping of a pendulum subsystem was selected as a control parameter while the natural frequency was about $1/2$ that of the main system.

The purpose of this chapter is to study possible dynamical phenomena of a coupled oscillator-pendulum system for realistic data, and to present a method of semi-active reduction of dangerous vibrations, mainly the chaotic oscillation and rotation. First, we propose to use the magnetorheological damper (MR), which is installed between the oscillator and the ground to provide controllable damping for the system. The effectiveness of the proposed smart base isolation system is studied numerically and experimentally. The second way to eliminate vibrations is to apply a nonlinear spring, mounted in the suspension of the autoparametric system. Experimental tests which confirm theoretical results are performed on a specially prepared two-degree-of-freedom physical model. The chaotic nature of the system is determined from real signals based on the reconstruction of the attractor and the recurrence plots. The results of this study present new possibilities for designers, in particular. It is found that magnetorheological fluids can be designed to be very effective for vibration control, and used in the autoparametric systems to produce controllable response.

2 Model of the Nonlinear Oscillator with an Attached Pendulum

The model of the autoparametric system is considered in two variants: with a classical linear spring and linear viscous damping, and with a nonlinear spring and magnetorheological (MR) damping. The nonlinear spring is introduced by a structural modification while the MR damper behaviour can be modified on-line from viscous, if the system is not activated, to mixed viscous with a dry friction component, when the damper is activated. All considered variants are presented in this section.

2.1 Differential Equations of Motion of a Model with a Viscous Damper

First, let's make a study of an autoparametric pendulum-like system with a typical viscous damper. The considered mechanical model, presented in Fig.1, consists of two subsystems. The main one, a nonlinear oscillator (I), is composed of the mass m_1 and a linear or nonlinear spring and a viscous damper. The second subsystem is an attached pendulum (II) composed of two masses m_p and m_2. The pendulum is attached at a pivot to the mass m_1. A length of the pendulum's arm is denoted by l, and a coefficient of angular damping in the pivot by c_φ. Stiffness coefficients of a linear and nonlinear part of the oscillator's supporting spring are expressed by k and k_1. The oscillator is excited by a classical linear spring k_2 due to harmonic vertical motion of the base (kinematic excitation). Amplitude and frequency of excitation are denoted Q and ω, respectively.

The motion of the autoparametric model is described by two generalized coordinates namely the displacement of the oscillator in the vertical direction x, and the angle of the pendulum rotation φ. Damping of the model is studied in two variants: (a) as linear viscous and (b) nonlinear magnetorheological damping. The second case is presented in Sect. 2.2. Differential equations of motion are derived by the second kind of Lagrange equations and they are written as:

$$
\begin{aligned}
&(m_1+m_2+m_p)\ddot{x}+c\dot{x}+(k+k_2)x+k_1x^3 \\
&\quad+\left(m_2+\frac{1}{2}m_p\right)l\left(\ddot{\varphi}\sin\varphi+\dot{\varphi}^2\cos\varphi\right)=k_2Q\cos\omega t\,, \\
&\left(m_2+\frac{1}{3}m_p\right)l^2\ddot{\varphi}+c_\varphi\dot{\varphi}+\left(m_2+\frac{1}{2}m_p\right)l\left(\ddot{x}+g\right)\sin\varphi=0.
\end{aligned}
\tag{1}
$$

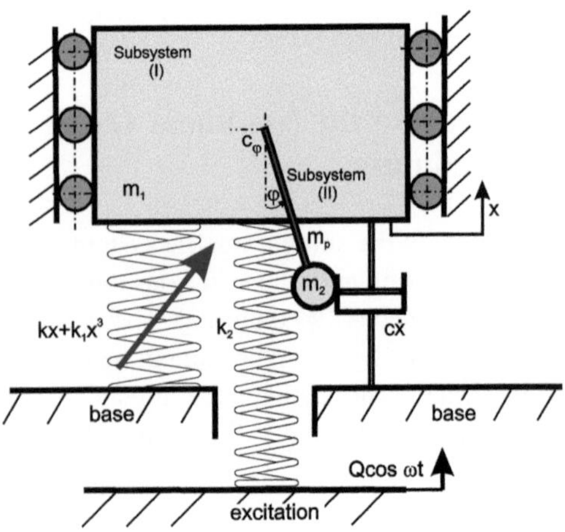

Fig. 1 Model of the autoparametric system

The first of the Eqs. (1) represents the oscillator's motion while the second – motion of the pendulum. If the pendulum does not move, it becomes a part of the oscillator mass, only. We call the solution of this state as the semi-trivial solution. Equations (1) are typical of this type of a two-degrees-of-freedom autoparametric system ([7], [36]). The nonlinear term k_1 of the supporting spring is a new element which appears in the model.

Introducing dimensionless time $\tau = \omega_0 t$, where $\omega_0 = \sqrt{(k+k_2)/(m_1+m_2+m_p)}$ is natural frequency of the oscillator, and dimensionless coordinates $X = x/x_{st}$ and $\varphi \equiv \varphi$, where $x_{st} = (m_1+m_2+m_p)g/(k+k_2)$ is the static displacement of the linear oscillator, we express the governing equations of motion in the dimensionless form:

$$\begin{aligned} \ddot{X} + \alpha_1\dot{X} + X + \gamma X^3 + \mu\lambda(\ddot{\varphi}\sin\varphi + \dot{\varphi}^2\cos\varphi) &= q\cos\vartheta\tau\,, \\ \ddot{\varphi} + \alpha_2\dot{\varphi} + \lambda\left(\ddot{X}+1\right)\sin\varphi &= 0\,. \end{aligned} \tag{2}$$

Now, the natural frequency of the linear oscillator in Eq. (2) is reduced to one, while the dimensionless parameters take definitions:

$$\begin{aligned} \alpha_1 &= \frac{c}{(m_1+m_2+m_p)\,\omega_0}, \quad \alpha_2 = \frac{c_\varphi}{\left(m_2+\frac{1}{3}m_p\right)l^2\omega_0}, \\ \vartheta &= \frac{\omega}{\omega_0}, \quad \mu = \frac{\left(m_2+\frac{1}{3}m_p\right)l^2}{(m_1+m_2+m_p)x_{st}^2}, \quad \lambda = \frac{(m_2+\frac{1}{2}m_p)x_{st}}{(m_2+\frac{1}{3}m_p)l}, \\ q &= \frac{k_2 Q}{(k+k_2)x_{st}}, \quad \gamma = \frac{k_1}{k+k_2}x_{st}^2\,. \end{aligned} \tag{3}$$

The autoparametric excitation in Eq. (2) is produced by coupling inertia term $\ddot{X}$ in the pendulum's equation. If the oscillator motion X is periodic, then the inertia term plays a role of a periodically changing coefficient. Because this phenomenon relies on the internal coupling, therefore the system is called autoparametric. Of course the first equation is also coupled by the second and the first derivative of the pendulum coordinate φ. More detailed information can be found in [16].

2.2 *Application of Magnetorheological (MR) Damper*

Active and semi-active control provides an important new tool for a control engineer. Many structures, such as automotive vehicles, tall buildings, robotic manipulator arms and flexible spacecraft have already been designed using smart vibration isolation as a part of the total design. Magnetorheological fluids (MRF) are a class of smart materials whose rheological properties may be rapidly varied by supplying the magnetic field. When outside the influence of the magnetic field, particles float freely, causing the material to behave like any colloidal mixture. Under the influence of the magnetic field suspended magnetic particles interact to form a structure that resists shear deformation or flow. Therefore, magnetorheological fluid devices are the most promising for vibration control and for vibration isolation.

Magnetorheological dampers are highly nonlinear devices. Their force-velocity relationship is not easy to describe. MR fluid behaves in two distinct modes: off state and activated state. There are several models that describe highly nonlinear behaviour of magnetorheological dampers, generally divided into two categories: parametric and non-parametric ones. The parametric models are physics-based models with parameters that have real physical meaning and carry engineering units (Bouc-Wen, Spencer, Bingham, LuGre and Dahl models). The non-parametric models are mostly empirical e.g. based on Artificial Neural Networks (ANN) or fuzzy logic models [14].

The investigated nonlinear model of an autoparametric system with applied MR damper is presented in Fig.2(a). Here, we propose to use a smooth function of modified Bingham's model (Fig.2b) suggested in the paper [31]. The restoring force of the MR damper with respect to the input velocity is represented by

$$F_d = d \tanh\left(e\dot{X}\right) + c\dot{X}\,, \tag{4}$$

where d is the force related to the rheological behaviour coefficient, produced by the fluid, $\dot{x}$ is velocity of the MR damper piston. In Eq. (4) e is a constant. In our study based on experimental tests, we assumed this value equals ten. In dimensionless form equation (4) is expressed as:

$$F_d = \alpha_3 \tanh\left(e\dot{X}\right) + \alpha_1\dot{X}\,, \tag{5}$$

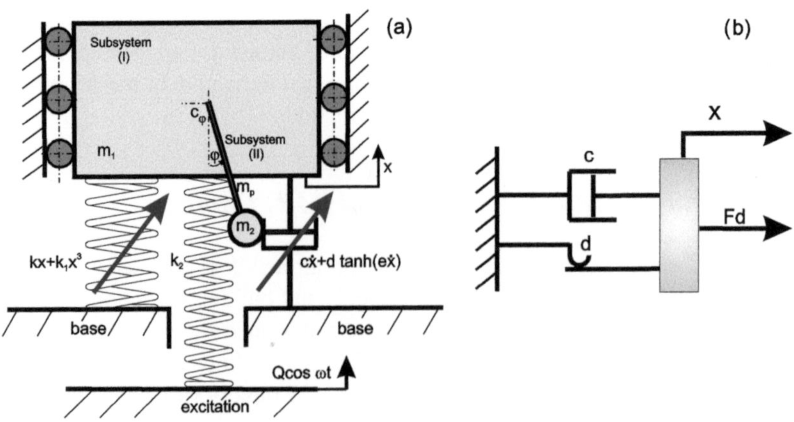

Fig. 2 Physical model of an autoparametric system with MR damper **a**, Bingham model of a controllable fluid damper **b**

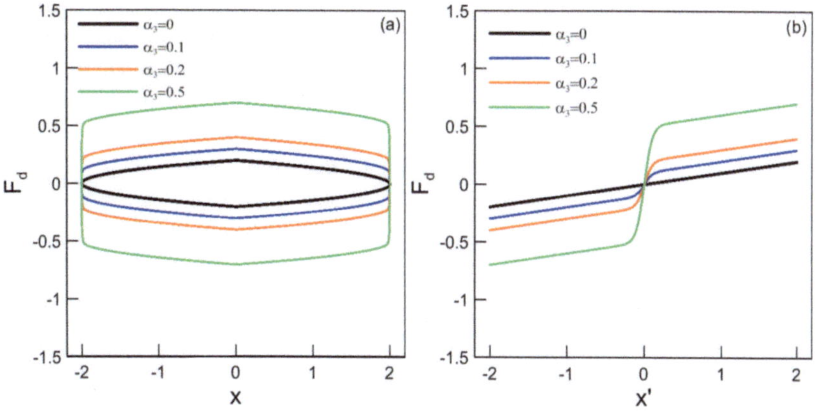

Fig. 3 Nonlinear damping force characteristics for different parameters α_3: **a** damping force versus displacement, **b** damping force versus velocity; viscous damping $\alpha_1 = 0.1$

where α_3 is a dimensionless coefficient of MR damping defined as:

$$\alpha_3 = \frac{d}{(m_1 + m_2 + m_p)\,\omega_0^2 x_{st}}. \tag{6}$$

This model consists of a combination of viscous damping (α_1) and Coulomb friction (α_3) as shown in Fig.3. The Bingham model can be used successfully when the width of hysteretic loop, Fig.3(a), in real characteristics of MR dampers is relatively narrow [15]. However, if the hysteretic loop is wider, it is necessary to construct a more complicated model.

The equations of motion of the system in Fig.2 that include the magnetorheological damper, take the form:

$$\ddot{X} + \alpha_1 \dot{X} + \alpha_3 \tanh(e\dot{X}) + X + \gamma X^3 + \mu\lambda(\ddot{\varphi}\sin\varphi + \dot{\varphi}^2\cos\varphi) = q\cos\vartheta\tau,$$
$$\ddot{\varphi} + \alpha_2\dot{\varphi} + \lambda\left(\ddot{X} + 1\right)\sin\varphi = 0. \tag{7}$$

If the parameter α_3 in Eq. (5) equals zero, we get a classical linear viscous damped model as in Eq. (2). Since in Eq. (5) hyperbolic tangential function occurs, which is very difficult to solve analytically, the strict solutions of an autoparametric pendulum-like system with MR damper are analysed by numerical methods.

3 Approximate Analytical Solutions and Their Stability

Understanding of the dynamics of the autoparametric pendulum-like system requires a study of its mathematical model and then finding analytical solutions.

Because the system is strongly nonlinear the solution can be found by approximate methods and only under assumed conditions. We often have to approximate nonlinear terms by polynomials that are valid for limited values of amplitude. On the basis of the analytical solution, bifurcations and stability analysis can be done giving insight into the system dynamics and allowing to find the parameters influence. This section deals with analytical results and their numerical validation.

3.1 Harmonic Balance Method

Because the equations of motions (2) include nonlinear terms, it is difficult to find their strict solutions. Therefore, Harmonic Balance Method (HBM) is used to find the approximated solution in the neighbourhood of the principal parametric resonance. The harmonic balance is a method effectively used to calculate the steady-state response of different dynamical systems. The method assumes that the solution can be represented by a linear combination of harmonic functions. Classical HBM assumes constant amplitudes and phases that represent steady state of the system. Here, to get more general solutions, we assume that amplitudes and phases are slowly varying functions of time. Around the principal parametric resonance the mass m_1 vibrates with frequency ϑ, which is equal to the excitation frequency, while the pendulum oscillates with frequency $\vartheta/2$ (1:2 subharmonic response). On the basis of this assumption we seek solutions in the form:

$$x(\tau) = A(\tau)\cos(\vartheta\tau + \phi_1),$$
$$\varphi(\tau) = B(\tau)\cos(\frac{\vartheta}{2}\tau + \phi_2). \tag{8}$$

Taking into account small oscillations of the pendulum around the equilibrium point $\varphi \approx 0$, the nonlinear terms $\sin\varphi$ and $\cos\varphi$ are expanded in Taylor's series taking into account terms up to the third-order

$$\sin\varphi = \varphi - \frac{\varphi^3}{6} + 0(\varphi)^5, \quad \cos\varphi = 1 - \frac{\varphi^2}{2} + 0(\varphi)^4. \tag{9}$$

Amplitudes $A(\tau)$, $B(\tau)$ and phases $\phi_1(\tau)$, $\phi_2(\tau)$ describe oscillator and pendulum motions, respectively. In further analysis the abbreviated notation $A(\tau) \equiv A$, $B(\tau) \equiv B$, $\phi_1(\tau) \equiv \phi_1$, $\phi_2(\tau) \equiv \phi_2$ is used. Introducing Eqs. (8) and (9) into (2), and next balancing coefficients of corresponding sine and cosine terms, and neglecting higher harmonics and small terms of a higher order, we get a set of the first order approximate differential equations:

$$\alpha_1 \dot{A} + \mu\lambda\frac{\vartheta}{2}\dot{B}\sin(\phi_1 - 2\phi_2)\left(2B - \frac{1}{3}B^3\right) - 2\vartheta A\dot{\phi}_1$$
$$-2\frac{\vartheta}{2}\mu\lambda B^2\dot{\phi}_2\cos(\phi_1 - 2\phi_2) + \frac{1}{6}\left(\frac{\vartheta}{2}\right)\mu\lambda B^4\dot{\phi}_2\cos(\phi_1 - 2\phi_2)$$
$$+A(1-\vartheta^2) + \frac{3}{4}\gamma A^3 - \left(\frac{\vartheta}{2}\right)^2\mu\lambda B^2\cos(\phi_1 - 2\phi_2)$$
$$+\frac{1}{12}\left(\frac{\vartheta}{2}\right)^2\mu\lambda B^4\cos(\phi_1 - 2\phi_2) - q\cos(\phi_1) = 0,$$
$$-2\vartheta\dot{A} + \mu\lambda\frac{\vartheta}{2}\dot{B}\cos(\phi_1 - 2\phi_2)\left(-2B + \frac{1}{3}B^3\right) - \alpha_1 A\dot{\phi}_1$$
$$-2\frac{\vartheta}{2}\mu\lambda B^2\dot{\phi}_2\sin(\phi_1 - 2\phi_2) + \frac{1}{6}\left(\frac{\vartheta}{2}\right)\mu\lambda B^4\dot{\phi}_2\sin(\phi_1 - 2\phi_2)$$
$$-\alpha_1\vartheta A - \left(\frac{\vartheta}{2}\right)^2\mu\lambda B^2\sin(\phi_1 - 2\phi_2)$$
$$+\frac{1}{12}\left(\frac{\vartheta}{2}\right)^2\mu\lambda B^4\sin(\phi_1 - 2\phi_2) - q\sin(\phi_1) = 0,$$
$$\alpha_2\dot{B} - \dot{\phi}_1\cos(\phi_1 - 2\phi_2)\left(\vartheta\lambda AB - \frac{1}{8}\lambda\vartheta AB^3\right)$$
$$-2\frac{\vartheta}{2}B\dot{\phi}_2 - B\left(\left(\frac{\vartheta}{2}\right)^2 - \lambda + \frac{1}{8}\lambda B^2\right)$$
$$-\left(\frac{1}{2}\lambda\vartheta^2 AB - \frac{1}{16}\lambda\vartheta^2 AB^3\right)\cos(\phi_1 - 2\phi_2) = 0,$$
$$-2\frac{\vartheta}{2}\dot{B} + \dot{\phi}_1\sin(\phi_1 - 2\phi_2)\left(\lambda\vartheta AB - \frac{1}{8}\lambda\vartheta AB^3\right) - \alpha_2 B\dot{\phi}_2$$
$$-\alpha_2 B\left(\frac{\vartheta}{2}\right) + \sin(\phi_1 - 2\phi_2)\left(\frac{1}{2}\lambda\vartheta^2 AB - \frac{1}{16}\lambda\vartheta^2 AB^3\right) = 0. \quad (10)$$

The above, so-called modulation equations, describe a change of amplitude and phase of the oscillator and the pendulum. For the steady state, amplitudes and phases are constant, thus their first order derivatives are equal zero

$$\dot{A} = 0,\ \dot{B} = 0,\ \dot{\phi}_1 = 0,\ \dot{\phi}_2 = 0, \quad (11)$$

and for small oscillations of the pendulum we may assume

$$B^4/48 \approx 0, \quad B^3/16 \approx 0. \quad (12)$$

Introducing (11), into (10) and taking into consideration (12) we receive a following set of algebraic equations:

$$(1-\vartheta^2)A+\frac{3}{4}\gamma A^3-\mu\lambda\left(\frac{\vartheta}{2}\right)^2 B^2\cos(2\phi_2-\phi_1)=q\cos\phi_1\,,$$

$$-\vartheta\alpha_1 A+\mu\lambda\left(\frac{\vartheta}{2}\right)^2 B^2\sin(2\phi_2-\phi_1)=q\sin\phi_1\,,$$

$$\left(\frac{\vartheta}{2}\right)^2-\lambda+\frac{\lambda}{8}B^2+A\frac{\lambda\vartheta^2}{2}\cos(2\phi_2-\phi_1)=0\,,$$

$$\alpha_2\left(\frac{\vartheta}{2}\right)+A\frac{\lambda\vartheta^2}{2}\sin(2\phi_2-\phi_1)=0\,. \tag{13}$$

Assumptions (12) are adopted in accordance with the work [28] in which analytical results both with neglected and without neglected parts are compared. The obtained results show that the influence of the terms (12) is very small.

Solutions of the nonlinear algebraic equations (13) represent amplitudes and phases of the model response in the steady state. It is not possible to receive the strict solutions of these equations in analytical form. However, if the nonlinearity of a spring is neglected, $\gamma=0$, the solution can be found. After some algebraic manipulations we get the following equations:

$$\tan\phi_1=\frac{4\vartheta\left(4A^2\alpha_1+B^2\alpha_2\mu\right)}{16A^2\left(\vartheta^2-1\right)-B^2\mu\left(\left(B^2-8\right)\lambda+2\vartheta^2\right)}\,,$$

$$\tan(2\phi_2-\phi_1)=\frac{\alpha_2\left(\frac{\vartheta}{2}\right)}{\left(\frac{\vartheta}{2}\right)^2-\lambda+\frac{\lambda B^2}{8}}\,,$$

$$A^2=\frac{B^4}{16\vartheta^4}+\frac{B^2(4\lambda\vartheta^2-16\lambda^2)}{16\lambda^2\vartheta^4}+\frac{64\lambda^2-32\vartheta^2\lambda+4\vartheta^4+16\vartheta^2\alpha_2^2}{16\lambda^2\vartheta^4}\,,$$

$$A^2=\frac{B^4\left(-\lambda^2\mu^2\vartheta^4+2\lambda\mu\left(\vartheta^2-1\right)\right)}{16\left(1+\left(-2+\alpha_1^2\right)\vartheta^2+\vartheta^4\right)}$$
$$+\frac{B^2\left(16\lambda\mu\left(1-\vartheta^2\right)+4\mu\vartheta^2\left(\vartheta^2-1-2\alpha_1\alpha_2\right)\right)}{16\left(1+\left(-2+\alpha_1^2\right)\vartheta^2+\vartheta^4\right)}+\frac{q^2}{1+\vartheta^2(-2+\alpha_1^2)+\vartheta^4}$$

$$B^4\left[\frac{-\lambda^2\mu^2\vartheta^4+2\lambda\mu\left(-1+\vartheta^2\right)}{16\left(1+\left(-2+\alpha_1^2\right)\vartheta^2+\vartheta^4\right)}-\frac{1}{16\vartheta^4}\right]$$
$$+B^2\left[\frac{16\lambda\mu\left(1-\vartheta^2\right)+4\mu\vartheta^2\left(\vartheta^2-1-2\alpha_1\alpha_2\right)}{16\left(1+\left(-2+\alpha_1^2\right)\vartheta^2+\vartheta^4\right)}+\frac{16\lambda^2-4\lambda\vartheta^2}{16\lambda^2\vartheta^4}\right]$$
$$+\frac{q^2}{1+\vartheta^2\left(-2+\alpha_1^2\right)+\vartheta^4}-\frac{64\lambda^2-32\lambda\vartheta^2+4\left(4\alpha_2^2\vartheta^2+\vartheta^4\right)}{16\lambda^2\vartheta^4}=0\,, \tag{14}$$

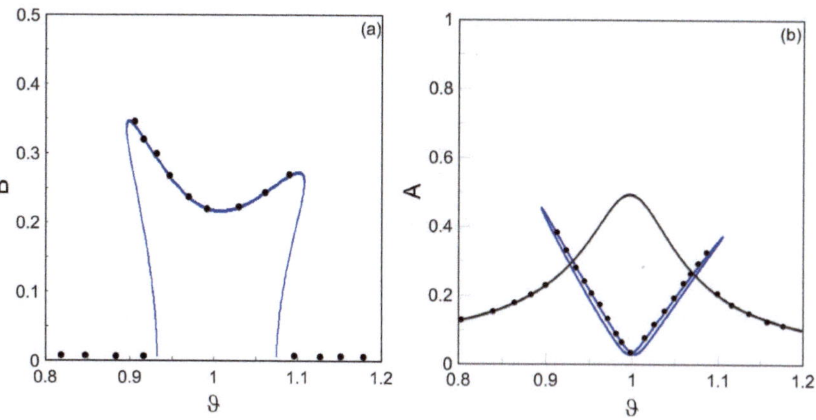

Fig. 4 Analytical resonance curve of the pendulum **a** and the oscillator **b**

which allow amplitudes and phases determination. For a full nonlinear case it is necessary to solve a set of nonlinear algebraic equations (13) numerically.

Sample resonance curves of the pendulum and the oscillator representing solutions of Eqs. (14) for data: $\alpha_1 = 0.1$, $\alpha_2 = 0.01$, $\mu = 15.2$, $\lambda = 0.25$, $q = 0.05$ are presented in Fig.4. The black dots denote the numerical verifications. It should be noted that harmonic balance method is valid for relatively small angles of the pendulum in the neighbourhood of the main parametric resonance ($\vartheta \approx 1$). The case of resonance with a fixed, not oscillating pendulum is presented by black line in Fig.4(b), while blue line corresponds to the case with a swinging pendulum. Close to $\vartheta \approx 1$, the dynamical elimination of oscillator's vibration caused by the swinging pendulum is clearly visible.

If we assume that damping of the pendulum is equal to zero, $\alpha_2 = 0$ in Eq.(13), we can find the condition of the full elimination of the oscillator's vibrations

$$\vartheta^* = \sqrt{2\lambda + \frac{\sqrt{2}\sqrt{2\lambda^2\mu^2 - \mu q}}{\mu}}, \tag{15}$$

where ϑ^* is the frequency of the full vibrations absorption. For such a case the amplitude of the pendulum is expressed

$$B^* = \sqrt{2}\sqrt{2 - \frac{\sqrt{2}\sqrt{\lambda^2\mu\,(2\lambda^2\mu - q)}}{\lambda^2\mu}}. \tag{16}$$

This state can be achieved only if the damping of the pendulum equals zero. We can notice that nonlinear characteristic of the spring does not influence this condition. If the pendulum does not swing, which corresponds to the semi-trivial solution, we can determine the resonance curve of the nonlinear one degree of freedom system with additional masses. Putting B=0 and assuming linear spring, $\gamma = 0$ in Eq. (13), after rearrangements, we get the equation:

$$A^* = \frac{q}{\sqrt{1+(-2+\alpha_1^2)\vartheta^2+\vartheta^4}} \tag{17}$$

which is the amplitude of the linear oscillator (the black line in Fig. 4(b)).

3.2 *Stability of Analytical Solutions*

Stability analysis of the harmonic solutions is carried out by using the approximate equations (10). Determining derivatives $\dot{A}$, $\dot{\phi}_1$, $\dot{B}$, $\dot{\phi}_2$ from Eqs. (10) we get the amplitude modulation equations, which can be written in the shortened form:

$$\begin{aligned}\dot{A} &= f_1(A,\ \phi_1,\ B,\ \phi_2)\\ \dot{\phi}_1 &= f_2(A,\ \phi_1,\ B,\ \phi_2)\\ \dot{B} &= f_3(A,\ \phi_1,\ B,\ \phi_2)\\ \dot{\phi}_2 &= f_4(A,\ \phi_1,\ B,\ \phi_2)\end{aligned} \tag{18}$$

where

$$f_1 = \frac{W_{\dot{A}}}{W},\ f_2 = \frac{W_{\dot{\phi}_1}}{W},\ f_3 = \frac{W_{\dot{B}}}{W},\ f_4 = \frac{W_{\dot{\phi}_2}}{W}.$$

Individual determinants in Eq. (18) are expressed in the form:

$$W = \begin{vmatrix} a_{11} & a_{12} & a_{13} & a_{14}\\ a_{21} & a_{22} & a_{23} & a_{24}\\ a_{31} & a_{32} & a_{33} & a_{34}\\ a_{41} & a_{42} & a_{43} & a_{44}\end{vmatrix},\ W_{\dot{A}} = \begin{vmatrix} a_0 & a_{12} & a_{13} & a_{14}\\ b_0 & a_{22} & a_{23} & a_{24}\\ c_0 & a_{32} & a_{33} & a_{34}\\ d_0 & a_{42} & a_{43} & a_{44}\end{vmatrix},\ W_{\dot{B}} = \begin{vmatrix} a_{11} & a_0 & a_{13} & a_{14}\\ a_{21} & b_0 & a_{23} & a_{24}\\ a_{31} & c_0 & a_{33} & a_{34}\\ a_{41} & d_0 & a_{43} & a_{44}\end{vmatrix},$$

$$W_{\dot{\phi}_1} = \begin{vmatrix} a_{11} & a_{12} & a_0 & a_{14}\\ a_{21} & a_{22} & b_0 & a_{24}\\ a_{31} & a_{32} & c_0 & a_{34}\\ a_{41} & a_{42} & d_0 & a_{44}\end{vmatrix},\ W_{\dot{\phi}_2} = \begin{vmatrix} a_{11} & a_{12} & a_{13} & a_0\\ a_{21} & a_{22} & a_{23} & b_0\\ a_{31} & a_{32} & a_{33} & c_0\\ a_{41} & a_{42} & a_{43} & d_0\end{vmatrix} \tag{19}$$

Coefficients included in the determinants (19) are defined as:

$$a_{11} = \alpha_1,\ a_{12} = -2\vartheta A,\ a_{13} = \mu\lambda\frac{\vartheta}{2}B\sin\Omega\left(\frac{B^2}{3} - 2\right),\ a_{14} = -\mu\lambda\vartheta B^2\cos\Omega,$$

$$a_{21} = -2\vartheta,\ a_{22} = -\alpha_1 A,\ a_{23} = \mu\lambda\frac{\vartheta}{2}B\cos\Omega\left(\frac{B^2}{3} - 2\right),\ a_{24} = \mu\lambda\vartheta B^2\sin\Omega,$$

$$a_{31} = 0,\ a_{32} = \lambda\vartheta AB\cos\Omega\left(\frac{B^2}{8} - 1\right),\ a_{33} = \alpha_2,\ a_{34} = -\vartheta B,$$

$$a_{41} = 0,\ a_{42} = \lambda\vartheta AB\sin\Omega\left(\frac{B^2}{8} - 1\right),\ a_{43} = -\vartheta,\ a_{44} = -\alpha_2 B,$$

$$a_0 = q\cos\phi_1 - A + \vartheta^2 A - \frac{3}{4}\gamma A^3 + \mu\lambda B^2\left(\frac{\vartheta}{2}\right)^2\cos\Omega,$$

$$b_0 = q\sin\phi_1 + \alpha_1\vartheta A - \mu\lambda B^2\left(\frac{\vartheta}{2}\right)^2\sin\Omega,$$

$$c_0 = B\left(\frac{\vartheta}{2}\right)^2 - \lambda B + \frac{\lambda B^3}{8} + \frac{1}{2}\lambda\vartheta^2 AB\cos\Omega,$$

$$d_0 = \alpha_2 B\frac{\vartheta}{2} + \frac{1}{2}\lambda\vartheta^2 AB\sin\Omega, \tag{20}$$

where $\Omega = 2\phi_2 - \phi_1$. Perturbing the analysed solutions, $A + \delta A$, $\phi_1 + \delta\phi_1$, $B + \delta B$, $\phi_2 + \delta\phi_2$, and next substituting them to (18), then subtracting from unperturbed equations and taking into account a linear part of their power series expansions, we get a set of linear differential equations in variations δA, $\delta\phi_1$, δB, $\delta\phi_2$:

$$\begin{bmatrix} \delta\dot{A} \\ \delta\dot{\phi}_1 \\ \delta\dot{B} \\ \delta\dot{\phi}_2 \end{bmatrix} = [J]\begin{bmatrix} \delta A \\ \delta\phi_1 \\ \delta B \\ \delta\phi_2 \end{bmatrix}, \tag{21}$$

where Jacobian $[J]$ takes the form:

$$[J] = \begin{bmatrix} \frac{\partial f_1}{\partial A} & \frac{\partial f_1}{\partial\phi_1} & \frac{\partial f_1}{\partial B} & \frac{\partial f_1}{\partial\phi_2} \\ \frac{\partial f_2}{\partial A} & \frac{\partial f_2}{\partial\phi_1} & \frac{\partial f_2}{\partial B} & \frac{\partial f_2}{\partial\phi_2} \\ \frac{\partial f_3}{\partial A} & \frac{\partial f_3}{\partial\phi_1} & \frac{\partial f_3}{\partial B} & \frac{\partial f_3}{\partial\phi_2} \\ \frac{\partial f_4}{\partial A} & \frac{\partial f_4}{\partial\phi_1} & \frac{\partial f_4}{\partial B} & \frac{\partial f_4}{\partial\phi_2} \end{bmatrix}. \tag{22}$$

The stability of the approximate solutions depends on the eigenvalues of the Jacobian (22). If at least one of the roots has a positive real part, the solution becomes unstable.

3.3 Model with the Inverted Pendulum

In this paragraph an autoparametric system with the inverted pendulum is presented. An inverted pendulum is a state when the pendulum swings or stands above its pivot point. Usually the pendulum hangs downwards, the inverted pendulum motion is inherently unstable. But for certain conditions the upper equilibrium point of a pendulum can become stable, especially if the pivot point is moving horizontally or vertically with a properly selected frequency and amplitude. The equations of motion for the system can be derived in a similar way as in Sect. 2.1 by using the second kind of Lagrange equations. Taking a new coordinate ψ pointed in Fig.5 we get differential equations of motion:

$$\begin{aligned}(m_1+m_2+m_p)\ddot{x}+c\dot{x}+kx+k_1x^3+\left(m_2+\frac{1}{2}m_p\right)l\left(\ddot{\psi}\sin\psi+\dot{\psi}^2\cos\psi\right)\\ =k_2Q\cos\omega t\,,\\ \left(m_2+\frac{1}{3}m_p\right)l^2\ddot{\psi}+c_\psi\dot{\psi}+\left(m_2+\frac{1}{2}m_p\right)l\left(\ddot{x}-g\right)\sin\psi=0\,.\end{aligned} \tag{23}$$

Introducing dimensionless time $\tau=\omega_0 t$, the equations of motion can be written in dimensionless form by using dimensionless parameters expressed earlier by (3). This yields:

$$\begin{aligned}\ddot{X}+\alpha_1\dot{X}+X+\gamma X^3+\mu\lambda\left(\ddot{\psi}\sin\psi+\dot{\psi}^2\cos\psi\right)=q\cos\vartheta\tau\,,\\ \ddot{\psi}+\alpha_2\dot{\psi}+\lambda\left(\ddot{X}-1\right)\sin\psi=0\,.\end{aligned} \tag{24}$$

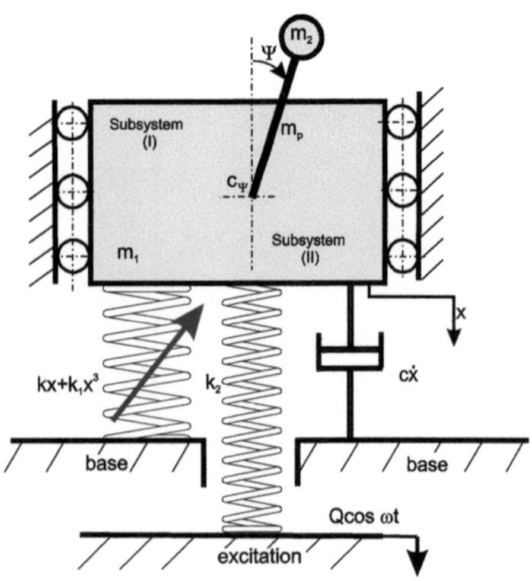

Fig. 5 Model of a system with the inverted pendulum

The equations of motion (24) are similar to those for a pendulum swinging about the lower equilibrium with one exception of the sign of the term related to parameter λ in the second equation. Analytical solutions of equations (24) are sought in a similar way to those presented in Sect. 3.1 by using the harmonic balance method.

Assuming small oscillations near the upper position the solutions are sought in the form:

$$
\begin{aligned}
x(\tau) &= C(\tau)\cos\left(\vartheta\tau+\phi_3(\tau)\right), \\
\psi(\tau) &= D(\tau)\cos\left(\frac{\vartheta}{2}\tau+\phi_4(\tau)\right),
\end{aligned}
\tag{25}
$$

where $C(\tau)$, $\phi_3(\tau)$ and $D(\tau)$, $\phi_4(\tau)$ are amplitude and phase of the oscillator and the pendulum, respectively. Making similar simplifications as in Sect. 3.1 we get a set of the nonlinear algebraic equations which allow to determine amplitudes and phases in the steady state around the upper position:

$$
\begin{aligned}
(1-\vartheta^2)C+\frac{3}{4}\gamma C^3-\mu\lambda\left(\frac{\vartheta}{2}\right)^2 D^2\cos\left(2\phi_4-\phi_3\right) &= q\cos\phi_3, \\
-\vartheta\alpha_1 C+\mu\lambda\left(\frac{\vartheta}{2}\right)^2 D^2\sin\left(2\phi_4-\phi_3\right) &= q\sin\phi_3, \\
\left(\frac{\vartheta}{2}\right)^2+\lambda-\frac{\lambda}{8}D^2+C\frac{\lambda\vartheta^2}{2}\cos\left(2\phi_4-\phi_3\right) &= 0, \\
\alpha_2\left(\frac{\vartheta}{2}\right)+C\frac{\lambda\vartheta^2}{2}\sin\left(2\phi_4-\phi_3\right) &= 0.
\end{aligned}
\tag{26}
$$

The analytical solution of the above equations is possible to get only for a system with a linear spring, $\gamma=0$. This leads to the resonance curve equation of the pendulum:

$$
\begin{aligned}
& D^4\left[-\frac{1}{16\vartheta^4}+\frac{-\lambda^2\mu^2\vartheta^4+2\lambda\mu\left(1-\vartheta^2\right)}{16\left(1+\left(-2+\alpha_1^2\right)\vartheta^2+\vartheta^4\right)}\right] \\
&+D^2\left[\frac{16\lambda^2-4\lambda\vartheta^2}{16\lambda^2\vartheta^4}+\frac{16\lambda\mu\left(\vartheta^2-1\right)+4\mu\vartheta^2\left(-1-2\alpha_1\alpha_2+\vartheta^2\right)}{16\left(1+\left(-2+\alpha_1^2\right)\vartheta^2+\vartheta^4\right)}\right] \\
&+\frac{q^2}{1+\vartheta^2\left(-2+\alpha_1^2\right)+\vartheta^4}-\frac{64\lambda^2+32\lambda\vartheta^2+4\left(4\alpha_2^2\vartheta^2+\vartheta^4\right)}{16\lambda^2\vartheta^4}=0
\end{aligned}
\tag{27}
$$

and two equivalent equations for oscillator's motion

$$
C^2=\frac{D^4}{16\vartheta^4}+\frac{D^2(16\lambda^2+4\lambda\vartheta^2)}{16\lambda^2\vartheta^4}+\frac{64\lambda^2+32\vartheta^2\lambda+4\vartheta^4+16\vartheta^2\alpha_2^2}{16\lambda^2\vartheta^4} \tag{28}
$$

or

$$C^2 = \frac{D^4\left[-\lambda^2\mu^2\vartheta^4 + 2\lambda\mu\left(1-\vartheta^2\right)\right]}{16\left(1+\left(-2+\alpha_1^2\right)\vartheta^2+\vartheta^4\right)} + \frac{q^2}{1+\vartheta^2(-2+\alpha_1^2)+\vartheta^4}$$
$$+\frac{D^2\left[16\lambda\mu\left(\vartheta^2-1\right)+4\mu\vartheta^2\left(\vartheta^2-1-2\alpha_1\alpha_2\right)\right]}{16\left[1+\left(-2+\alpha_1^2\right)\vartheta^2+\vartheta^4\right]}. \quad (29)$$

The phase angles ϕ_3 and ϕ_4 are defined as

$$\tan(2\phi_4-\phi_3) = \frac{\alpha_2\left(\frac{\vartheta}{2}\right)}{\lambda-\frac{\lambda D^2}{8}+\left(\frac{\vartheta}{2}\right)^2},$$
$$\tan\phi_3 = \frac{4\vartheta\left(4C^2\alpha_1+D^2\alpha_2\mu\right)}{D^2\mu\left(\lambda\left(D^2-8\right)-2\vartheta^2+16C^2\left(\vartheta^2-1\right)\right)}. \quad (30)$$

Differential equations of motion of the system with MR damper and the inverted pendulum are expressed as:

$$\ddot{X}+\alpha_1\dot{X}+\alpha_3\tanh\left(e\dot{X}\right)+X+\gamma X^3+\mu\lambda(\ddot{\psi}\sin\psi+\dot{\psi}^2\cos\psi)$$
$$=q\cos\vartheta\tau$$
$$\ddot{\psi}+\alpha_2\dot{\psi}+\lambda\left(\ddot{X}-1\right)\sin\psi=0 \quad (31)$$

These equations include strongly nonlinear terms therefore it is difficult to find their solutions analytically. The equations are solved and analysed numerically.

4 Regular and Chaotic Dynamics Near the Main Parametric Resonance

Analytical methods allow to find stable and unstable periodic solutions for specific resonance regions. In spite of the fact that the results are valid only for limited values of parameters and for limited amplitudes, they are very valuable. The determined resonance regions and zones in which unstable solutions occur give the base for further search by numerical methods. Chaotic vibrations, large regular oscillations or rotation of the pendulum are investigated by advanced numerical methods.

4.1 Instability Region and Parameters Influence

The autoparametric system with an attached pendulum exhibits the resonance behaviour near the frequency $\vartheta \approx 1$. Approximate analytical resonance solutions are presented in Sect. 3. We may take advantage of this phenomenon to use the pendulum as a vibration dynamical absorber. However, the detailed checkup of the solutions stability shows that for a certain set of parameters the instability region occurs inside the resonance region. A similar result is observed by Song et al. [28] for a system with pendulum and by Nayfeh and Zavodney [24] for a

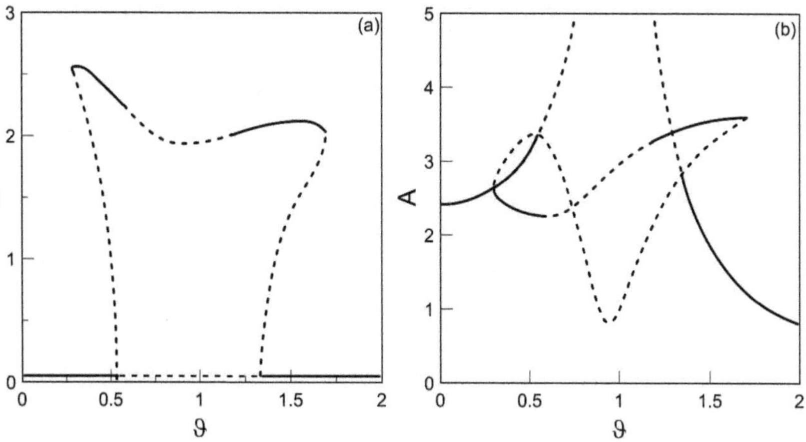

Fig. 6 Analytical resonance curves of a pendulum **a** and an oscillator **b** with marked unstable regions (*dashed line*)

continuous autoparametric model. This unwanted area is usually located close to the minimum of the pendulum's resonance curve. Such instability may lead to rapid increase of pendulum oscillations, finally resulting in rotation or irregular response. This phenomenon is not desirable if the pendulum is to be a dynamical absorber. Nevertheless, we can take advantage of this instability to use the pendulum as a generator producing energy from see waves, for instance [40], [41]. Therefore the detailed study of the stability and possible system's response is the main purpose of this section.

Sample resonance curves of the pendulum and the oscillator obtained on the basis of Eq. (14) for data: $\alpha_1 = 0.261354$, $\alpha_2 = 0.1$, $\alpha_3 = 0$, $\mu = 17.2279$, $\lambda = 0.127213$, $q = 2.45094$ and $\gamma = 0$ are presented in Fig.6(a) and (b), respectively. The solution stability analysis is carried out taking into account Eqs. (18)–(22).

This resonance is manifested by vibrations of the oscillator with frequency ϑ and the pendulum with frequency $\vartheta/2$. Figure 6 presents amplitudes of analytical solutions, trivial one equals zero and nontrivial is greater than zero. The solid line points stable, while the dashed line represents unstable solutions. The solution is unstable if at least one real part of eigenvalues of the Jacobian matrix (22) is negative.

Inside the resonance zone the motion of the oscillator is reduced by the pendulum oscillations which play a role of a dynamical absorber. Out of this region, this phenomenon fades out, and then the oscillator's amplitudes reach high values. As can be seen in Fig.6, there exists an unstable region in the middle of the resonance, located around the amplitude 'well'. If the system works in this area, the periodic motion becomes quasi-periodic, chaotic or the pendulum may rotate [28]. Transition to this region can lead to unexpected increase of amplitude and eventually to destruction of the system. Therefore, it is very important to avoid such situations. Additionally, near the unstable area it is possible to observe the shift of the pendulum vibration

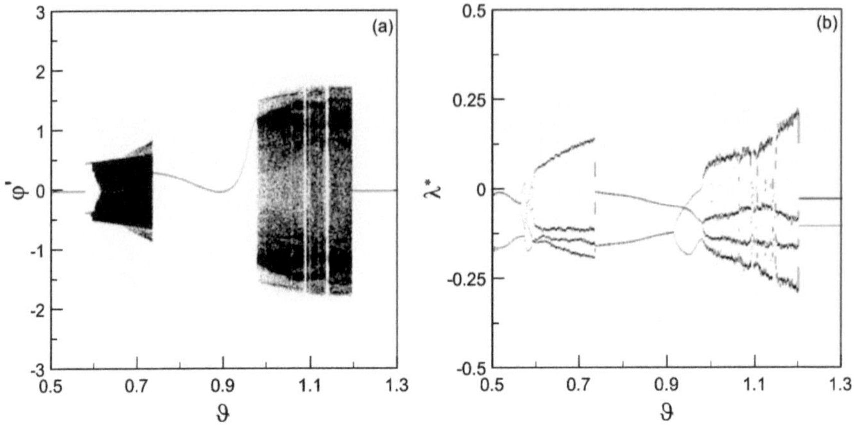

Fig. 7 Bifurcation diagram of angular velocity $\dot{\varphi}$ versus excitation frequency ϑ **a** and corresponding Lyapunov exponent **b** for $\alpha_1 = 0.261354$, $\alpha_2 = 0.1$

centre [37]. If the system is to work in this region, one should know how to reduce or control its response.

Figure 7(a) presents a numerical bifurcation diagram and the corresponding Lyapunov exponents (Fig.7b) near the main parametric resonance. If the frequency of excitation takes value from the range $\vartheta \approx 0.6$ to 0.72 or $\vartheta \approx 0.98$ to 1.19 then chaotic motions appear. The chaotic attractors for four selected frequencies of excitation ϑ are presented in Fig.8. It is interesting that at the beginning of the chaotic zone, $\vartheta \approx 0.6$ to 0.62, the chaotic motion is represented by a strange attractor which only consists of swings (Figs. 8 a and b). We call this motion *chaotic swings*. For the frequency $\vartheta \approx 0.63$ to 0.72 and $\vartheta \approx 0.98$ to 1.19, chaotic motion is composed of both rotation and swings of the pendulum, Fig.8(c) and (d). If the frequency of excitation is located between 0.73 and 0.97, then the pendulum performs the full rotation, negative or positive, depending on the initial conditions.

Considering that the system with an attached pendulum can be designed as a dynamical absorber, it is necessary to recognise the reason of the occurrence of the instability region and to find the influence of system parameters on this phenomenon. The set of parameters should be chosen carefully, to eliminate this instability, but also not to reduce a range of active pendulum operation. The first possible intuitive solution is to increase the system damping. Figures 9 and 10 present influence of damping on the pendulum amplitudes. Both graphs are plotted on the basis of Eq. (14). From Fig.9 it results that we can reduce the instability area by increasing damping of the oscillator (parameter α_1).

The value of that damping coefficient $\alpha_1 \approx 0.6$ causes total vanishing of the instability region. The influence of this parameter is clearly visible on the surface cross-sections in Fig.9(b). The disadvantage of such a solution is that the right side of the resonance curve (at the higher frequency direction) is totally reduced. Also the increase of the pendulum damping in the pivot, parameter α_2, causes reduction in

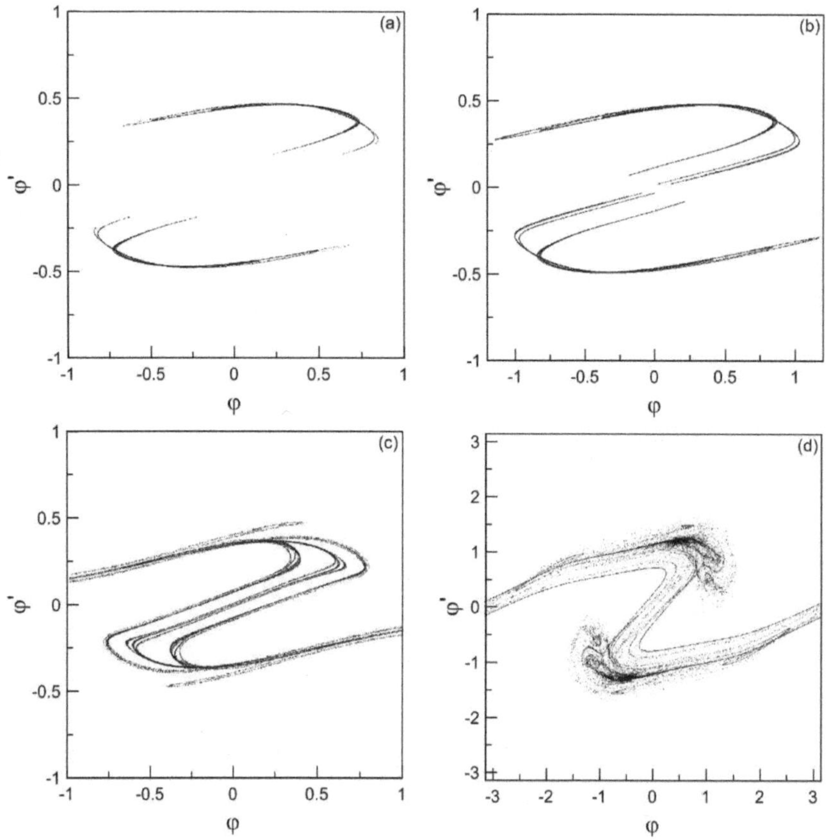

Fig. 8 Chaotic attractors for frequency $\vartheta = 0.6$ **a**, $\vartheta = 0.61$ **b**, $\vartheta = 0.7$ **c** and $\vartheta = 1$ **d**; $\alpha_1 = 0.261354$, $\alpha_2 = 0.1$

both the right and the left sides of the resonance curve, which essentially decreases effectiveness of the pendulum as a dynamical absorber (Fig.10). For damping in the pivot of $\alpha_2 \approx 0.5$, this instability does not exist at all, however, a very narrow region of absorption is the price we pay for this solution.

Figure 11 presents the influence of the excitation amplitude q on the unstable region. If the value of excitation amplitude increases, then the instability effect is still present, and its width increases, too. Then pendulum's oscillations increase dangerously.

Because autoparametric systems are very sensitive to the system's parameters, therefore, instability region can also be sensitive to some of them.

The dynamics of the system strongly depends on the values of the nonlinear terms which couple the main structure (the oscillator) and the pendulum. The response of the system is very sensitive to a change of parameters λ and μ. Parameter λ couples the oscillator and the pendulum motion (Eq.7). Parameter μ appears only in the

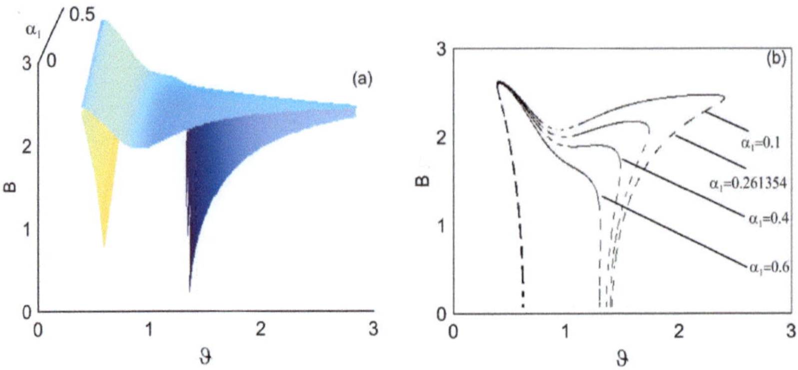

Fig. 9 Amplitude versus excitation frequency for chosen coefficients of oscillator's damping, 3D graph **a** and cross-sections **b**

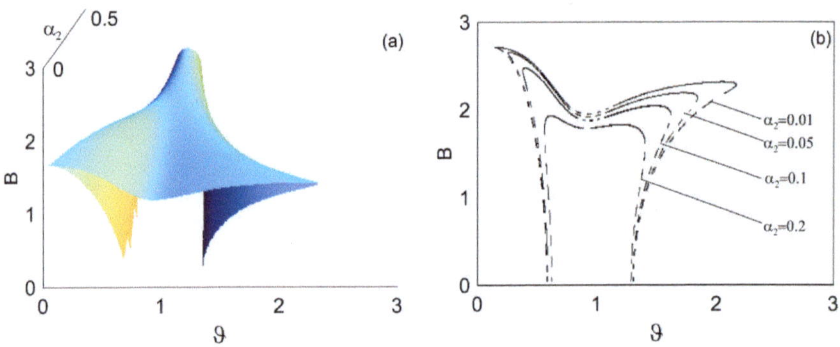

Fig. 10 Amplitude versus excitation frequency for chosen coefficients of pendulum's damping, 3D graph **a** and cross-sections **b**

oscillator's equation and plays a role of a gain in the product $\mu\lambda$. The influence of both parameters μ and λ on the pendulum oscillation around the principal parametric resonance is presented in Figs.12 and 13, respectively. These plots allow one to determine a proper selection of the pendulum parameters, according to the primary structure (the oscillator). However, a modification of the parameter λ or μ while the system vibrates in order to control the response online, could be rather difficult.

Another proposal to change the system dynamics is to include nonlinear stiffness (parameter γ) or magnetorheological damping (parameter α_3) to constitute a semi-active suspension. Because it is not possible to solve strictly analytically such a nonlinear model, the influence of the supporting spring stiffness and MR damping is analysed in next sections by numerical methods, and then verified experimentally.

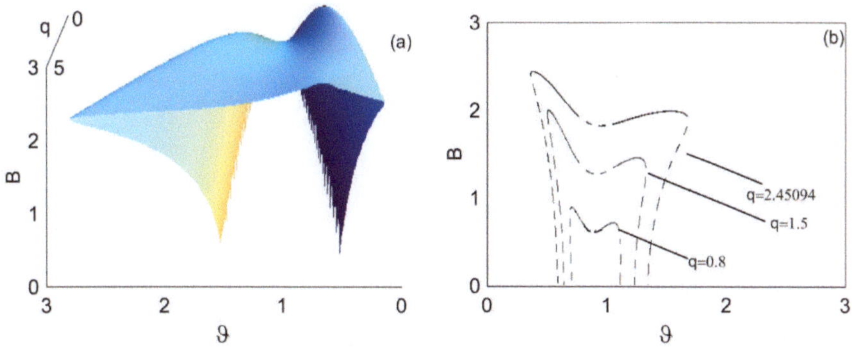

Fig. 11 Pendulum's amplitude versus excitation frequency for different excitation amplitudes, 3D graph **a** and cross-sections **b**

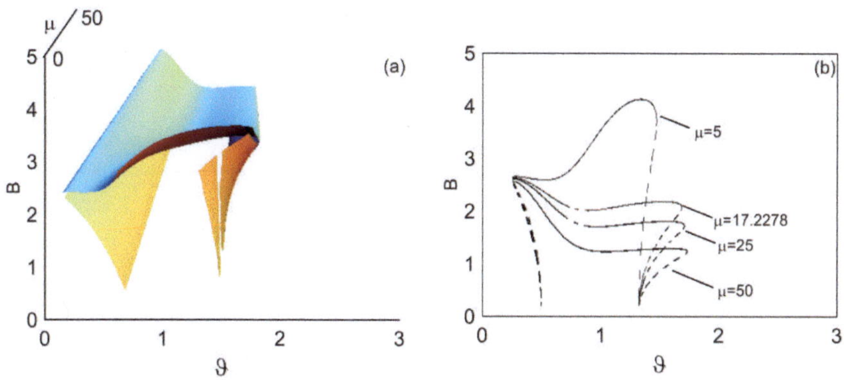

Fig. 12 Pendulum's amplitude versus excitation frequency for different μ parameter, 3D graph **a** and cross-sections **b**

4.2 Upside-Down Pendulum

Stabilising a pendulum at the upright unstable position has become an interesting object for physicists. Such stabilisation was first done using feed forward vertical vibration of the pivot of a pendulum by Stephenson [30] at the beginning of 20th century and it was analysed and demonstrated by Acheson [1], [2]. The inverted pendulum is related to rocket or missile guidance, where thrust is actuated at the bottom of a tall vehicle. If the support of the pendulum is caused to oscillate in an appropriate manner by an external applied force, then a configuration of stable equilibrium exists in which the pendulum remains upside-down. This solution characterises by oscillations around the inverted position after its slight displacing.

The obtained numerical results are based on equations of motions (24). In order to determine the ranges in which the pendulum can maintain the inverted equilibrium position ($\psi \approx 0$), two bifurcation diagrams are done. In Fig.14(a) the bifurcation

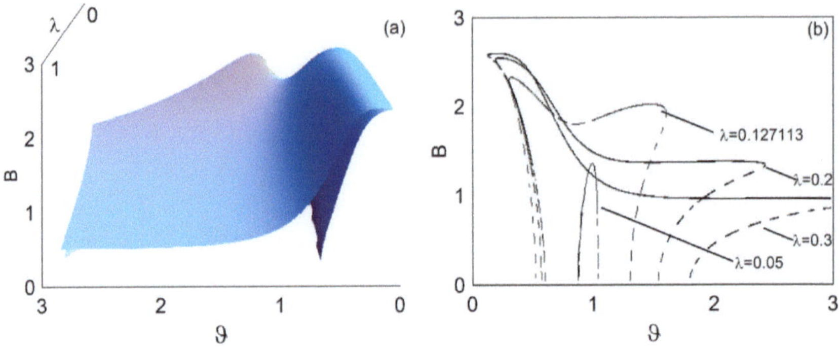

Fig. 13 Pendulum's amplitude versus excitation frequency for different λ parameter, 3D graph **a** and cross-sections **b**

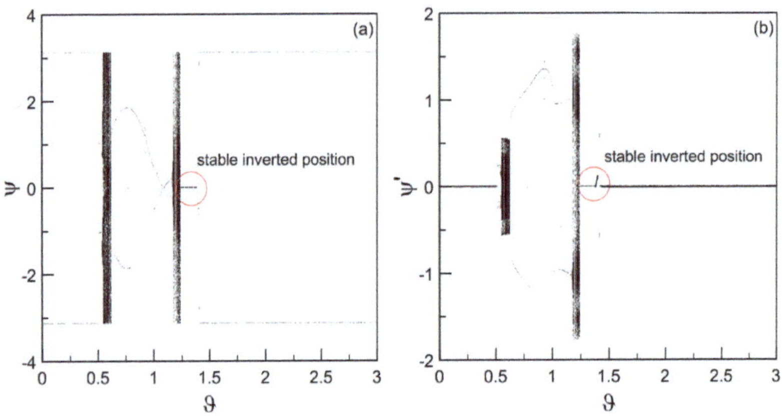

Fig. 14 Bifurcation diagrams of angle ψ **a** and angular velocity $\dot{\psi}$ **b** versus excitation frequency presenting stable inverted pendulum position

diagram of an angle of the pendulum ψ versus excitation frequency ϑ is presented, while in Fig.14(b) the equivalent bifurcation diagram of angular velocity $\dot{\psi}$ is plotted. Numerical simulations are done for data: $\alpha_1 = 0.26$, $\alpha_2 = 0.1$, $\alpha_3 = 0$, $q = 3$, $\gamma = 0$, $\mu = 17.2$, $\lambda = 0.12$ and initial conditions $\psi(0) = 0.1$, $\dot{\psi}(0) = 0.1$, $x(0) = 0$ and $\dot{x}(0) = 0$. Analysing the results in both diagrams we can precisely determine the ranges of excitation frequency where the pendulum stands up and stays at rest, or executes small vibrations in the inverted position.

A region in which a pendulum is upside down is marked by a circle. The transition to the vertical position occurs when the frequency of excitation is located between $\vartheta \approx 1.25 - 1.38$. Figure 15(a) shows time history of the pendulum coming to rest in the upper position.

However, for the range of $\vartheta \approx 1.39 - 1.40$ a very narrow region of swinging upside down pendulum exists (Fig.14). This swinging in the inverted position (see

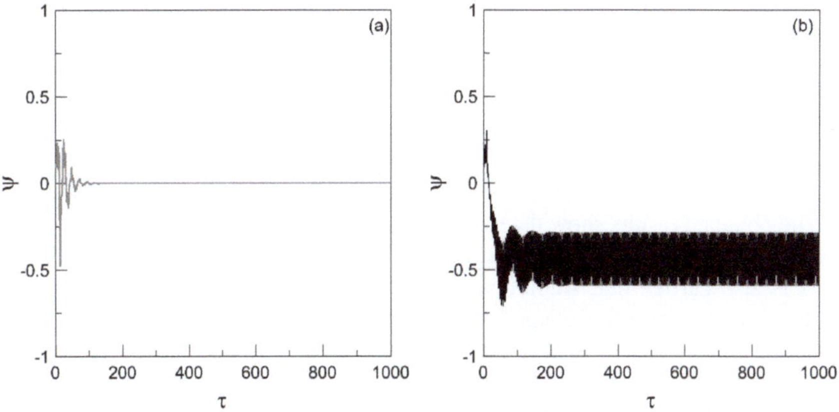

Fig. 15 Time responses of the pendulum for $\vartheta = 1.3$ **a** and $\vartheta = 1.4$ **b**

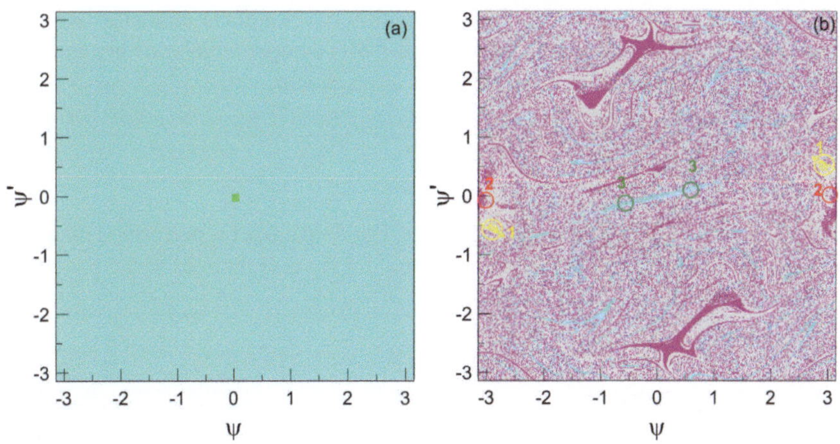

Fig. 16 Basins of attraction of the pendulum for $\vartheta = 1.3$ **a** and $\vartheta = 1.4$ **b**

time history in Fig.15b) may occur in two forms. These forms are characterised by a positive or negative shift of the vibration centre. The direction of this centre shift depends on initial conditions put on the pendulum.

Between frequency $\vartheta \approx 1.18 - 1.25$ the pendulum executes irregular motion composed of rotation and swinging. If the frequency of excitation is between $\vartheta \approx 0.62 - 1.17$, then the pendulum rotates in negative or positive directions. After crossing the value $\vartheta \approx 1.41$, the pendulum goes down and stops there. The same situation occurs if frequency of excitation does not exceed $\vartheta \approx 0.5$.

A very interesting situation is observed for $\vartheta = 1.3$ where the pendulum always goes up to the inverted vertical position independently of its initial conditions. This situation is clearly shown in Fig.16(a) presenting only one attractor at the upper equilibrium position with the homogenous basin of attraction in light blue colour.

This solution suddenly alters with a slight change of frequency to $\vartheta = 1.4$. For this data there are three different solutions dependent on initial conditions, three double attractors in Fig.16(b). Attractor No.1 is responsible for chaotic swings in the lower pendulum position (yellow colour of attractor and grey colour of its basin of attraction). Attractor No.2 (in red with pink basins of attraction) denotes the rest of the pendulum at a lower position at $\psi = \pi$ and, attractor No. 3 (in blue with light blue basins) corresponds to the swinging pendulum near a lower position, $\psi \approx \pi$.

The obtained results show the sensitivity of the system both on initial conditions and parameters. Therefore, in next sections, the study of the random initial conditions is also carried out. Additionally, the received results show the need for the application of semi-active element to control the system response.

5 Experimental Setup of the System with Active MR Damper

Results of theoretical analysis are tested on a real system which has been designed and then produced to satisfy all required conditions. Nonlinear elements like nonlinear springs with different characteristics and MR damper with hardware and software for its activation and control are taken into account.

5.1 *Laboratory Rig*

The experiment of the studied two-degree-of-freedom model is performed on a specially prepared test stand (Fig.17). The laboratory rig is composed of the pendulum

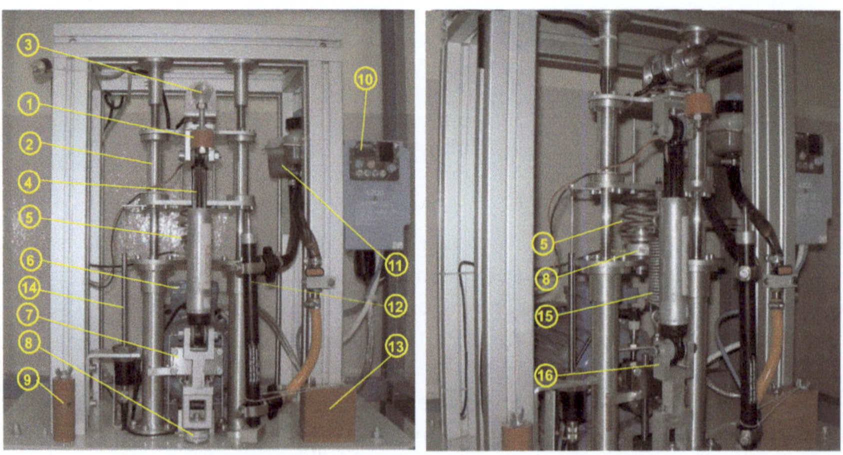

Fig. 17 General view of experimental setup and its main components

(1) with possibility of full rotation and the mass of oscillator (2) together with additional masses (13). The mass moment of inertia of the pendulum may be set for required dynamical conditions by changing its mass (9) or length. The pivot of the pendulum is connected with encoder MHK 40 (3). Motion of the system is generated by mechanisms: a motor (6), an inverter (10) and a system changing rotation of the DC motor into translational motion (7). Amplitude of the kinematical excitation is fixed by a pitch of the drive shaft. The spring (5) which connects mass of oscillator with the foundation is considered as linear or nonlinear with different stiffness characteristics. The spring (15) transmits motion from the motor to the oscillator. An additional strain gauge (8) is mounted under the spring connecting the oscillator's mass with the base. The angle of rotation φ of the pendulum and the displacement x of the oscillator are measured by the encoder and special proximity detector (14). Velocity and acceleration of the pendulum and the oscillator are calculated from signals received by their numerical derivation. The application of additional sensors also allows measuring the damping force and the force transmitted on the foundation.

Dynamics of the system is investigated for a classical linear viscous damper (12) which is controlled by a hydraulic valve connected to an oil tank (11), or a nonlinear composite MR damper RD 1097-01 (4) with a suitable control system. The damper (MR or viscous one) is mounted in a special holder (16) and fixed on the strain gauge KMM30 (8) which allows the damping force measurement. This MR damper may be effectively used to avoid unwanted situations and control the system motion. National Instrument card NI-DAQPad 6015 with DasyLab version 9 and Measurement Studio development environments, with features such as DAQ Assistant, and a single programming interface for all device functions are used for data acquisition and for the system control.

5.2 Characteristics of the Magnetorheological Damper and Nonlinear Springs

Magnetorheological dampers are specific class components that may be used for a semi-active suspension design. After the application of the magnetic field, the fluid used in a MR damper changes in a few milliseconds from liquid to semi-solid state. So, the result is an infinitely variable, controllable damper, capable of large damping forces. MR dampers offer an attractive solution to energy absorption in mechanical systems and structures. There are three main types of MR dampers: the mono tube, the twin tube, the double-ended MR damper. Of the three types, the mono tube is the most common since it can be installed in any orientation and is compact in size.

The used mono tube magnetorheological composite damper RD 1097-01 (Fig.18a), manufactured by Lord Corporation, may perform maximal damping force $F_{max} = 100\,\mathrm{N}$ and is suitable for low force, light duty suspension and isolation applications. Friction of the damper is controlled by the increase in yield strength of the MR fluid in response to magnetic field strength. The MR damper is mounted vertically between the mass of oscillator and the base. The key functional parameters

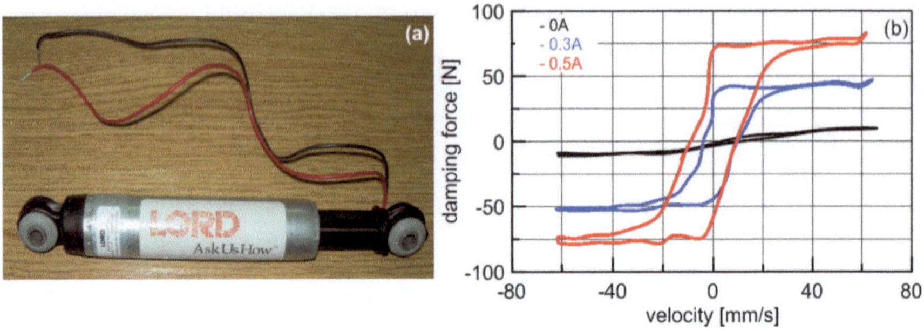

Fig. 18 Magnetorheological composite damper RD 1097-01**a** and its characteristics - damping forces versus velocity of a piston **b**

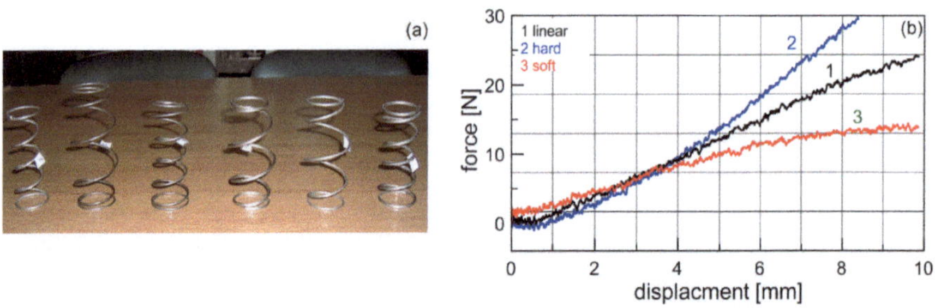

Fig. 19 Nonlinear oscillator's springs **a** and sample of hard and soft stiffness characteristics **b**

of the damper RD-1097-01 listed by the manufacturer are: maximum force 100 N, stroke 25 mm, response time less than 25 ms. The force in the passive-off mode (0 A current) is about 9 N. Its construction is different from typical linear stroke dampers filled with MR fluids and operating in the valve mode. The main difference lies in the presence of an absorbent matrix saturated with an MR fluid in this damper. More information about technical specification is provided on website www.lord.com.

In Fig.18(b) experimental characteristics of MR damper RD 1097-01 for three different current intensity ($0\,A$, $0.3\,A$, $0.5\,A$) are shown. The same damper is used in [27] to control cables vibrations and in [20] for the active suspension of the body.

In the laboratory rig, the spring which connects the oscillator and the base is considered in two variants, linear or nonlinear, with different soft or hard stiffness characteristics. Nonlinearity of springs have been reached by designing a special shape of springs e.g. barrel shape and spiral hourglass helical shape (Fig.19a). In Fig.19(b) sample characteristic of nonlinear springs used in experimental system are presented. The spring corresponding to the linear characteristics is denoted as No.1 ($k = 103$ N/m), the hard characteristics as No.2 ($k = 10^3$ N/m, $k_1 = 5 \times 10^6$ N/m^3) and the soft characteristic as No.3 ($k = 2.5 \times 10^3$ N/m, $k_1 = -9 \times 10^6$N/m^3).

6 Dynamics of an Autoparametric System with MR Damper

To improve dynamics of the studied autoparametric system and to control its response online, a magnetorheological damper, considered as a semi-active device, mounted in the suspension is proposed. Variable MR force gives possibility to control the system. However, to get the desired response, it is necessary to know the influence of MR damping when the system operates in regular or chaotic zones. Therefore, intensive numerical and experimental analysis by a study of time series, phase diagrams, Poincaré maps, Layapunov exponents, bifurcation diagrams, basins of attraction and two-parameter plots supports the possible use of the semi-active damper.

6.1 Regular Oscillations

As it is reported in Sect.4, for fixed values of parameters, without any control, the pendulum can generate various types of motion, starting from simple periodic oscillations to complex chaotic response. The motion depends on the initial conditions and the system parameters. The objective of this part is to create an open-loop control strategy of the autoparametric system by application of the MR damper. The open-loop control may be implemented easier and cost-effective comparing with a closed-loop control. A schematic diagram of the control by the MR damper is shown in Fig.20.

First, let us analyse the influence of two types of damping, the viscous and the magnetorheological ones, on free vibrations of the pendulum and the oscillator. We assume that the viscous damping of the pendulum is fixed $\alpha_2 = 0.01$, the other parameters are: $q = 0$, $\gamma = 0$ (linear spring), $\mu = 14.6863$, $\lambda = 0.12342$, and initial conditions $\varphi(0) = 0.1$, $\dot{\varphi}(0) = 0$, $x(0) = 0$, $\dot{x}(0) = 0$. The influence of viscous damping ($\alpha_1 = 0.01$, $\alpha_3 = 0$) on free vibrations is presented in Fig.21(a) and (b), respectively, while MR damping for $\alpha_3 = 0.1$ and $\alpha_1 = 0$ is presented in Fig.21(c) and (d). Analysing time series of the responses we note that the ratio of pendulum and oscillator frequencies is equal to $1/2$. Qualitative analysis shows that response of the pendulum is generally almost the same, but the oscillator's response is different. The oscillator's response in Fig.21(d) is evidently influenced by nonlinear part of MR damping.

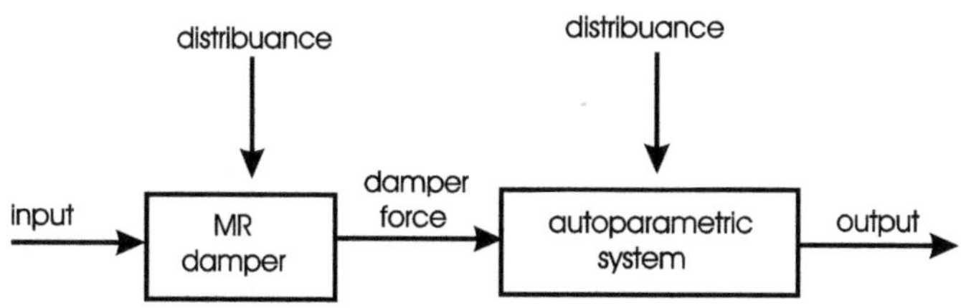

Fig. 20 Block diagram of open-loop control with the application of MR damper

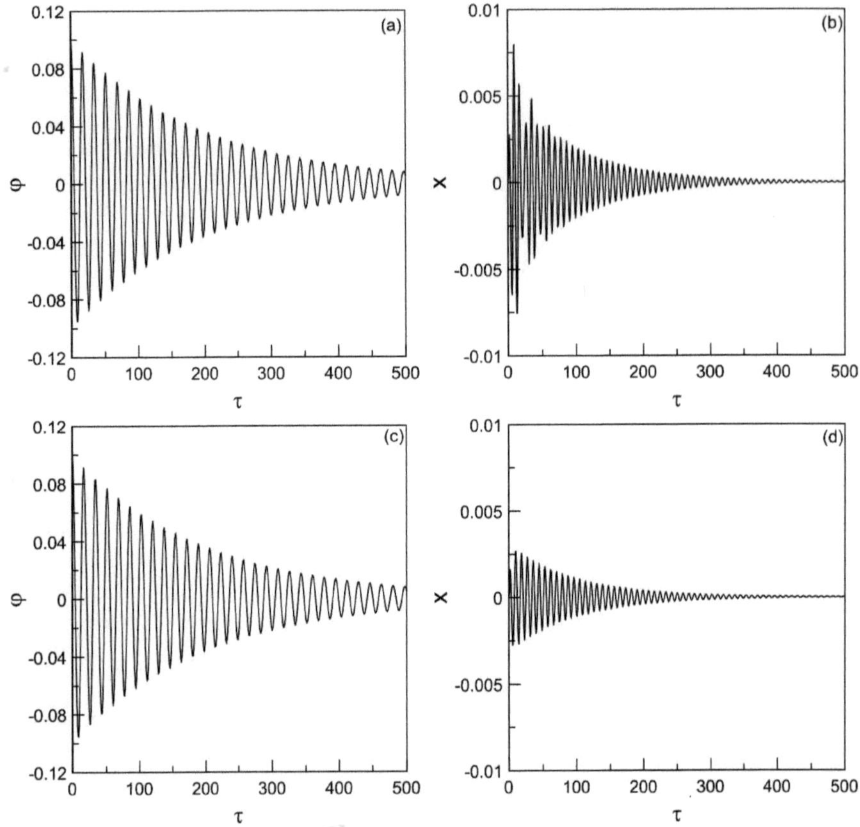

Fig. 21 Time histories of free vibrations of the pendulum and the oscillator without MR damping $\alpha_3 = 0$ **a**, **b** and with activated MR damping $\alpha_3 = 0.1$ **c**, **d**

However, the magnetorheological damper can be used to reduce amplitude of the forced pendulum motions. Sample results demonstrating this situation are shown in Fig.22(a). In this case numerical investigations are done for the value of parameters determined on the basis of experimental setup. The parameters take dimensionless values: $\alpha_1 = 0.3054$, $\alpha_2 = 0.1$, $\mu = 14.6863$, $\lambda = 0.1342$, $q = 2.3239$, $\gamma = 0$. To reduce amplitude of the pendulum (Fig.22a) and oscillator (Fig.22b), the MR damping is activated at time $\tau = 1000$, Fig.22(c), by setting value of $\alpha_3 = 1.0$. After the activation of the MR damper, the angle of rotation of the pendulum φ is slightly reduced by a significant reduction of the oscillator's vertical displacement x. The result shows that MR damping has direct influence on effective reduction of oscillator's motion and indirectly reduces oscillations of the pendulum. When the MR damper is switched on to the level of $\alpha_3 = 1.0$, then the damping force increases about twice (Fig.22d), and vibrations are successfully suppressed.

Nevertheless, we have to remember that if the initial value of MR damping is fixed on a high level while the system is operating, a rapid drop in damping value may destabilise the system. Such a situation is presented in Fig.23(a). The rapid drop in MR damping from $\alpha_3 = 0.8$ to $\alpha_3 = 0$ (Fig.23c) results in large oscillations of the pendulum (Fig.23a). The necessary conditions of this increase are different than zero initial conditions imposed for the pendulum, which is always satisfied in practice. The activated MR damper suppresses the pendulum motion to small oscillations. The result indicates that MR damper may be used to prevent a loss of stability of the pendulum motion. Time histories of oscillator's response, the dynamic force in activated and inactivated MR damper are presented in Fig.23(b), Fig.23(d).

To check the influence of magnetorheological damping we observe the system response in bifurcation diagrams in Fig.24. The excitation frequency is fixed taking the value $\vartheta = 0.57$. Fig.24(a) presents the angle of pendulum's motion φ versus varying α_3, while initial condition of the pendulum is fixed $\varphi(0) = 0.1$ with the

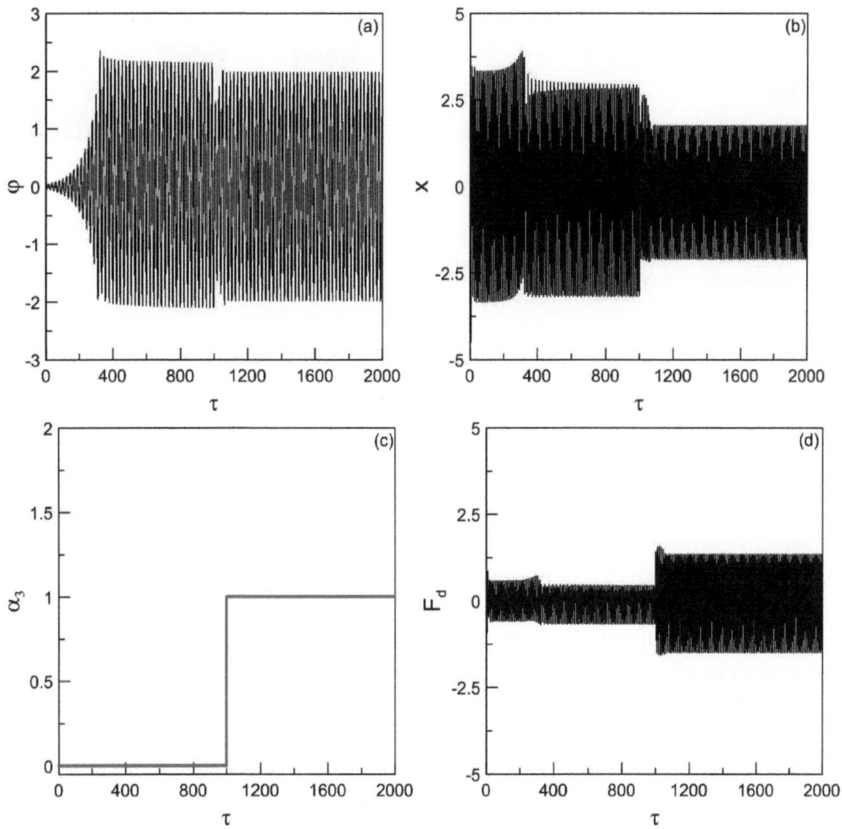

Fig. 22 Regular motion of the pendulum **a** and oscillator **b** for excitation frequency $\vartheta = 0.57$ and time histories of MR damper parameter α_3 **c** and damping force F_d **d**

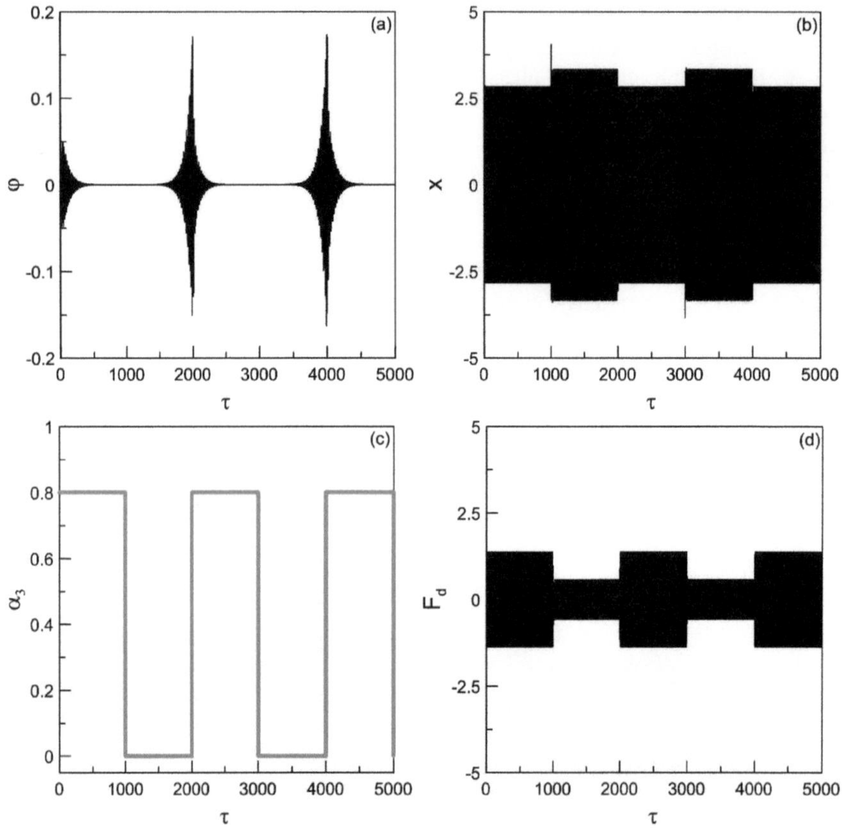

Fig. 23 Regular motion of the pendulum **a** and oscillator **b** for excitation frequency $\vartheta = 0.57$ and time histories of parameter α_3 **c** and damping force F_d **d**

others equal zero. For a wide range of MR damping the response is periodic, but after exceeding a certain critical value ($\alpha_3 = 0.4$, Fig.24a), the pendulum motion vanishes. However, remembering that in nonlinear systems dynamic response depends strongly on initial conditions, the bifurcation analysis is repeated for different randomly chosen initial conditions. Figure 24(b) presents solutions obtained for ten various initial conditions for each α_3 parameter. The total range of the pendulum swinging is much wider now, exceeding $\alpha_3 = 1.2$, that is three times greater than the result obtained in bifurcation diagram for one fixed initial condition.

This example emphasises a very important aspect of the existence of possible multiple solutions in nonlinear structures. This observation has practical meaning in engineering and physical problems. Bearing in mind that for different initial conditions MR damping may lead to a few steady solutions, their basins of attraction also have to be determined. The basin of attraction is a set of points (domain of initial conditions) in the phase space which tend to the selected attractor. In Fig.25(a)

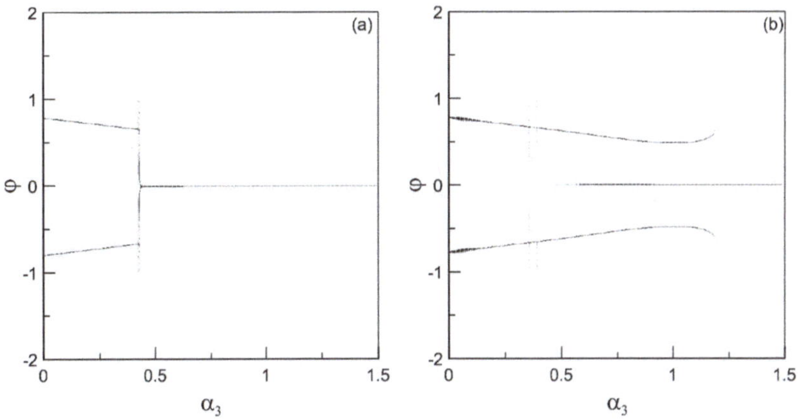

Fig. 24 Bifurcation diagrams of angle φ vs. α_3 for fixed initial condition $\varphi(0) = 0.1$ **a** and for ten randomly selected initial conditions **b**

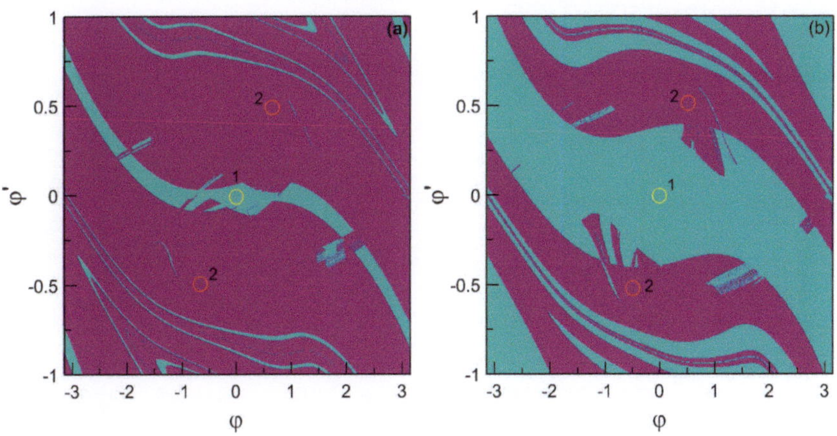

Fig. 25 The basins of attraction of the pendulum for $\vartheta = 0.57$ and $\alpha_3 = 0.5$ **a** and $\alpha_3 = 1.0$ **b**

we observe two possible solutions, represented by attractors No.1 and No.2 with relevant basins of attraction and marked by two different colours.

The two points No.2 (in red) denote the double point attractor with basin of attraction in pink colour, responsible for swinging of the pendulum (Fig.26b). The attractor No.1 with its basin of attraction in light green colour denotes semi-trivial solution, with the oscillator vibrating only and without swinging of the pendulum. If MR damping increases, up to $\alpha_3 = 1$ (Fig.25b), then the area of semi-trivial solutions is much larger. The usage of a pendulum as a dynamical absorber is then limited. Figure 26 presents time histories from numerical simulations for $\vartheta = 0.57$, $\alpha_3 = 0.5$ and for two different initial conditions. The result in Fig.26(a) represents

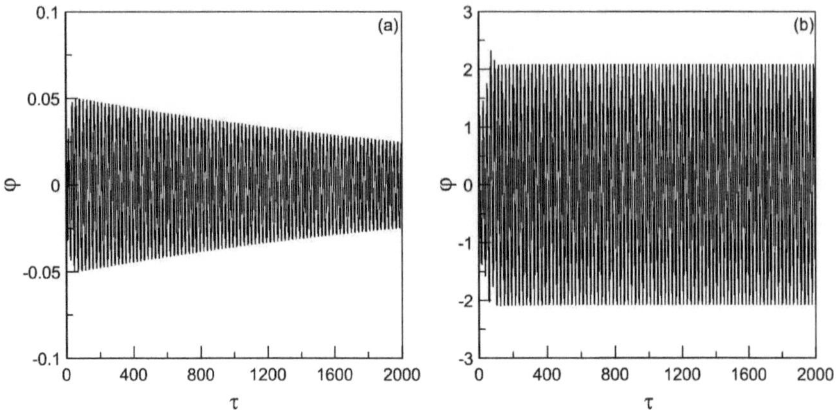

Fig. 26 Numerical time responses of the pendulum for $\vartheta = 0.57$, $\alpha_3 = 0.5$ and initial conditions $\varphi(0) = 0.1$ **a** and $\varphi(0) = 2$ **b**

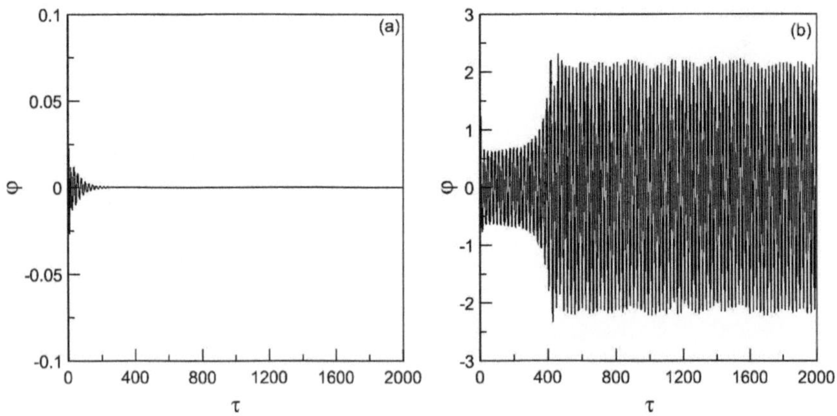

Fig. 27 Experimental time responses of the pendulum for $\vartheta = 0.57$, $\alpha_3 = 0.5$ and initial conditions $\varphi(0) = 0.1$ **a** and $\varphi(0) = 2$ **b**

attractor No.1, the pendulum motion vanishes, while in Fig.26(b) nontrivial solution with the swinging pendulum is shown.

Experimental validation of the numerical results taken from model is given in Fig.27(a) - vanishing pendulum's motion, and Fig.27(b) - periodic motion. As can be seen, numerical results are in good agreement with the experimental test in this example. However, some disturbances in the experimental response can be seen, too. Identification and analysis of experimental signals is presented later in Sect. 6.3.1.

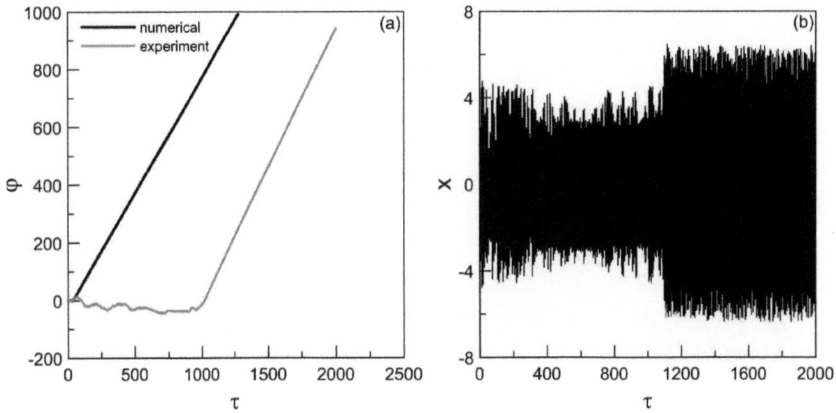

Fig. 28 Numerical and experimental rotation of the pendulum **a** and experimental vibrations of the oscillator **b** for $\varphi(0) = 0.1$, $\dot{\varphi}(0) = 0$, $x(0) = 0$, $\dot{x}(0) = 0$ and $\vartheta = 0.8$

6.2 Influence of MR Damping on Pendulum's Rotation

The parameters of an autoparametric system can be tuned in such a way that a small perturbation of initial conditions transits its response to dangerous motion, like a full rotation of the pendulum or chaotic dynamics. If the pendulum plays a role of a dynamical absorber, this kind of motion is unwanted. But sometimes we may want to generate the pendulum's rotation, for example to produce energy. The concept of energy extraction is based on the conversion of vertical motion of the oscillator into rotational motion of the pendulum by means of a parametric pendulum operating in full rotational mode [13]. The existence of rotational attractors through the parameter space and a robust method proposed for parameter identification of an experimental pendulum rig is given in [40]. In addition, closed form analytical expressions for the primary bifurcations leading to the existence of oscillations and rotations are presented in [41].

We propose to use a MR damper as a tool which fast and easily may prevent dangerous dynamics or in specific situations can generate rotation of the pendulum. The rotation is defined as a case when the pendulum amplitude exceeds $\phi = \pm\pi$ [6]. In Fig.28(a) we observe numerical time history (black line) and its experimental verification (grey line) in case of full rotation of the pendulum. Parameters $\alpha_1 = 0.3054$, $\alpha_2 = 0.1$, $\alpha_3 = 0$, $\mu = 14.6863$, $\lambda = 0.1342$, $q = 2.3239$, $\gamma = 0$, are used in numerical and experimental tests. Both motions are very similar, and rotations have the same direction. The clockwise angle is taken as positive. Figure 28(b) presents oscillator's vibration during rotations of the pendulum obtained experimentally. It may be observed that after transition to the rotation ($\tau >\sim 1000$), the amplitude of oscillator motion has become much higher. Therefore, if the pendulum is to play the role as a dynamical absorber, this kind of motion is unwanted.

The autoparametric system with pendulums belongs to strongly nonlinear models, in which, for the fixed system parameters, several different solutions depending

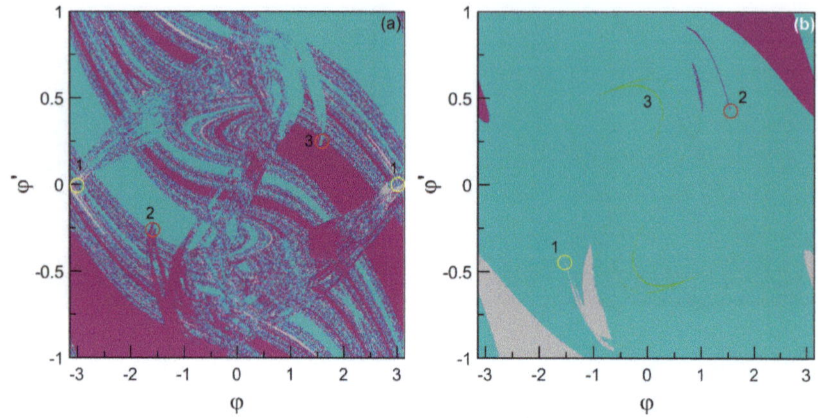

Fig. 29 Basins of attraction of the pendulum for $\vartheta = 0.8$, **a** $\alpha_3 = 0$ and **b** $\alpha_3 = 0.4$

on initial conditions may coexist. In order to verify the number of solutions, including full rotation, the basins of attraction are calculated for $\varphi \in [-\pi, +\pi]$. In Fig.29(a) we observe basins of attraction for initial conditions of pendulum for $\vartheta = 0.8$ and $\alpha_3 = 0$. Analysing the result we conclude that there are three stable solutions, two one-point and one double-point attractors. For each attractor, its basins of attraction are stated by a set of initial conditions approaching that attractor, in long-time behaviour. The attractor No.1 is responsible for the upper stable position of the pendulum. The corresponding solution of the inverted stable position is presented in Fig.30(a). Due to a small size of these regions (grey colour in Fig.29a) it is necessary to impose precise initial conditions to get the required attractor corresponding to the stable upper pendulum's position. Moreover, the basins have a strong fractal structure, therefore the weakly damped system is very sensitive to a change of initial conditions. Increasing MR damping till $\alpha_3 = 0.4$, we may eliminate the fractal nature and settle the position of basins, Fig.29(b). Now positive and negative rotations of the pendulum exist for a small range but precisely located initial conditions of the pendulum (pink and grey colours in Fig.29b). However, a new attractor No.3 (green colour) which has fractal structure, arises in this case. This strange chaotic attractor has a very wide and well-established basins of attraction marked by light green colour.

Trials of experimental verification of the numerical result of the stable inverted pendulum position presented in Fig.29(a) and Fig.30(a), lead to the rotation of the pendulum, (Fig.30(b). In a real object while operating, the external disturbances or even small changes of parameters e.g. damping, lead to interferences that move the system to the other dominating basin, the pendulum's rotation in this case.

The solutions that represent rotations (attractors No.2 and No.3 in Fig.29a), are reached numerically in Fig.31(a) and experimentally in Fig.31(b). For the initial condition $\varphi(0) = 1$ (light blue colour of basins of attraction) both numerical and experimental results show positive, clockwise rotation of the pendulum. The

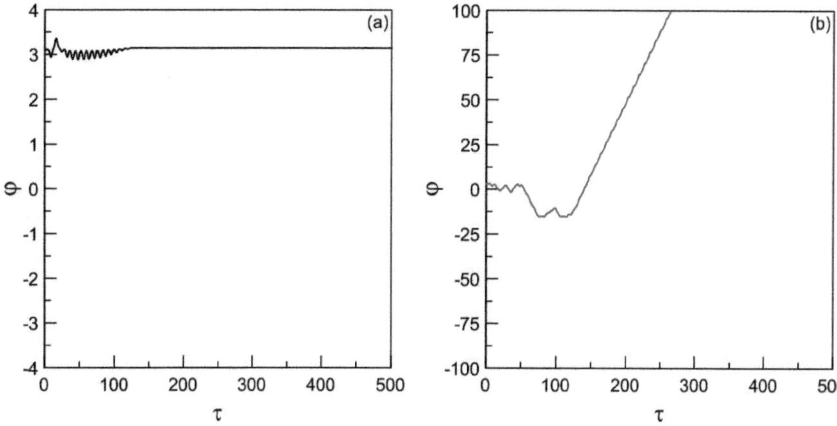

Fig. 30 Time history of the pendulum for $\vartheta = 0.8$, $\alpha_3 = 0$ and $\phi(0) = 3.13$, **a** numerical, **b** experimental

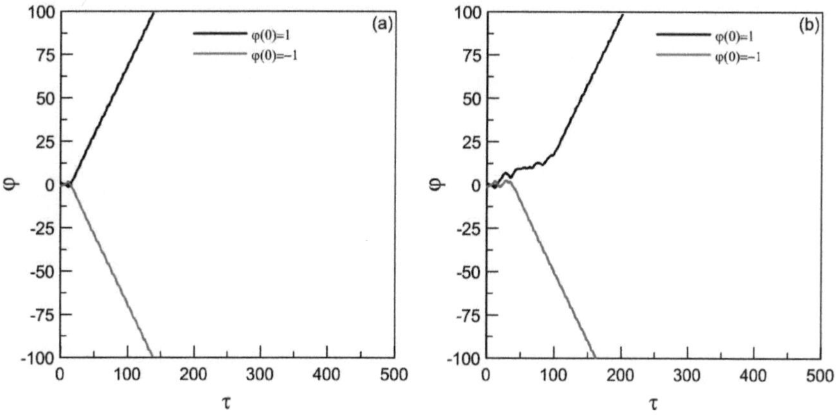

Fig. 31 Time histories of the pendulum for $\vartheta = 0.8$, $\alpha_3 = 0$ representing attractors No.2 and No.3, **a** numerical and **b** experimental results

negative rotation of the pendulum obtained for $\varphi(0) = -1$ (pink colour of basins of attraction) agrees with experimental tests, too.

The motion starting in the vicinity of fractal basin boundaries, e.g. Fig.(29a), behaves 'randomly' along the boundary for a long time, therefore it is difficult to predict which attractor is chosen. During this period of uncertainty the motion is irregular and it has the fractal structure [32].

Time histories of attractors presented in Fig.29(b) are plotted in Fig.32. The response of the pendulum in Fig.32(a) corresponds to the strange chaotic attractor No.3 in Fig.29(b). We see that chaotic motion is composed of both rotations and pendulum's swings. Clockwise and anti-clockwise rotations are received by putting

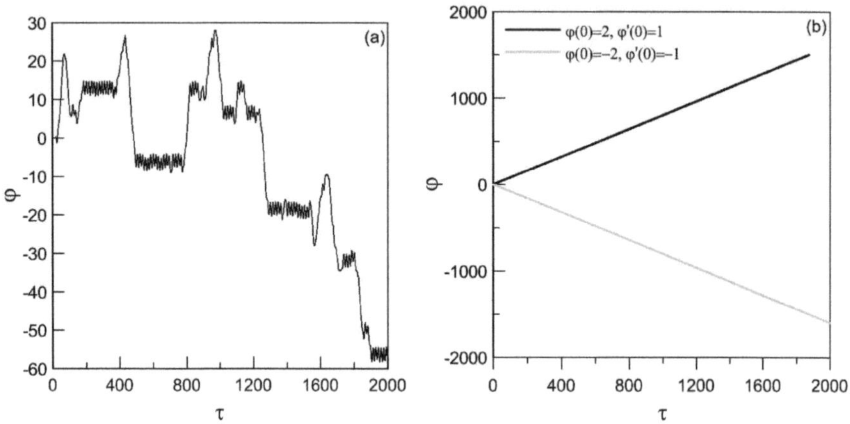

Fig. 32 The numerical time history of the pendulum for $\vartheta = 0.8$, $\alpha_3 = 0.4$, $\varphi(0) = 0.1$ and time responses for attractor **a** No.1 and **b** No.2

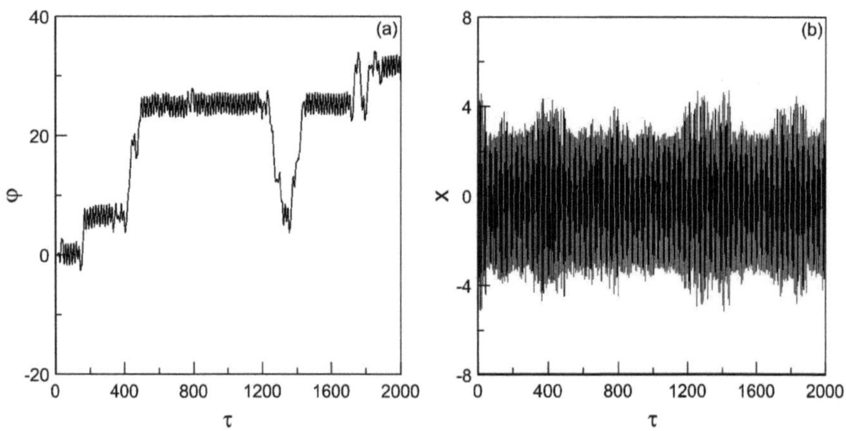

Fig. 33 The experimental time history of the pendulum **a** and the oscillator **b** for $\vartheta = 0.8$, $\alpha_3 = 0.4$ and initial conditions $\varphi(0) = 0.1$

initial conditions $\varphi = 1$ and $\varphi = -1$. The black line in Fig.32(b) denotes attractor No.2, the grey line means attractor No.1 from Fig.29(b).

The experimental validation of numerical results is presented in Fig.33. Experimental chaotic oscillations similarly to numerical results are composed of swinging and rotation (Fig.33a). Chaotic nature of oscillator vibrations can be deduced from Fig.33(b). However, to prove the oscillations' irregularity, methods of chaos identification have to be applied. A detailed description of time series analysis and methods of chaos identification are presented in the next section.

Figure 34 presents influence of the magnetorheological damping on the rotation of the pendulum. In Fig.34(a), starting from the full rotation of the pendulum, we

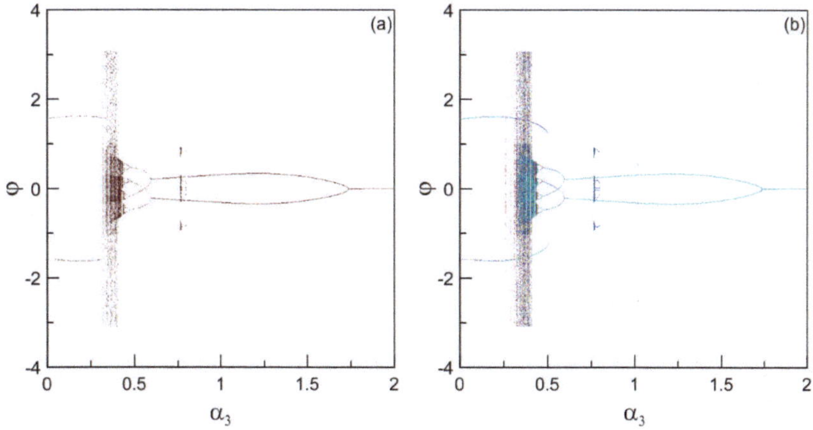

Fig. 34 Bifurcation diagrams φ versus α_3 parameter **a** for one initial condition $\varphi(0) = 0.1$ and **b** for ten random initial conditions of the pendulum, $\vartheta = 0.8$

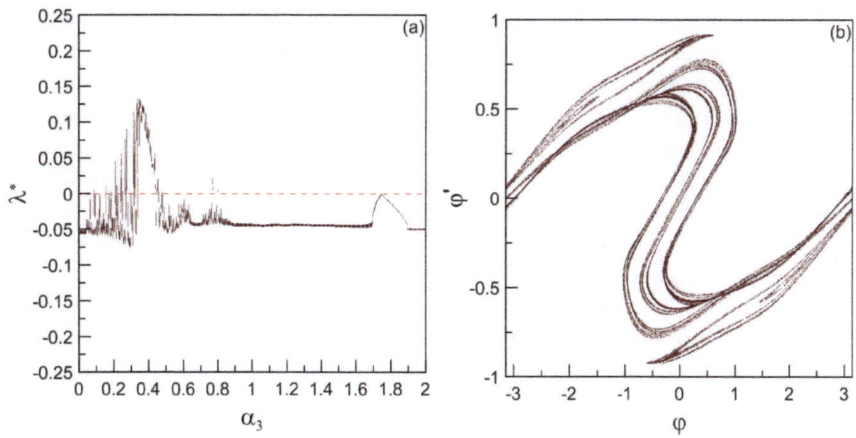

Fig. 35 The maximal Lyapunov exponent versus MR damping **a** and Poincaré map **b** for $\alpha_3 = 0.4$ and $\vartheta = 0.8$

observe transition of motion while MR damping is increased, and pendulum initial conditions are fixed $\varphi(0) = 0.1$ and $\dot{\varphi}(0) = 0$. For completeness the equivalent bifurcation diagram is plotted for ten randomly chosen initial conditions in Fig.34(b). If the magnetorheological damping increases, after crossing the bifurcation point $\alpha_3 \approx 0.36$, the rotation transits to chaotic motion. Calculated maximal Lyapunov exponent, shown in diagram Fig.35(a), as well as the Poincaré map in Fig.35(b) confirm chaotic nature of the experimental signal. Further increase of MR damping leads to the cascade of inverse period doubling bifurcations resulting in regular subharmonic response. Critical value, at which the pendulum motion disappears, is equal $\alpha_3 \approx 1.74$. It is interesting that in the neighbourhood of $\alpha_3 \approx 0.8$ a very narrow chaotic region appears (positive value of Lyapunov exponent in Fig. 35(a).

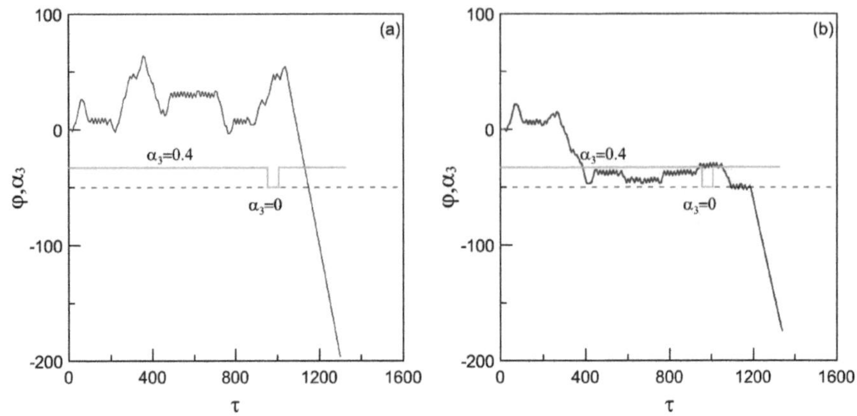

Fig. 36 The numerical **a** and experimental **b** time response of the pendulum for $\vartheta = 0.8$ with impulse deactivation of MR damping $\alpha_3 = 0.4$

The obtained results show that the application of nonlinear damper may be an effective method of the pendulum's rotation elimination.

However, for some parameters magnetorheological damping may transit the pendulum from rotation to chaotic motions. In Fig.36 we observe a change of the solution when MR damping is set to zero in selected time intervals. As we discussed in Fig.29(b) for MR damping of $\alpha_3 = 0.4$ there are three attractors No.1 and No.2 responsible for two kinds of rotation and No.3 for chaotic motion. If the MR damping is deactivated from $\alpha_3 = 0.4$ to $\alpha_3 = 0$, for a certain interval of time, short period lasting $\tau \approx 25$, at time $\tau \approx 1000$, then chaotic motion changes into negative rotations of the pendulum (Fig.36a). It means that solutions jump from attractor No.3 into attractor No.2. A very similar result was obtained by experimental tests in Fig.36(b).

As results from this paragraph, after proper tuning of the structure its response can be effectively transmitted from chaos to rotation. It is confirmed experimentally that a simple open-loop technique, allows easy control of the system response and, in case of emergency, reduction or escape from unwanted motion.

6.3 Chaotic Motion under MR Damping Influence

Chaotic dynamics which has been reached by numerical methods should be validated by experimental tests. Then the observed chaotic oscillations are subjected to the influence of a real MR damper. Such a study shows real performance of the structure under MR damping influence. Experimental time series which include external noise, need a special treatment to extract required information.

6.3.1 Experimental Verification of Chaotic Motion

The presence of chaos in physical systems is very common and is a key feature of nonlinear systems. It is extensively demonstrated in mechanical engineering

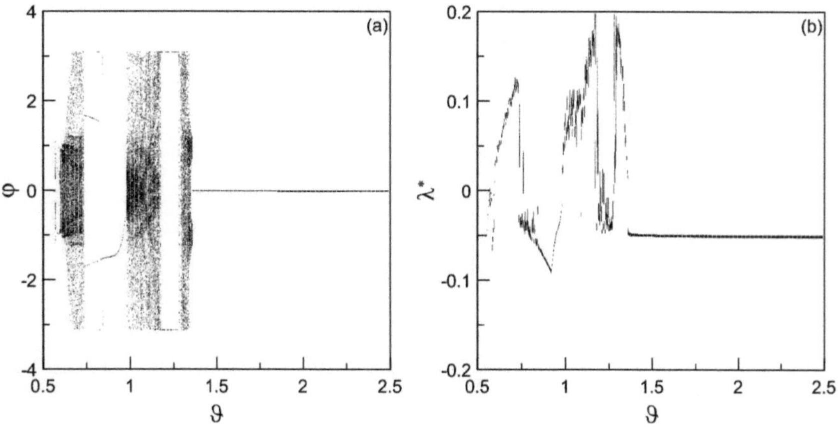

Fig. 37 Bifurcation diagram **a** and corresponding maximal Lyapunov exponent **b**, $\alpha_3 = 0$

applications, too. The regions of chaotic behaviour of the considered autoparametric system are presented in Sect. 4.1. Chaos is a state where small variations in initial conditions produce different results, in such a way that the long-term behaviour of chaotic systems cannot be predicted. The trajectory remains in a bounded phase space but it is unstable in the Lyapunov's sense. A small disturbance of the trajectory leads to its exponential divergence.

Chaotic motion can be useful in selected practical applications like percussion drill systems, for example. In practice, however, it is often desired to avoid chaos to improve a system performance. Therefore the structure and its parameters have to be properly designed or, in emergency, changed on-line during operational time. Chaotic oscillations of different pendulum-like structures, with one or multiple pendulums and nonlinear damping are investigated in a large number of papers (for example [33], [26]).

As it is pointed in Sect. 4.1 and in [28] the autoparametric system exhibits the instability region near the main parametric resonance. If a system works in this region then, rotation or chaotic oscillations of the pendulum occur [38]. Therefore, in such a situation efficient methods for chaotic motion elimination are desirable. One of the proposed method which can be applied to chaotic motion elimination is semi-active control by magnetorheological damping. This controllable damping has to be properly selected in accordance with very important parameters which influence the response of dynamical system i.e. the amplitude and frequency of excitation. Therefore, the first bifurcation diagram versus frequency of excitation (Fig.37a) and corresponding Lyapunow exponent (Fig.37b) are analysed for a system with classical viscous damping. Parameters: $\alpha_1 = 0.3054$, $\alpha_2 = 0.1$, $\mu = 14.6863$, $\lambda = 0.1342$, $q = 2.3239$, $\gamma = 0$ are taken in the bifurcation analysis.

Changing the bifurcation parameter ϑ, three separated regions with irregular motion are obtained in Fig.37(a). Chaotic motion in these zones is confirmed by

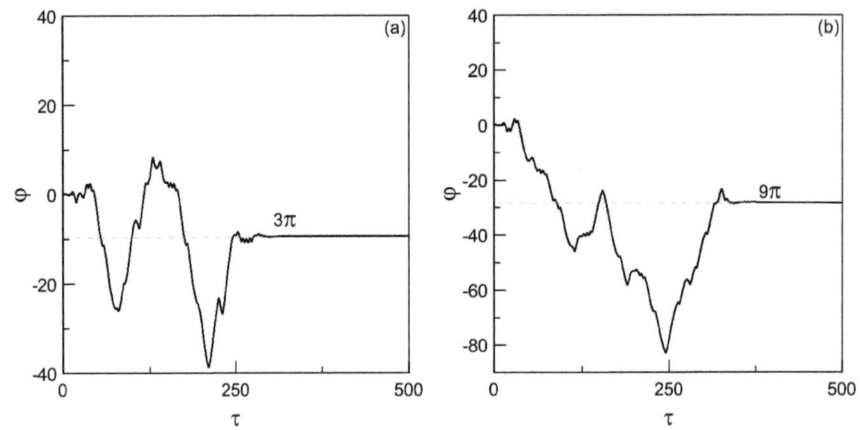

Fig. 38 Time response of the pendulum **a** numerical and **b** experimental for frequency $\vartheta = 1.25$

positive values of the maximal Lyapunov exponent (Fig.37b). Between the first and the second chaotic areas the rotation region is placed. Between the second and the third chaotic zones, the pendulum goes to stable inverted equilibrium position. Time history of this solution is presented in Fig.38(a). This result is confirmed by experiment in Fig.38(b). The strange chaotic attractors determined in the three chaotic regions are presented in Fig.39. Interestingly, that chaotic motion may occur in two forms [38]. The first form of chaotic motion, occurring more frequently is composed of irregular rotation and swings of the pendulum (Figs.39a, b and c).

The second form of chaos only consists of irregular swings (called 'chaotic swings'). The chaotic attractor of this kind of motion is split into two separate parts (Fig.39d). Comparing the attractors' set in Fig.39, we see that the pendulum motion reaches the highest velocity in the widest second chaotic region. The smallest velocity of the pendulum is obtained in the first chaotic region. It follows that the velocity of the pendulum during chaotic motion does not depend only on the frequency of excitation. The time histories of experimental chaotic motion are shown in the next part where the chaos is detected directly from experimental time series by using methods of nonlinear time series analysis.

Some behaviours e.g. chaotic oscillations composed of swings are very difficult to be detected by experiment due to a very small existence zone. It is caused by dynamically changed damping in the pendulum pivot and in the oscillator. Slightly dynamically varied parameters may sometimes eliminate a very narrow region of chaotic swings, which is discovered numerically. As shown in the previous analysis, the response of the autoparametric systems depends on the initial conditions. Therefore, the influence of MR damping on chaotic motion is analysed simultaneously, by the bifurcation diagrams calculated for one fixed or for ten randomly selected

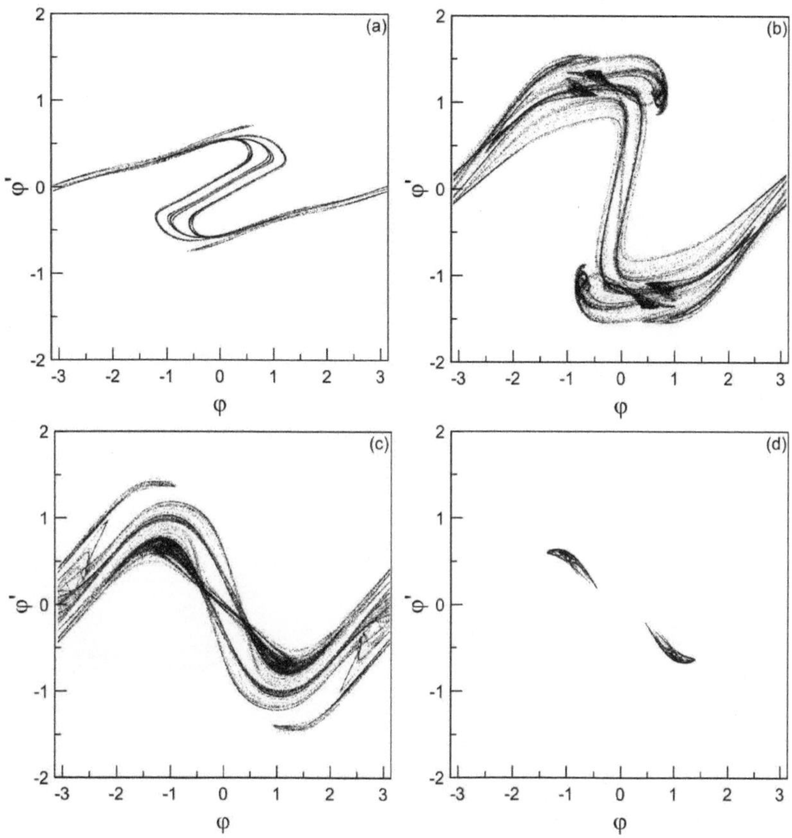

Fig. 39 Poincaré maps for the first **a** $\vartheta = 0.7$, the second **b** $\vartheta = 1.1$ and the third **c** $\vartheta = 1.32$, **d** $\vartheta = 1.35$, chaotic regions; $\alpha_3 = 0$

initial conditions. In Fig.40 the first chaotic area is taken into consideration. MR damping less than $\alpha_3 \approx 0.25$, practically does not cause qualitative changes, which is demonstrated in Fig.41 by the comparison of basins of attraction. There is chaotic motion or rotation of the pendulum.

High value of $\alpha_3 > 0.25$ parameter leads to chaos exhibited only by chaotic swings. However, apart from chaotic swings, by changing initial conditions regular rotation of pendulum is reached (Fig.40b). Magnetorheological damping over value $\alpha_3 \approx 0.32$ eliminates both the rotation and chaos phenomena. After the inverse period doubling bifurcation the system oscillates periodically. Regular motion exists in a very wide range. Maximal value of α_3 which allows the pendulum for periodic oscillation is equal $\alpha_3 = 1.8$.

In Figs.41(a) and (b) basins of attraction of the pendulum are calculated for $\alpha_3 = 0$ and for relatively small MR damping $\alpha_3 = 0.3$. The attractor No.1 (in green)

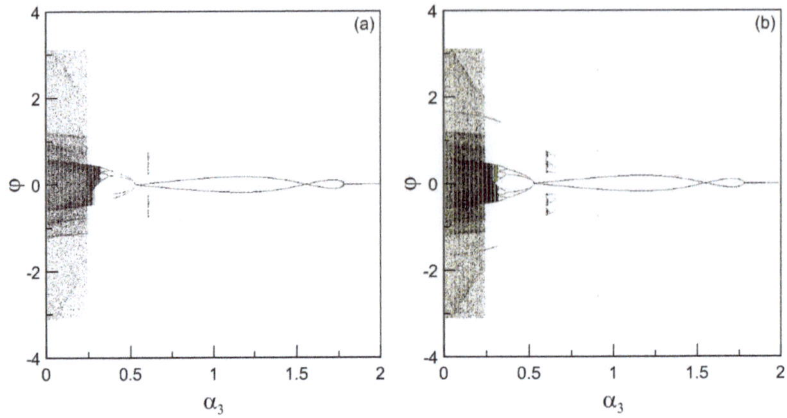

Fig. 40 Bifurcation diagrams for **a** one fixed initial condition $\varphi(0) = 0.1$ and **b** ten random initial conditions of the pendulum, $\varphi = 0.7$

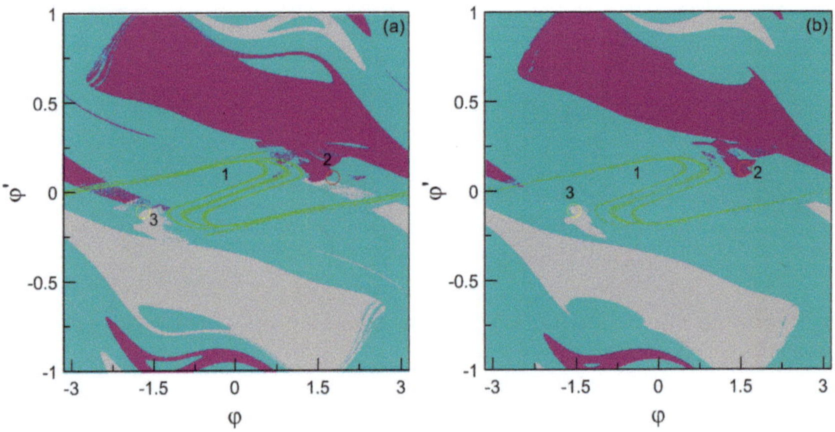

Fig. 41 Basin of attraction for $\vartheta = 0.7$ and **a** $\alpha_3 = 0$, **b** $\alpha_3 = 0.2$

denotes a strange chaotic attractor composed of chaotic rotation and swings of the pendulum, the light green colour represents its basin of attraction.

Attractors No.2 and No.3 correspond to negative and positive rotation of the pendulum, respectively. Introducing MR damping the areas of rotation existence are slightly reduced. In Fig.41(b), pink and grey colours cover smaller areas in comparison to Fig.41(a). In the second chaotic region (found in Fig.37) magnetorheological damping eliminates chaotic motion when the value of $\alpha_3 \approx 0.31$ is exceeded (Figs.42a and b). A very interesting effect was observed for a range of MR coefficient $\alpha_3 \in (0.32 - 0.38)$. After the initial conditions perturbation the pendulum jumps from the lower equilibrium ($\varphi(0) = 0.1$) to the periodic swings around the upper equilibrium point. (Fig.43a). Despite numerous attempts this result is not

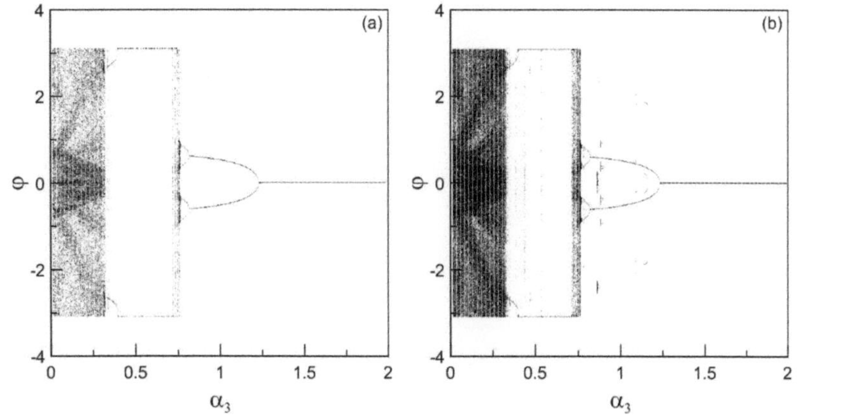

Fig. 42 Bifurcation diagrams for **a** one fixed initial condition $\varphi(0) = 0.1$ and **b** ten random initial conditions; $\vartheta = 1.1$

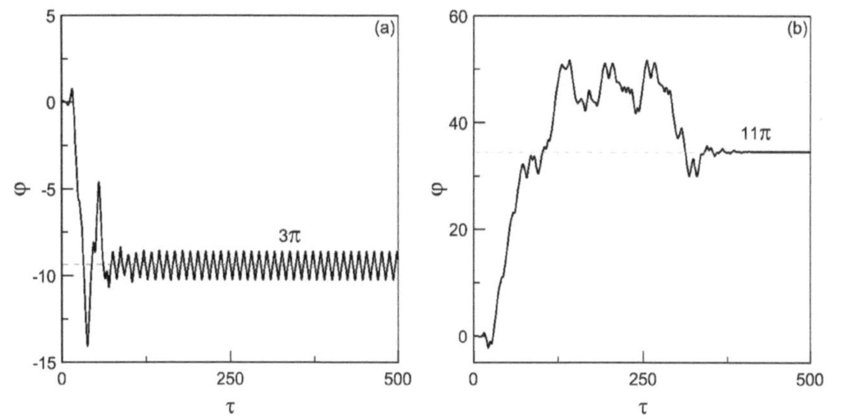

Fig. 43 Time history of the pendulum for $\vartheta = 1.1$, **a** $\alpha_3 = 0.35$ and **b** $\alpha_3 = 0.5$

confirmed by experimental tests. For $\alpha_3 \in (0.39 - 0.73)$ the pendulum also jumps into the upper position, but after transient time it goes to the upper equilibrium point (Fig.43b). This effect is observed in the experiment.

For a narrow zone of $\alpha_3 \in (0.74 - 0.78)$ chaotic motion consists of rotation and oscillation of the pendulum. After crossing the value $\alpha_3 \approx 0.78$, irregular motion transits into regular one and then for $\alpha_3 > 1.22$ the pendulum does not oscillate.

The influence of MR damping on the smallest, the third chaotic region is presented in Fig.44. We observe that the parameter greater than $\alpha_3 \approx 0.24$ eliminates irregular motion. If $\alpha_3 \in (0.14 - 0.23)$, then chaotic motion consists of pendulum swings only (Fig.45a), for $\alpha_3 = 0.1$ in chaotic response also rotation is included (Fig.45b).

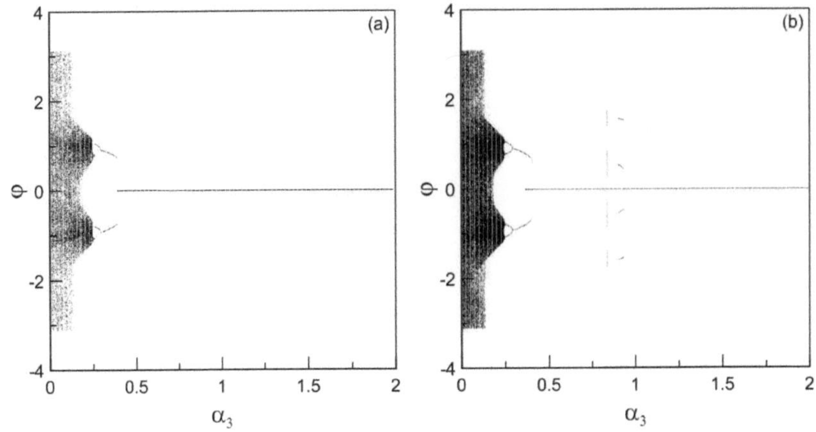

Fig. 44 Bifurcation diagrams for **a** one fixed initial condition $\varphi(0) = 0.1$ (a) and **b** ten random initial conditions for $\vartheta = 1.32$

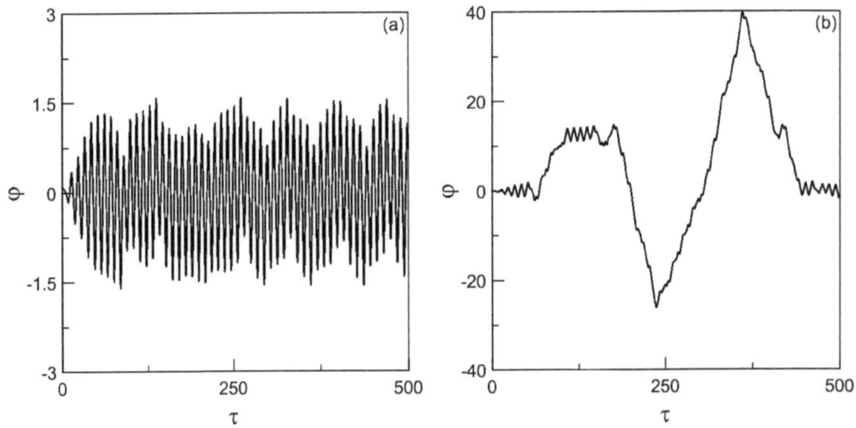

Fig. 45 Time response of the pendulum showing two forms of chaotic behaviour for $\vartheta = 1.32$, **a** $\alpha_3 = 0.2$, **b** $\alpha_3 = 0.1$

For MR damping larger than $\alpha_3 \approx 0.37$ pendulum's oscillations decay and only the oscillator vibrates. In conclusion to this analysis we may note that the reduction of the pendulum motion is the most difficult in the first chaotic region, a very large value of MR parameter is required at $\alpha_3 \approx 1.8$.

The avoidance of chaotic motion is less difficult in the second area, because MR damping at $\alpha_3 \approx 0.78$ has to be given. In Fig.46 the evolution of attractors in all chaotic regions is presented. We observe on Poincaré maps that in the second and third chaotic regions (maps in the second and the third row in Fig.46) motion is the most developed which is demonstrated by the most complex fractal structure.

To investigate how external excitation influences the system, two parameter plots, amplitude q versus excitation frequency ϑ are calculated in Fig.47. The response of

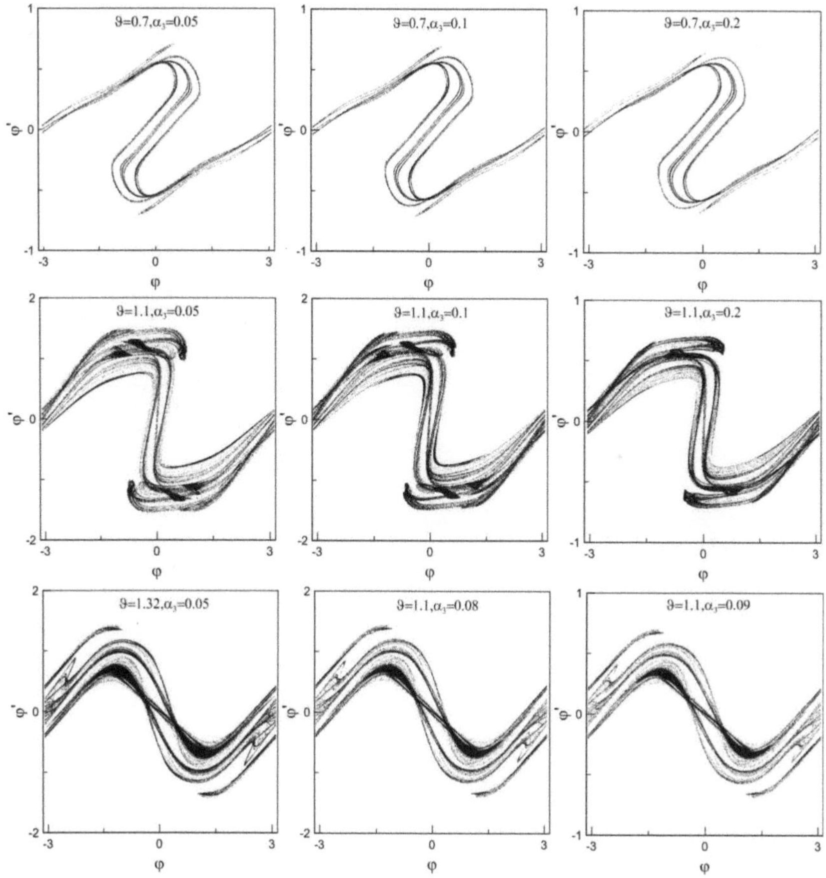

Fig. 46 Evolution of strange attractors in the three chaotic areas under MR damping influence

the system is marked by colours related to regular and chaotic dynamics taking into account selected values of MR damping $\alpha_3 = 0$, $\alpha_3 = 0.5$, $\alpha_3 = 0.8$, $\alpha_3 = 1.0$ respectively, presented in Fig. 47(a)–(d). In this way both small and large MR damping on chaotic behaviour is analysed. The calculations are performed in such a way that the first 500 solution periods are excluded from the time series. For each value of the varied parameter the same initial conditions, $\varphi(0) = 0.1$, $\dot{\varphi}(0) = 0$, $x(0) = 0$, $\dot{x}(0) = 0$, are taken. Thus, the results for different system parameter values may be directly compared. The dark blue, light blue and green colours indicate chaotic regions estimated on the basis of positive value of the maximal Lyapunov exponent. The dark blue colour is plotted if the value of Lyapunov exponent is between $+0.001$ and $+0.1$, the green colour between $+0.1$ and $+0.2$, the light blue colour between $+0.2$ and $+0.3$. White colour defines periodic motion, oscillation, rotation or regions where the pendulum goes to the lower equilibrium state. As we may see, chaotic response occurs near the main parametric resonance $\vartheta \approx 1$. Introducing MR damping we move up chaotic tongues towards the axis of amplitude of excitation q.

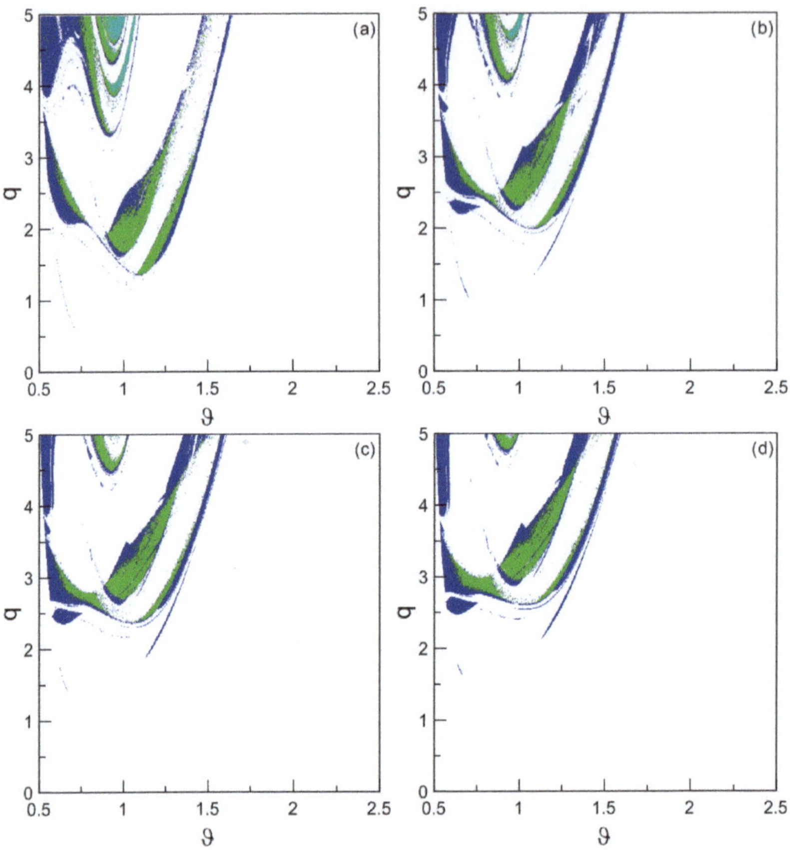

Fig. 47 Two-parameter space plots for different sets of MR damping, **a** $\alpha_3 = 0$, **b** $\alpha_3 = 0.5$, **c** $\alpha_3 = 0.8$ and **d** $\alpha_3 = 1.0$

Additionally, small chaotic zones are reduced and part of chaotic areas are divided into smaller domains (Fig.47d). The results of the two-parameter space plots are tested by the bifurcation diagrams (cross-checks). For example Fig.48(a) is a cross-check of Fig.47(a) for the value of $q = 5$ and $\alpha_3 = 0$, the colour of chaotic region corresponds to the value of the maximal Lyapunov exponent (see Fig.48b), where all Lyapunov exponents are plotted).

In Fig.49 we observe two-parameter space diagrams α_3 versus q. These diagrams present the influence of MR damping on chaotic motion versus amplitude of excitation for two selected values of frequency excitation: $\vartheta = 0.8$, (Fig.49a), and $\vartheta = 0.1$ (Fig.49b). We can clearly observe that MR damping can both eliminate or give rise

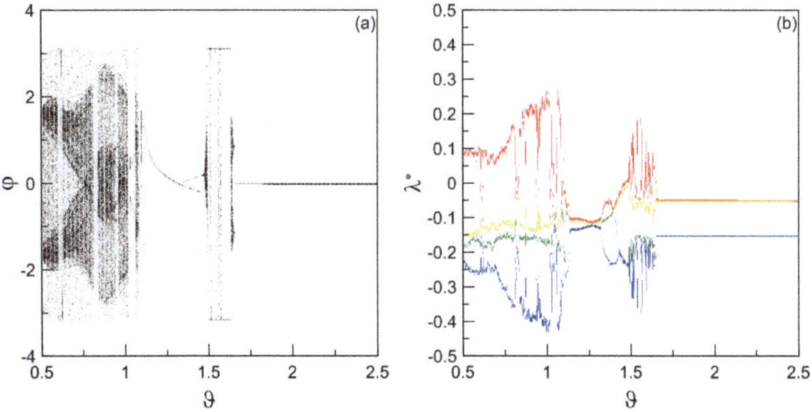

Fig. 48 The bifurcation diagram for **a** $q = 5$ and $\alpha_3 = 0$ and **b** corresponding Lyapunov exponents

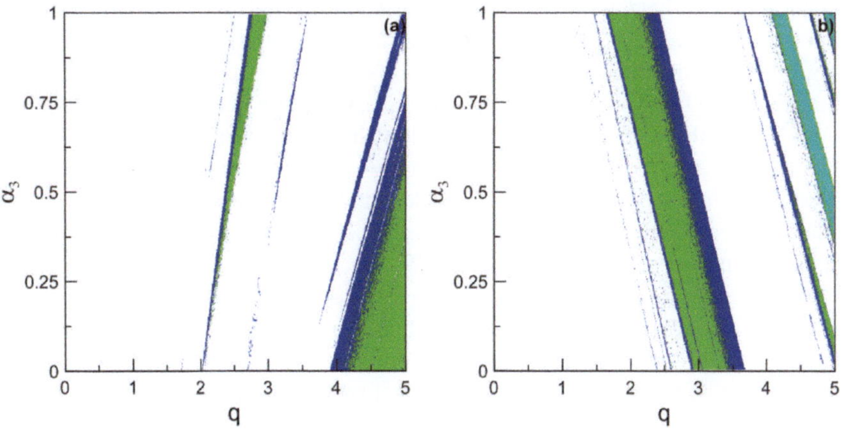

Fig. 49 Influence of MR damping on chaotic motion versus amplitude of excitation q for **a** $\vartheta = 0.8$ and **b** $\vartheta = 1$

to chaotic motion. It means that the increase of MR damping may not guarantee suppression of chaotic oscillations.

Summarising results of the analysis presented above we note that a change of system dynamics must be preceded by an intensive numerical simulation with particular emphasis on the initial conditions and value of MR damping. Bearing it in mind we can avoid dangerous situations and control chaotic dynamics of the autoparametric system by using MR damper. The semi-active MR device allows also to control the system motion on-line by open or closed-loop control strategy.

6.3.2 Experimental Time Series Analysis: Chaotic and Regular Dynamics

Experimentally observed behaviour, regular or chaotic one, may differ from that obtained from the governing equations of motion. Therefore, showing that a mathematical model exhibits chaotic dynamics is not a proof that chaos is also present in the corresponding experimental system [19]. In order to unambiguously confirm the irregular motion in a dynamic system, the chaos should be identified directly from experimental data series. For this purpose, nonlinear time series analysis methods are applied. The standard procedure to perform nonlinear analysis is the phase space reconstruction. At first, we construct the delay vectors

$$\mathbf{s}(i) = \left(x_i, x_{i+d}, x_{i+2d}, \ldots, x_{i+(m-1)d}\right) , \tag{32}$$

where parameter m is the embedding dimension and parameter d is the embedding delay. Each unknown point of the phase space at time i is reconstructed by the delayed vector $\mathbf{s}(i)$ in an m-dimensional space called *the reconstructed phase space*. To reconstruct the attractor successfully by using Eq. (32), appropriate values for d and m have to be determined. For determining time delay d the mutual information between x_i and x_{i+d} is introduced as a suitable quantity [10]. To determine the proper embedding dimension m, we use the false nearest neighbour method (FNN) introduced in [18]. This method relies on the assumption that an attractor folds and unfolds smoothly with no sudden regulations in its structure. Therefore, all points that are close in the reconstructed embedding space have to stay sufficiently close also during forward iteration. Having calculated two parameters m and d, we are able to successfully reconstruct the attractor. The reconstructed attractor and the calculated largest Lyapunov exponent [39] are the key step towards establishing whether the experimentally analysed time series originate from a chaotic system.

One of the newer methods to identify chaotic motion in pendulum-like system is the recurrence plot (RP) method [17] introduced in [9] and developed by Marvan et al. in [22]. Recurrence plots are a useful tool to identify a structure in a time data set. A recurrence plot is a graph which shows all those time instants at which a state of the dynamical system recurs. In other words, the RPs method reveals all the times when the phase space trajectory visits roughly the same area in the phase space. A recurrence plot can be described by computing the matrix M_{ij}

$$M_{ij} = \theta\left(\varepsilon - \left|\mathbf{s}_i - \mathbf{s}_j\right|\right) , \tag{33}$$

where θ is the Heaviside step function, ε is a tolerance parameter (threshold), to be chosen. If the trajectory in the reconstructed phase space returns at time i into the neighbourhood of ε where it was j then $M_{ij} = 1$, otherwise $M_{ij} = 0$. This matrix is symmetric by construction. In Fig.50(a) we show the method of determining the proper embedding delay for the chaotic signal $\vartheta = 0.7$ presented in Fig.39(a) and in Fig.51. For the phase portrait reconstruction only one signal of angular velocity of the pendulum, shown in Fig.51(a), is used. This choice of a signal makes the analysis easier because in velocity domain a rotation of the pendulum is eliminated. The mutual information has the first minimum at the time step $d = 21$.

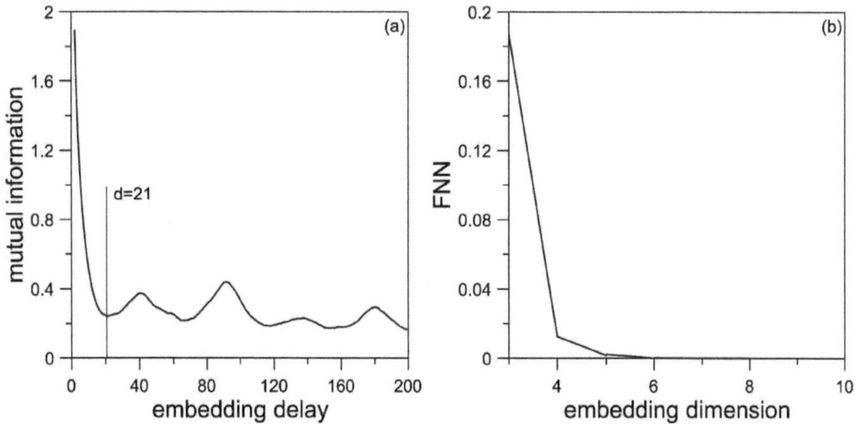

Fig. 50 The embedding delay **a** and minimal required embedding dimension **b**; chaotic motion $\vartheta = 0.7$

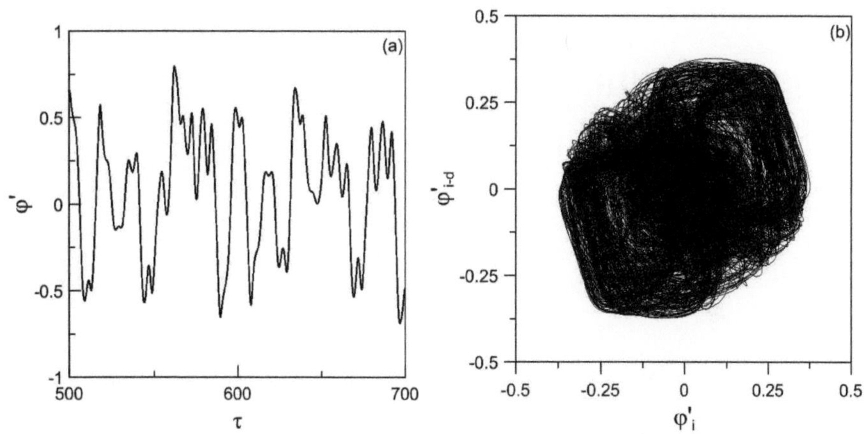

Fig. 51 Time history **a** and its reconstructed phase space **b** obtained with the optimal embedding parameters: $d = 21$, $m = 5$; chaotic motion, $\vartheta = 0.7$

The results obtained with the False Nearest Neighbour (FNN) method are presented in Fig.50(b). It can be observed that the fraction of FNN convincingly drops to zero for $m = 5$. This means that behaviour of the system requires a five-dimensional phase-space. The obtained recurrence diagram in (Fig.52b) includes a large number of short discontinuous lines. Their distance from the diagonal line is varied and does not exhibit any symmetry. This indicates that the analysed signal is irregular. The same conclusion can be drawn from the reconstructed attractor in Fig.52(a) and the phase space in Fig.51(b). These results clearly indicate that the motion is irregular.

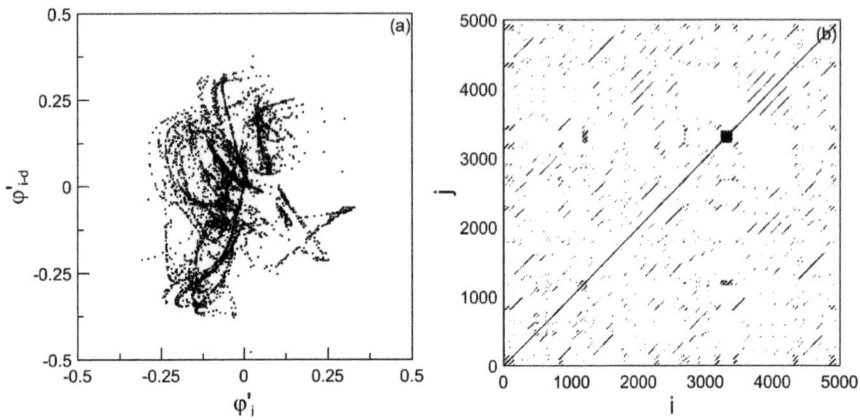

Fig. 52 Reconstructed attractor **a** and recurrence plot **b** obtained for embedding parameters: $d = 21$, $m = 5$ and $\varepsilon = 0.05$; for chaotic motion for $\vartheta = 0.7$

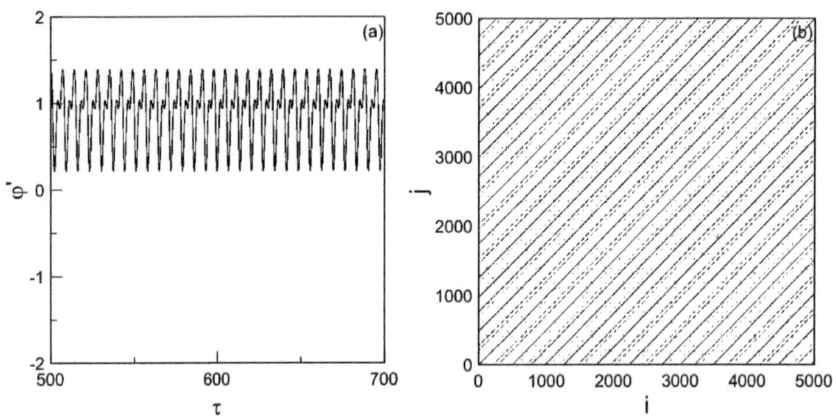

Fig. 53 Experimental time response **a** and recurrence plot **b** obtained with the optimal embedding parameters: $d = 9$, $m = 5$ and $\varepsilon = 0.01$; rotation of the pendulum $\vartheta = 0.9$

The recurrence analysis is repeated, for comparison, for the rotation of the pendulum. Again, the angular velocity presented in Fig.53(b) is used as a representative signal. For recurrence plot analysis 5 000 data points are taken. The long lines, parallel to the main diagonal, suggest that motion is regular. The experimental time series and the recurrence plot for the periodically swinging pendulum, for $\vartheta = 0.58$ are presented in Fig.54(b). In this case the patterns of the plot are reflected by long and non-interrupted diagonals. The vertical distance between these lines corresponds to the period of oscillation. This distance between individual lines is larger compared to those obtained in Fig.53(b) for rotation. It means that the period of the pendulum rotation is smaller than the swinging period. The reconstructed Poincaré maps for

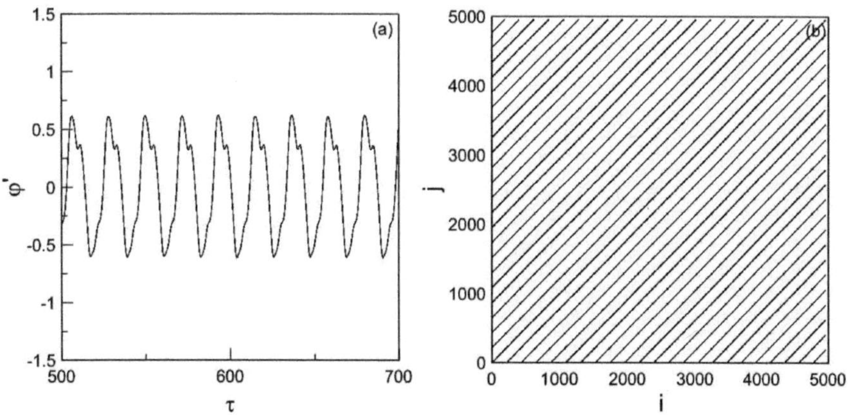

Fig. 54 Experimental time response **a** and its recurrence plot **b** obtained with the optimal embedding parameters: $d = 15$, $m = 4$ and $\varepsilon = 0.05$; regular motion $\vartheta = 0.58$

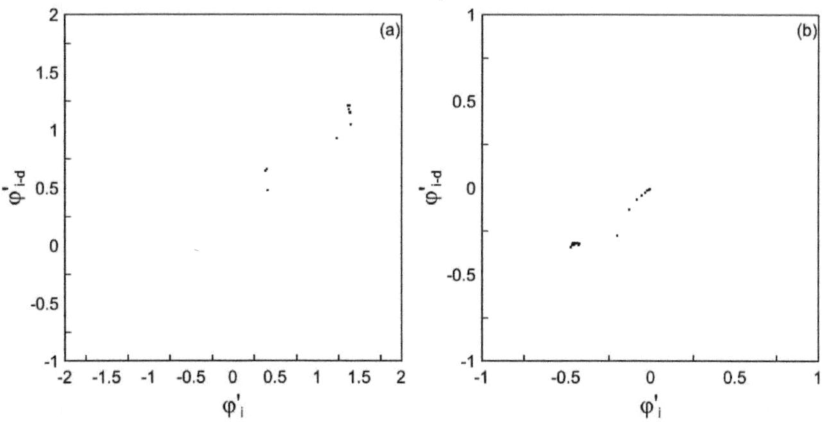

Fig. 55 Reconstructed Poincaré maps for **a** rotation $\vartheta = 0.9$ and **b** swings $\vartheta = 0.58$ of the pendulum

rotation and swinging of the pendulum are presented in Fig.55. The reconstructed attractors agree with the recurrence diagrams.

It should be noted that the analysis of real signals is more difficult, owing to various kinds of disturbances and an external noise. An inaccurate selection of delay parameters significantly influences the obtained results. It is very important to properly identify parameters from the real structure. In the considered autoparametric system the experimental results confirm those obtained from the mathematical model.

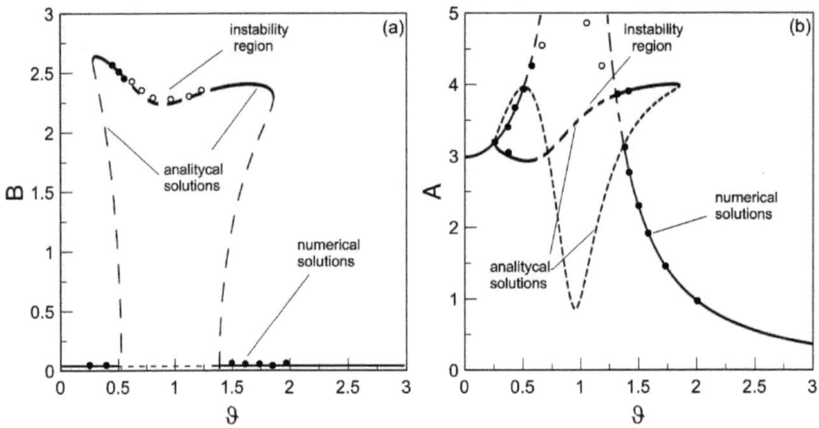

Fig. 56 Analytical resonance curve and its numerical verification with marked instability region for **a** the pendulum and **b** the oscillator

7 Influence of a Nonlinear System's Suspension on the Instability Regions

The natural presence of the nonlinear terms in the considered system with a pendulum can lead to a certain type of instability referred to autoparametric resonances. The instability may result in various forms of dynamic phenomena, such as amplitude jumps and an energy exchange between modes. This dynamics cannot be predicted by the classical theory of small linear oscillations. Apart from instability of trivial solutions, unstable nontrivial solutions may also appear. The instability region near the main parametric resonance of the autoparametric system is analysed in Sect. 4.1 and [28], [24]. An effective method of eliminating chaotic oscillations or rotation of the pendulum is to apply magnetorheological damping (Sect. 6.3). An alternative method of reducing unwanted vibrations is a change of stiffness in suspension of the autoparametric system.

Figure 56 presents the resonance curve with the instability region near the main parametric resonance obtained from Eq. (14) for parameters: $\alpha_1 = 0.26$, $\alpha_2 = 0.1$, $\mu = 17.2$, $\lambda = 0.12$, $q = 3$ and $\alpha_3 = 0$. The stability is determined analytically by the eigenvalues of the Jacobian matrix (22). The solid line denotes stable, trivial and nontrivial solutions, the dashed line represents the unstable solution. Black and white circles mean selected points of the numerical verification of analytical results, black for stable and white for unstable solutions.

The unstable region is also clearly visible in bifurcation diagram in Fig.57(a) and the corresponding plot of Lyapunov exponent in Fig.57(b). The lines between the two chaotic regions (dark areas) denote rotation of the pendulum.

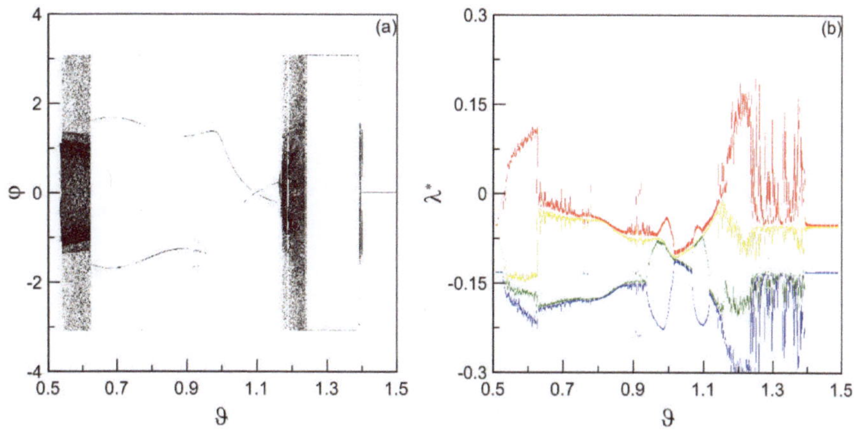

Fig. 57 Bifurcation diagram for linear spring **a** and corresponding Lyapunov exponent **b**, obtained for the initial condition $\varphi(0) = 0.1$

To reduce or shift this unstable area we can apply semi-active MR damper (as it is shown in Sect. 6.3) or to make a structural modification in the suspension. We propose to consider additional structural nonlinearity in the system's suspension considering that spring may have a nonlinear part represented by parameter γ in Eq.(7).

To check whether the introduction of nonlinearity is a promising method to improve dynamical absorption and to eliminate the unstable region the parameter γ is taken as the bifurcation parameter. The bifurcation diagrams for varying frequency of excitation ϑ for hard nonlinear spring, $\gamma > 0$ is shown in Fig.58. Stiffness causes mainly an increase in the second chaotic area and additionally, for nonlinearity $\gamma = 0.02$ (Fig.58b), the stable inverted pendulum position vanishes. The first chaotic zone is shifted towards the second. The advantage of this effect, caused by suspension stiffening, is a shift of the first chaotic zone out of possible working region. The obtained results show that hard stiffness of a spring can be used for avoiding unwanted dynamics. It should be remembered that the introduction of hard or soft nonlinearity may lead to new attractors or areas of chaos.

The influence of a soft nonlinear spring is demonstrated in Fig.59. The soft nonlinearity slightly moves the resonance tongue to the left side. Additionally, near the frequency $\vartheta \approx 0.8$, for $\gamma = -0.002$, a new bifurcation point arises. Therefore, it is very important to check new attractors existence.

The two-parameter space plots of the pendulum for different stiffness are presented in Fig.60. The blue colour denotes chaotic motion estimated by a positive value of the maximal Lyapunov exponent. Comparing the space plots for $\gamma = 0.01$ and $\gamma = 0.02$ (Fig.60) we observe that for greater value of hard characteristics γ the first chaotic tongue is smaller and is moved to the right side, but the second zone is merged with the small chaotic island. The same effect was observed in bifurcation diagram (Fig.58).

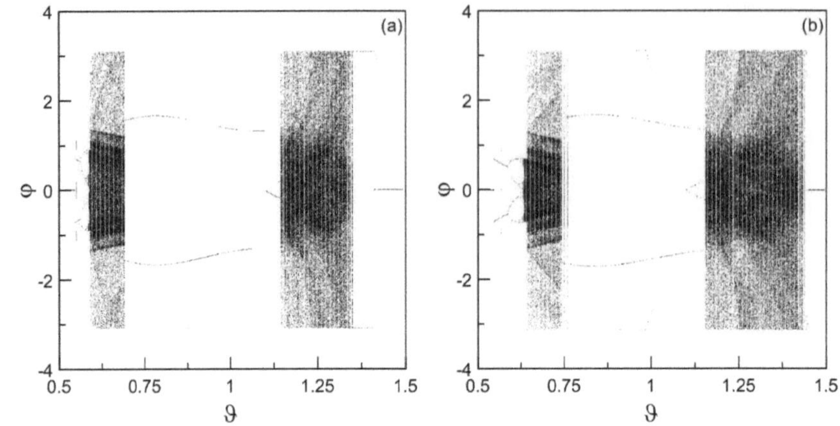

Fig. 58 Bifurcation diagram for hard nonlinear spring **a** $\gamma = 0.01$ and **b** $\gamma = 0.02$, obtained for the initial condition $\varphi(0) = 0.1$

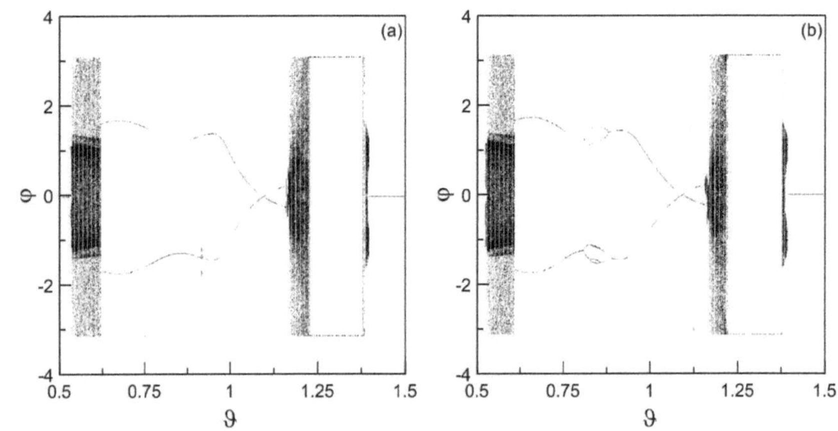

Fig. 59 Bifurcation diagram for soft nonlinear spring **a** $\gamma = -0.001$ and **b** $\gamma = -0.002$, obtained for the initial condition $\varphi(0) = 0.1$

The influence of the spring's stiff characteristics on the first chaotic region is presented in Fig. 61 by varying the parameter γ while the system operates in the chaotic zone. Figure 61(a) shows results for the fixed initial condition $\varphi(0) = 0.1$ and Fig.61(b) for ten randomly selected initial conditions. We observe that parameter $\gamma \approx 0.013$ eliminates chaotic motion, which is confirmed for all taken initial conditions.

The influence of γ in the second chaotic zone is presented in Fig.62, for one or ten random initial conditions respectively. The nonlinearity value of $\gamma \approx 0.045$ suppresses the chaotic behaviour. But a zone where the pendulum swings is very

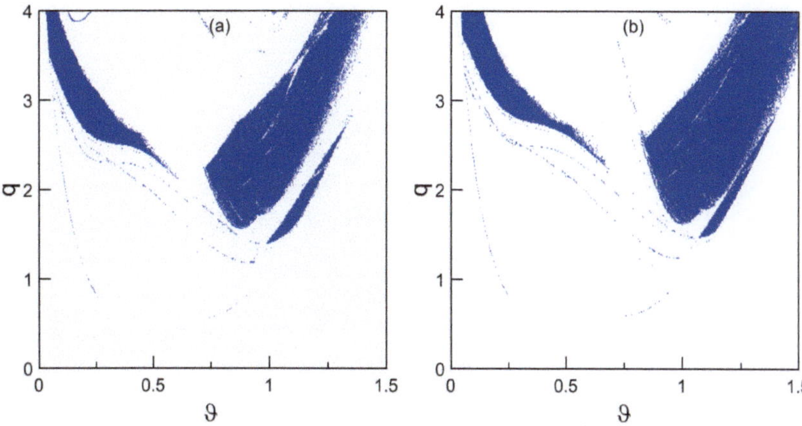

Fig. 60 Two-parameter space plot of the pendulum obtained for stiffness **a** $\gamma = 0.01$ and **b** $\gamma = 0.02$ and the initial condition $\varphi(0) = 0.1$

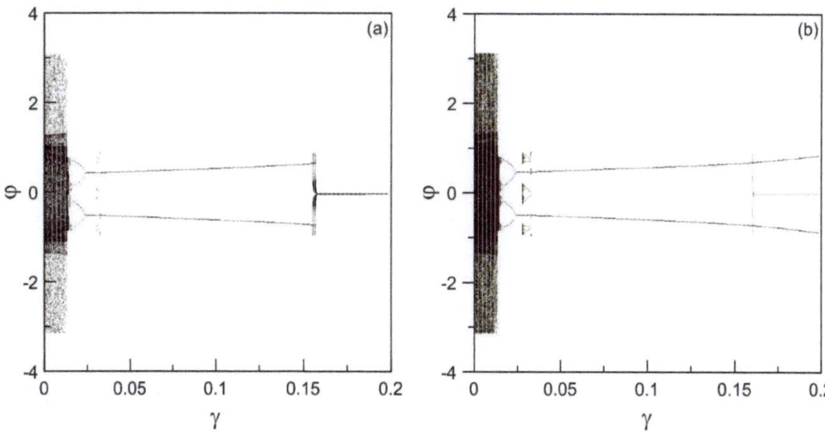

Fig. 61 Bifurcation diagram φ versus parameter γ for **a** one initial condition $\varphi(0) = 0.1$ and **b** ten random initial conditions (b), $\vartheta = 0.6$

narrow, $\gamma \in\sim (0.04 - 0.05)$, and for $\gamma \in\sim (0.05 - 0.08)$ it stops in the inverted position. For random initial conditions the rotation of the pendulum may be got as well (the line emerging from a dark area in Fig.62(b).

The evolution of attractors under the influence of stiff characteristics of a spring in the first and the second chaotic region is presented in Fig.63. In the second chaotic region the nonlinearity strongly affects the shape of the attractor. The high nonlinearity reduces pendulum's angular velocity.

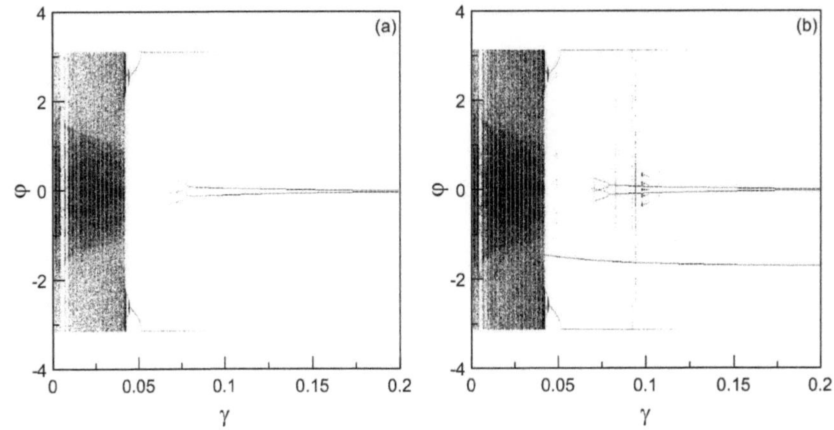

Fig. 62 Bifurcation diagram φ versus parameter γ for **a** one initial condition $\varphi(0) = 0.1$ and **b** ten random initial conditions, $\vartheta = 1.2$

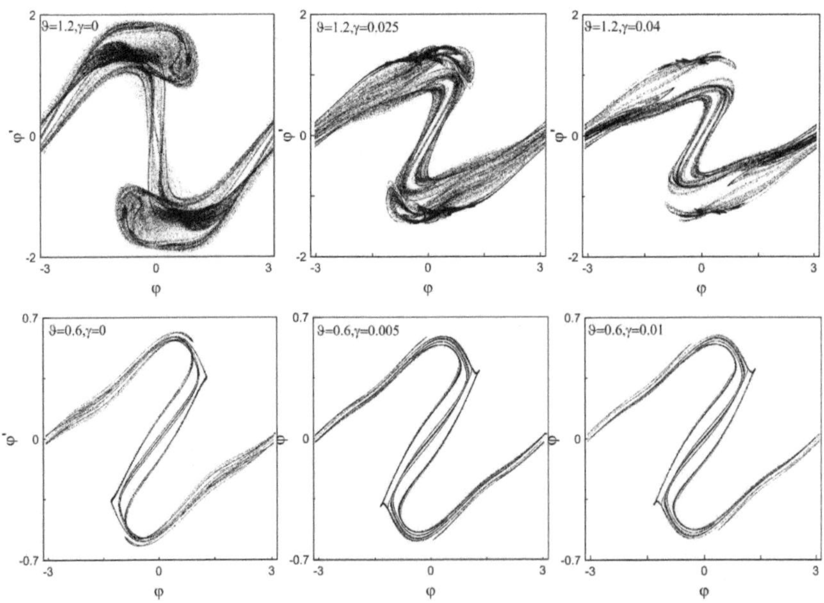

Fig. 63 Evolution of attractor under the influence of nonlinear spring in two selected chaotic regions (*row 1 and 2 respectively*)

It is very interesting that in the analysed range of γ from 0 to 0.2 hard nonlinearity does not eliminate rotation of the pendulum. As shown in the bifurcation diagrams, the response of the autoparametric system with nonlinear spring depends on initial conditions.

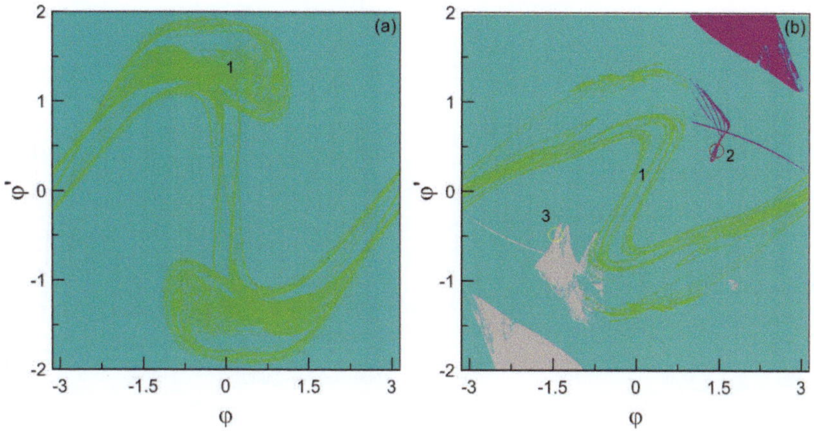

Fig. 64 Basin of attraction for frequency $\vartheta = 1.2$ and nonlinearity **a** $\gamma = 0$ and **b** $\gamma = 0.04$

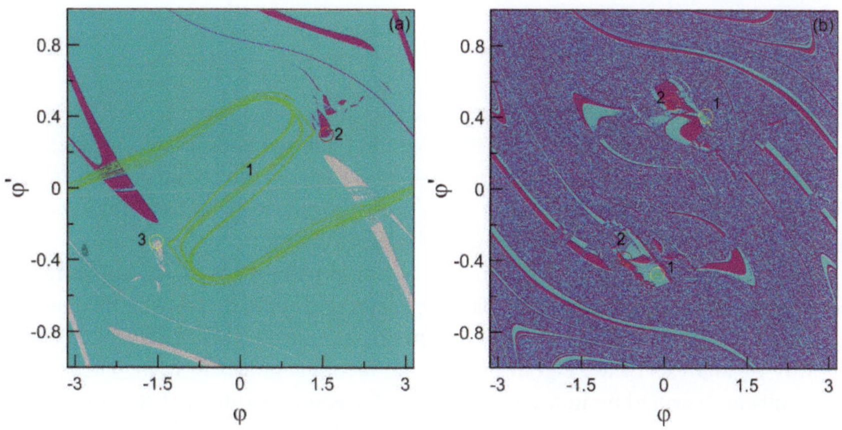

Fig. 65 Basin of attraction for frequency $\vartheta = 0.6$ and nonlinearity **a** $\gamma = 0$ and **b** $\gamma = 0.015$

The increase in nonlinearity reduces the zone of chaotic motion (attractor No.1 in Fig.64 and Fig.65), but it forms the areas of positive and negative rotation of the pendulum (attractor No.2 - positive, and No.3 - negative rotation). A change of the patterns of basins of attraction is also evident in the presented basins of attraction.

It is worth pointing out another interesting phenomenon which may appear if the system works near the resonance region. The pendulum's vibration centre is shifted from the lower neutral position. Depending on the initial conditions the centre of the angle φ may be shifted in the positive (Fig.66a) or negative (Fig.66b) direction. The two possible shifts are symmetric around the lower equilibrium position of the pendulum. The shift of the pendulum's vibration is detected in the real setup, too.

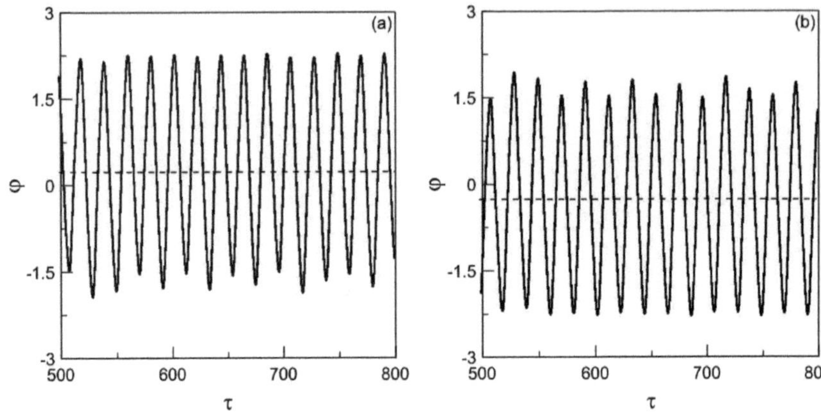

Fig. 66 Time response of the pendulum for frequency $\vartheta = 0.6$ and nonlinearity $\gamma = 0.015$ and the initial condition **a** $\varphi(0) = 0.1$ and **b** $\varphi(0) = -0.1$

8 Conclusions and Remarks

This chapter delivers a vibration analysis of a linear or nonlinear oscillator with an attached pendulum constituting the so-called autoparametric system. Near the parametric resonance zone the pendulum does not stay in the equilibrium point, but oscillates. This phenomenon, under certain conditions, may be used for vibration suppression of the oscillator treated as the main structure.

To determine the influence of the most essential parameters on the autoparametric system response, the harmonic balance method is applied. The obtained analytical solutions allow to find the main parametric resonance curves and to check the stability by analysis of the eigenvalues of the Jacobian matrix system. The analytical results are verified by numerical calculations and experiment. They are in good agreement if vibration amplitudes do not exceed large values. Therefore for larger values of amplitude mainly numerical methods and experiment are used. The considered autoparametric system exhibits, for some sets of parameters, the instability region near the principal parametric resonance. This phenomenon may lead to rotation of the pendulum or its chaotic motion - chaotic swings or irregular motions composed both of swings and rotation. The dynamic response is examined by constructing parameter plots which determine regular or irregular motions. Bifurcation diagrams, Lyapunov exponents, time histories and basins of attractions are used to check the nature of motion in regular and chaotic regions.

To eliminate unwanted vibrations two efficient methods are proposed. The first method is to use a magnetorheological damper as a semi-active device. Activation of MR damping reduces amplitudes in the resonance region. However, the area of effective dynamical vibration absorption, generated by a swinging pendulum, is also reduced. Therefore, a careful selection of the system parameters is required in this approach. The second method is to apply a nonlinear spring in the system suspension.

The nonlinear spring with hard stiffness characteristics shifts the instability region to the right side of the resonance zone. But this new nonlinear structural element involves additional attractors which exist for various initial conditions.

The received results show that semi-active suspension of the autoparametric system, based on MR damper and the nonlinear spring, freely moves the chaotic regions and also reduces or eliminates zones of unwanted pendulum's rotation. This concept allows the system, after proper tuning, to maintain on a regular or chaotic attractor, or if necessary to change one attractor into another. Moreover, by applying simple open-loop control, it is possible to fit on-line the structure response to the frequency and amplitude of external excitation. The rotation, swings and chaotic motion, taken from the experimental system are investigated on the basis of nonlinear time series analysis. The experimental results confirm regular and chaotic nature of system's dynamics.

The next step of the investigations is to use a shape memory alloy (SMA) spring together with a magnetorheological (MR) damper and apply a closed-loop control to get a smart dynamical absorber.

References

1. Acheson, D.J.: A pendulum theorem. Proc. R. Soc. Lond. A (443), 239–245 (1993)
2. Acheson, D.J.: Upside-down pendulums. Nature 336, 215–216 (1993)
3. Anxin, G., Xu, Y.L., Hui, L.: Dynamic performance of cable-stayed bridge tower with multi-stage pendulum mass damper under wind excitations - I: Analytical. Earthquake Engineering and Engineering Vibration 6(3), 295–306 (2007)
4. Areemit, N., Warnttchai, P.: Vibration suppression of a 90-m tall steel stack by using tuned mass damper. In: The Eighth East Asia-Pacific Conference on Structural Engineering and Constrution. December 5-7, Nanyang Technological University, Singapore (2001); paper 1316
5. Bajaj, K., Chang, S.I., Johnson: Amplitude modulated dynamics of a resonantly excited autoparametric two degree-of -freedom system. Nonlinear Dynamics 5, 433–457 (1994)
6. Bishop, S.R., Cliford, M.J.: Rotating periodic orbits of the parametrically excited pendulum. Physical Letter A 201, 191–196 (1995)
7. Cartmell, M.P., Lawson, J.: Performance enhancement of an autoparametric vibration absorber by means of computer control. Journal of Sound and Vibration 177(2), 173–195 (1994)
8. Cartmell, M.P., Roberts, J.W.: Simultaneous combination resonances in an autoparametrically resonant system. Journal of Sound and Vibration 123(1), 81–100 (1988)
9. Eckmann, J.P., Kampshort, S.O., Ruelle, D.: Recurrence plot of dynamical systems. Euro-physics Letters 4, 973–977 (1987)
10. Fraser, A.M., Swinney, H.L.: Independent coordinates for strange attractors from mutual information. Physical Review A 33, 1134–1140 (1986)
11. Hatwal, H., Mallik, A.K., Ghos, A.: Forced nonlinear oscillations of an autoparametric System - Part1: Periodic Responses. Journal of Applied Mechanics, Transactions of the American Society of Mechanical Engineers 50, 657–662 (1983)
12. Hatwal, H., Mallik, A.K., Ghosh, A.: Forced nonlinear oscillations of an autoparametric system. Journal of Applied Mechanics, Transactions of the American Society of Mechanical Engineers, Part 2. Chaotic responses 50, 663–668 (1983)

13. Horton, B.W., Wiercigroch, M.: Effects of heave excitation on rotations of a pendulum for wave energy extraction. In: IUTAM Bookseries, IUTAM Symposium on Fluid-Structure Interaction in Ocean Engineering, vol. 8, pp. 117–128 (2008)
14. Jumani, S.S.: An invertible open-loop nonlinear dynamic temperature dependent MR damper model. Master of Science (2010)
15. Kamath, G.M., Hurt, M.K., Wereley, N.M.: Analysis and testing of Bingham plastic behavior in semi-active electrorheological fluid dampers. Smart Materials and Structures 5(5), 576–590 (1996)
16. Kecik, K.: Regular and chaotic vibrations of a nonlinear mechanical system with a pendulum. PhD Thesis, Lublin University of Technology (2009)
17. Kecik, K., Warminski, J.: Analysis of chaotic and regular motions of an autoparametric system by recurrence plots applications. Vibrations in Physical Systems 24, 221–226 (2010)
18. Kennel, M.B., Brown, R., Abarbanel, H.D.I.: Determining embedding dimension for phase space reconstruction using a geometrical construction. Physics Review A 45, 3403–3411 (1992)
19. Kodba, S., Perc, M., Marhl, M.: Detecting chaos from a time series. European Journal of Physics 26, 205–215 (2005)
20. Kromulski, J., Kazimierczak, J.: Damping of vibrations with using magnetorheological fluid devices. Journal of Research and Applications in Agricultural Engineering 51(3), 47–49 (2006)
21. Lee, W.K., Hsu, C.S.: A global analysis of an harmonically excited spring-pendulum system with internal resonance. Journal of Sound and Vibration 171, 335–359 (1994)
22. Marwan, N., Romano, M.C., Thiel, M., Kurths, J.: Recurrence plots for the analysis of complex systems. Physics Reports 438, 237–329 (2007)
23. Nagasaka, I., Ishida, Y., Ishii, T., Okada, T., Koyoma, T.: Vibration suppresion of helicopter blades by pendulum absorbers. Analytical and experimental investigations in case of rigid -body mode. Transactions of the Japan Society of Mechanical Engineers 73(725), 129–137 (2007)
24. Nayfeh, A.H., Zavodney, L.D.: Experimental observation of amplitude and phase modulated responses of two internally coupled oscillators to a harmonic excitation. Journal of Applied Mechanics 10, 706–710 (1988)
25. Sado, D.: Energy Transfer in Nonlinearly coupled systems with two degrees of freedom. In: Oficyna Wydawnicza Politechniki Warszawskiej, Prace Naukowe, Mechanika, Warszawa, p. 166 (1997)
26. Sanjuan, M.A.F.: The effect of nonlinear damping on the universal escape oscillator. International Journal of Bifurcation and Chaos 9, 735–744 (1999)
27. Sapinski, B., Snanima, J., Maslanka, M., Ros, M.: Factility for testing magnetorheological damping system for cable vibrations. Mechanics 25(3), 135–142 (2006)
28. Song, Y., Sato, H., Iwata, Y., Komatsuzaki, T.: The response of a dynamic vibration absorber system with a parametrically excited pendulum. Journal of Sound and Vibration 259, 747–759 (2003)
29. Spencer, B.F., Sain, M.K.: Controlling Buildings: A new frontier in feedback. Special Issue of the IEEE Control Systems Magazine on Emerging Technology 17(6), 19–35 (1997)
30. Stephenson, A.: On a new type of dynamical stability. Manchester Memoirs 8, 1–10 (1908)

31. Tang, D., Gavin, H., Dwell, E.: Study of airfoil gust response alleviation using on electro-magnetic dry friction damper. Part1: Theory. Journal of Sound and Vibration 269, 853–874
32. Tel, T., Grulz, M.: Chaotic Dynamics. An introduction based on classical mechanics. Cambridge University Press, Cambridge (2006)
33. Thomas, K.I., Ambika, G.: Occurrence of stable modes in a pendulum with cubic damping. Pramana-Journal of Physics 59, 445–456 (2002)
34. Tondl, A., Ruijgork, T., Verhulst, F., Nabergoj, R.: Autoparametric resonance in mechanical system. Cambridge University Press, New York (2000)
35. Vyas, A., Bajaj, K.: Dynamics of autoparametric vibration absorbers using multiple pendulums. Journal od Sound and Vibration 246, 115–135
36. Warminski, J., Kecik, K.: Autoparametric vibration of a nonlinear systems with pendulum. Mathematical Problems in Engineering. Article ID 80705 (2005)
37. Warminski, J., Kecik, K.: Instabilities in the main parametric resonance area of mechanical system with a pendulum. Journal of Sound Vibration 332, 612–628 (2009)
38. Warminski, J., Kecik, K.: Regular and chaotic motions of an autoparametric real pendulum system with the use of a MR damper. In: Modeling, Simulation and Control of Nonlinear Engineering Dynamical Systems, pp. 267–276. Springer, Heidelberg (2009)
39. Wolf, A., Swift, J.B., Swinney, H.L., Vastano, J.A.: Determining Lyapunov exponent from a time series. Physica D 16, 285–317 (1985)
40. Xu, X., Wiercigroch, M., Cartmell, M.P.: Rotating orbits of a parametrically excited pendulum. Chaos, Solitons and Fractals 23(5), 1537–1548 (2005)
41. Xu, X., Wiercigroch, M.: Approximate analytical solutions for oscillatory and rotational motion of a parametric pendulum. Nonlinear Dynamics 47, 311–320 (2006)
42. Yabuno, K., Endo, Y., Aoshima, N.: Stabilization of 1/3-order subharmonic resonance using an autoparametric vibration absorber. Journal of Vibration and Acoustics 121, 309–315 (1999)

On the Dynamics of Pedestrians-Induced Lateral Vibrations of Footbridges

Stefano Lenci and Laura Marcheggiani

Abstract. This chapter is concerned with the problem of the large horizontal oscillations induced on slender footbridges by the motion of pedestrians, a phenomenon which involves the synchronization between the motion of walkers and that of the bridge deck. We initially review the extensive technical and scientific literature, and then we focus on two models to detect numerically and analytically the phenomenon. A continuous-time bridge-pedestrians model initially developed by Strogatz et al. is improved to better understand some aspects of the underlying mechanical phenomena. We perform extensive parametric investigations by means of many numerical simulations. This permits to highlight the parameters which mainly affect the trigger and the development of the phenomenon of synchronous lateral excitations, thus allowing a good understanding of the physical event and an evaluation of the engineering reliability of the model. Then, in order to obtain analytical instead of numerical predictions, a nonlinear discrete-time model based on an appropriate 1D map is considered. It is able to provide a reliable value of the number of pedestrians which trigger the synchronization, thus predicting the onset of instability which is also the onset of crowd synchronization. From a dynamical system point of view, the main result is that the model highlights how the phenomenon can be seen as a perturbation of a classical pitch-fork bifurcation, which is then shown to be the underlying dynamical event.

Keywords: Synchronization-induced lateral vibrations, continuous- and discrete-time models, pedestrians-bridge interactions, Millennium Bridge.

Stefano Lenci · Laura Marcheggiani
Department of Architecture, Buildings and Structures, Polytechnic University of Marche, via Brecce Bianche, Ancona, 60131, Italy
e-mail: {lenci,l.marcheggiani}@univpm.it

J. Warminski et al. (Eds.): Nonlinear Dynamic Phenomena in Mechanics, SMIA 181, pp. 63–114.
springerlink.com

1 Introduction and Literature Review

In the last 10–20 years there has been a growing attention toward slender footbridges by engineers and architects, also a consequence of renowned structures, such as the London Millennium Bridge, built by worldwide reputation designers. Often, footbridges are playing a central role in the urban renewal demanded by modern society. It is clear that the beauty and elegance in the architectural concept must spring from simplicity in the structural design, in order to achieve a sort of natural harmony between the fairness of the structure, the physical configuration of the local landscape and the social attractiveness and usefulness of the footbridge. Because of all these social, aesthetic and technical requirements, the trend in footbridge design is towards an increasing flexibility and lightness.

This tendency is supported by the fact that modern materials, thanks to both their mechanical characteristics and their cost competitiveness, allow to consider a high stress level and thus to conceive slender structures with small cross section and large spans.

As a consequence, modern footbridges have small natural frequencies, and possibly have high sensitivity to dynamic loads induced by pedestrians. Excessive vibrations can be caused by resonance between pedestrian loading and one or more natural frequencies of the structure. The reason is that the range of footbridge natural frequencies, both vertical and lateral, often coincides with the dominant frequencies of the pedestrians-induced load [54]: 1.4–2.4 Hz for pedestrian vertical forcing and 0.7–1.2 Hz for pedestrian lateral forcing. It is obvious that if footbridges are designed for static loads only, they may be susceptible to vertical as well as horizontal vibrations, thus confirming the necessity to pay attention to dynamic aspects with a detailed analysis. As a matter of fact, very complex and partially unexpected dynamical phenomena may, and actually do, occur.

Several footbridges experiencing excessive lateral vibrations due to pedestrians-induced loading have been reported in the last years; the most famous is the London Millennium Bridge (a shallow suspension footbridge, Fig.1a) which experienced, on its opening day, strong horizontal vibrations due to the synchronization of the pedestrians motion with the natural modes of the structure [12]–[16]. Other bridges which suffered a similar problem are the Toda Park Bridge (a cable-stayed footbridge, best known as T-Bridge) in Japan [34], [17], the Maple Valley Bridge (a great suspension footbridge, best known as M-Bridge) in Japan [34], the Solferino Footbridge (a double arc steel structure, Fig.1b) in Paris [11] and the Alexandra Bridge in Ottawa [12]. In all these cases, the natural frequency of the principal lateral mode was mainly excited by pedestrians.

We can observe that the problem of the pedestrians-induced lateral vibrations occurred on a range of different structural types (suspension, cable-stayed and steel girder bridges) as well as on footbridges made of different materials (steel, composite steel-concrete and reinforced and pre-stressed concrete) [54]. It is therefore confirmed that a large enough crowd of pedestrians can induce strong lateral vibrations on a footbridge of any structural form as well as of any material, although this requires the lateral mode to have a low enough natural frequency [12], approximately below 1.3 Hz and close to the range of 0.7–1.2 Hz typical of the pedestrian lateral motion [54].

The theory behind pedestrians-induced lateral vibrations on footbridges is that of synchronous lateral excitation [54], [12], [47]–[14]. People walking in a crowd exhibit a random level of synchrony, and large enough crowds might produce lateral forcing causing a bridge to vibrate laterally. This forcing is the sum of all lateral forces applied to the bridge by the footsteps of each pedestrian. Even if the bridge vibrations are initially very small, pedestrians tend spontaneously to walk in synchrony with the bridge. This ensures an interaction with the bridge for each step of the pedestrians. This instinctive behaviour causes the synchronized pedestrians' footfall forces to be applied at the frequency of the vibration and with a phase such as to increase the motion of the bridge, with each process pumping the other in a sort of *positive* feedback loop [3]; the increase in oscillations amplitude causes pedestrians to increase their lateral footfall forcing and their level of synchrony, by following the movements of the deck in order to balance themselves [12]–[16], [31]: the more the bridge moves, the more the crowd pushes it to move further.

It is recognized from observations that for potentially susceptible spans there exists a critical number of pedestrians that will cause the vibrations to increase suddenly to unacceptable levels. The nature of this problem, in terms of dynamic response of the bridge, is clearly nonlinear (as it is for example confirmed by tests performed on the London Millennium Bridge [12]): the oscillations are small until a critical number of walking pedestrians N_c and then, due to the synchronization, they increase rapidly until a final threshold. This number N_c is of practical engineering interest and its reliable prediction is the final aim of all the theoretical studies.

Several research papers have recently investigated this topic, even if a standard and generally accepted model of pedestrians-induced lateral dynamic loading and of dynamical interaction with the bridge, especially in case of crowding, is missing and

Fig. 1 London Millennium Bridge **a** and Solferino Bridge in Paris **b**: two examples of footbridges undergoing the phenomenon of synchronization induced lateral oscillations

still under research. Zivanovic et al. [54] have performed a comprehensive review of the existing literature on the topic until 2003, while an updated review can be found in [51].

Early studies on pedestrians-induced vibrations of footbridges were carried out at the end of the seventies by Blanchard et al. [6], Matsumoto et al. [29], and Wheeler [53], but they concerned only the measurement and modelling of the vertical component of pedestrian load on a motionless surface. It is worthy to note that Matsumoto and co-workers first attempted to define the vertical dynamic load induced by a crowd and to investigate its stochastic nature, but their approach did not account for interaction with structure and could not describe synchronization.

To model pedestrians-induced lateral vibrations it is necessary to have some understanding of the mechanics which describes human walking, with special attention to the lateral component of the ground reaction force; therefore, some authors (Bauby et al. [4], Belli et al. [5], Vaughan [50], Hof et al. [22], [23], Macdonald [28], Trovato et al. [49]), especially in the area of biomechanical engineering, have recently investigated this topic developing and reorganizing theories on bipedal walking, forcing and frequency of human footfall during walking and balance control. The common observations we can draw from these researches are that: (*i*) locomotion is the translation of the centre of gravity along a pathway requiring the least expenditure of energy; (*ii*) walking biomechanical strategy is to minimize the risk of falling; (*iii*) bipedal locomotion is generated through global real-time entrainment of the neural system on the one hand, and the musculoskeletal system and environment on the other. All these aspects obviously influence the gait pattern and the ground reaction forces in the three directions: vertical, longitudinal horizontal and lateral/transversal horizontal, and are useful to understand what happens when we walk on a laterally moving surface.

Dallard et al. [12]–[16] have conducted a series of controlled crowd tests on the Millennium Bridge and consequently they have proposed a load model based on an empirical observation: during the transient, when bridge wobbling is growing up, the crowd force can be assumed linearly dependent on the lateral oscillation velocity of the deck; this means that pedestrians act like negative damping on the structure. Also a formula has been obtained to evaluate the critical number of pedestrians; it actually depends only on the modal damping of the bridge through a proportionality constant which is strictly related to the specific real case-study (the Millennium Bridge). This phenomenological approach assumes but does not explain the observed synchronization effect, and cannot predict the steady state amplitude for bridge motion, as it is due to not modelled nonlinearities.

Nakamura [33] has proposed an interactive forcing model analogous to the previous one, but which allows the schematization of the self-limiting nature of the synchronization phenomenon and the prediction of the steady state amplitude. Also this model is based on coefficients which have been estimated from experimental tests [34], [17] and cannot easily be generalized to other footbridges.

Newland [36] has approached the problem by referring to the interaction phenomenon between fluid flow and structures which is widely studied in wind engineering (vortex-induced oscillations) and commonly known as lock-in [9], [43].

His model includes the empirical assumption that the 40% of the pedestrians are synchronized with bridge lateral frequency, independently on the amplitude of the oscillations.

Fujino et al. [17] have adopted a model of harmonic forcing by empirically tuning a synchronization parameter for the lateral vibrations of the T-Bridge (according to their experimental data [17]). This model does not predict any sudden transition to a vibrating state of the bridge but assumes a continuous increase in the vibration amplitude as the number of pedestrians increases.

Roberts [38]–[40] has schematized the interaction between the pedestrians and the footbridge assuming that synchronization occurs when the pedestrians motion is larger than the bridge motion; from this critical condition, he obtains a limit number of pedestrians.

Laboratory tests involving a single pedestrian walking on platforms or treadmills forced to move laterally have been carried out by Dallard et al. [12] and McRobie et al. [31] soon after the occurrence of the Millennium Bridge problem.

Ricciardelli and Pizzimenti [41], [42] have recently performed a systematic experimental campaign aimed at characterizing dynamically the lateral force exerted by pedestrians on footbridges, both in case of still deck and in case of laterally moving deck; deterministic and stochastic lateral loading models for the static case have been provided and the bases for more sophisticated dynamic models including crowd-structure interaction have been put. The mechanism of crowd synchronization has been investigated only from the qualitative point of view, deferring its quantitative study and modelling after further future measurements.

Blekherman [7] has explained the excessive lateral vibrations on the Solferino Bridge in Paris on the basis of autoparametric resonance by using a double pendulum model; the process of possible synchronization of pedestrian loading with the relevant vibrational modes, which are nonlinearly coupled in the ratio 2:1 between their frequencies, depends on the achievement of parametric resonance.

Piccardo and Tubino [37] have performed an interesting extensive critical analysis of the excitation mechanisms identified in the literature and they have proposed a new forcing model based on experimental tests carried out on harmonically moving platforms [12], [42]: the force exerted by pedestrians is modelled as harmonic with an amplitude depending on the deck lateral displacement, and a simple criterion defining the limit pedestrian mass is introduced. They mainly ascribe to a mechanism of parametric excitation the lateral sway motion induced by crowds in very flexible, slowly damped footbridges, with a first lateral natural frequency around 0.5 Hz corresponding to a half of the first lateral walking frequency.

Very recently, Venuti et al. [52] have developed a first-order model based on the mass conservation equation, in order to macroscopically describe the dynamics of the crowd in the framework of hydrodynamic modelling: the crowd, considered as pedestrian flow, is assumed to behave like a continuous compressible fluid. The structural system is modelled by means of a generalized single degree of freedom model. The two-way interaction between the crowd and the structure is studied. This model permits to take into account the triggering of the lock-in and its self-limited nature, previously explained only by Strogatz et al. [47]. The effects of two different kinds of synchronization, i.e. between pedestrians and structure and among pedestrians, are introduced; the presence of different frequency components in the overall force exerted by pedestrians is considered. Some parameters, used in the formulation of the model, come from reasonable qualitative considerations about pedestrian behaviour and they would require specific experimental tests to be confirmed.

Bodgi et al. [8] have adopted a similar approach to simulate the mechanics of synchronous lateral excitation induced by pedestrians on footbridges.

Strogatz et al. [47], [1] have been the first, up to our knowledge, to mathematically describe and predict the simultaneous growth of bridge movement and crowd synchronization, an observation that was unexplained in previous models but that is confirmed by analyses of video footages [2] recorded during overcrowding conditions on lively footbridges [12], [17].

This continuous-time bridge-pedestrians model (called SAMEO in the following) is particularly interesting for its contribution to the physical-mathematical explanation of the underlying mechanical event, besides for the reasonable description of the phenomenon itself. Moreover the model is quite simple in its formulation and general in its application possibilities to any bridge at risk of synchronous lateral excitation; obviously it includes some important simplifications which could affect its predictions from a quantitative point of view. For this reason a consistent detailed analysis can be worthwhile in order to evaluate the engineering reliability of the SAMEO model, and is pursued in the first part of this chapter (Sect. 2).

Due to the large number of nonlinear equations, the analysis is numerical, and is performed by means of a self-made code; extensive parametric investigations are performed through wide numerical simulations and some modifications are introduced with the aim of improving the understanding and the description of the main aspects of the underlying mechanical phenomena. In particular the aim is to give a deeper insight into the synchronization phenomena.

To overcome the limits of a completely numerical analysis, which is accurate but not easy enough for 'immediate' understanding of the involved phenomena, the second main section of the chapter (Sect. 3) is dedicated to present and to analyze in detail a nonlinear discrete-time model which allows to obtain analytical instead of numerical predictions, and to give a dynamic interpretation of the synchronization phenomenon. In this case the approach is therefore analytical by using the classical tools of the discrete nonlinear dynamics.

Some concluding remarks (Sect. 4) about both models end this chapter, which is completely dedicated to the synchronous lateral excitation phenomenon in slender monodimensional civil structures; this is an example of both biological and mechanical synchronization which can be cause of instability.

2 A Continuous-Time Model: The SAMEO Model

The SAMEO model [47]–[14] is able to describe the synchronization of pedestrians motion with the lateral vibrations of footbridges, by adapting ideas originally developed to explain the collective synchronization of biological oscillators, such as neurons and fireflies, or other nonlinear systems able to phase-lock to an external periodic drive [44], [45]. The basic idea is just the observation that also human walking is governed by unconscious rhythmic biological signals, so an analogy for example with the rhythmic flashing of fireflies seems to be possible.

In the problem we are studying, the stimulus signal is given by the dynamic active feedback between the two involved systems: the bridge and the crowd.

Here lateral synchronization is assumed to involve only one vibrational mode of the structure; this hypothesis is true in *non-pathological* cases in which internal resonance conditions among different natural modes of vibration of the bridge itself do not occur. This is acceptable, e.g., in the London Millennium Bridge case: the analysis of its vertical forces and lateral oscillations shows no correlation between such quantities [12] and so internal resonance is not necessary to explain its lateral sway motion. Moreover the pedestrian lateral excitation is supposed to have a dominant frequency close to the bridge fundamental frequency, as it happens in most of the real observed cases of wobbling footbridges. As a consequence it is sufficient to perform a mono-modal analysis, by projecting the equation of motion of the footbridge on the relevant modal shape.

A footbridge span of length L, measured along the coordinate y, is modelled as a linear mono-dimensional damped dynamical system. To obtain the reduced order, single degree of freedom (SDOF) model, a modal analysis of the whole 3D structure is performed in order to identify the eigenfunction $\varphi(y)$ involving lateral displacements and corresponding to a natural frequency f_0 close to the range 0.7–1.2 Hz typical of the pedestrian lateral excitation. Hence, the dynamics is projected on $\varphi(y)$ and the equation of motion along the selected lateral mode (usually the first) is obtained (Fig.2):

$$M\ddot{X}(t) + B\dot{X}(t) + KX(t) = F_{ped}(t), \tag{1}$$

$$F_{ped}(t) = \int_0^L F_p(y,t)\,\varphi(y)\,dy \approx G\sum_{i=1}^{N} \sin\Theta_i(t). \tag{2}$$

The overdots denote differentiation with respect to time t. $X(t)$ is the generalized displacement (amplitude) of the relevant lateral mode and M, B, K are the modal mass, damping and stiffness, respectively. $F_{ped}(t)$ is the lateral modal force exerted on the bridge by the pedestrians, projection of the forces on the relevant modal shape, being $F_p(y,t)$ the crowd-induced force per unit length. G is its amplitude and $\Theta_i(t)$ is the phase in the walking cycle for each of the N pedestrians. It is assumed that $\Theta_i = 0$ when the pedestrian's left foot first touches the ground, and $\Theta_i = \pi$ when the right foot is on the ground, interpolating for phases between these events (Fig. 3). According to (2) the pedestrian load is approximated as sinusoidal with respect to the pedestrians phases: it can be thought as the first term in the Fourier series of the (obviously periodic) load function [48].

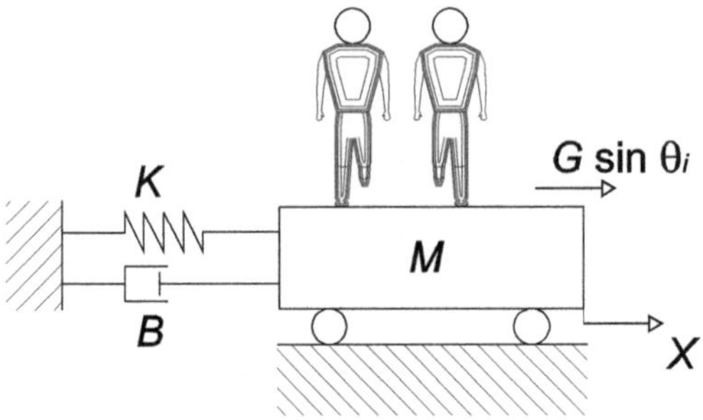

Fig. 2 Mechanical model of the single mode dynamics of the bridge

The dynamical pedestrians-bridge interaction is introduced by assuming that

$$\dot{\Theta}_i(t) = \Omega_i + C_i A(t)\sin(\Psi(t) - \Theta_i(t) + \alpha)\,. \tag{3}$$

Therefore, the pedestrians are considered as limit-cycle phase oscillators with a random distribution of native frequencies Ω_i. C_i measures the pedestrians' sensitivity to bridge lateral vibrations and can be determined experimentally, α is a constant phase lag. It is assumed $\alpha = \pi/2$ in order to match the worst resonance condition in which the instantaneous lateral excitation frequency is approximately equal to the natural frequency of the bridge relevant lateral mode, $\dot{\Theta}_i/\Omega_0 \cong 1$, $\Omega_0 = 2\pi f_0 = \sqrt{K/M}$ being the bridge natural (circular) frequency. $A(t)$ and $\Psi(t)$ are the bridge vibrations amplitude and phase (Fig. 3), which are defined by

$$X = A\sin\Psi,\ \dot{X} = \Omega_0 A\cos\Psi \ \rightarrow\ A = \sqrt{X^2 + \frac{\dot{X}^2}{\Omega_0^2}}\,. \tag{4}$$

The term $f = C_i A\sin(\Psi - \Theta_i + \alpha)$, added to Ω_i in (3), is chosen to be a function of the bridge motion amplitude, of the bridge phase and of the walker phase through a constant of proportionality. It is therefore evident that f has the effect of shifting walkers to a phase closer to that of the bridge, thus modelling the active dynamical bridge-pedestrians interaction and describing the natural tendency of the systems to synchronize. In fact, when the phase difference $(\Psi - \Theta_i + \alpha)$ is positive, i.e. the stimulus is ahead in the cycle (Θ_i lags $\Psi + \alpha$), f is globally positive and the pedestrian speeds up in an attempt to synchronize with the bridge. Conversely, when the phase difference $(\Psi - \Theta_i + \alpha)$ is negative, i.e. the stimulus is behind in the cycle (Θ_i leads $\Psi + \alpha$), f is globally negative and the pedestrian slows down his walking frequency in order to lock to the bridge.

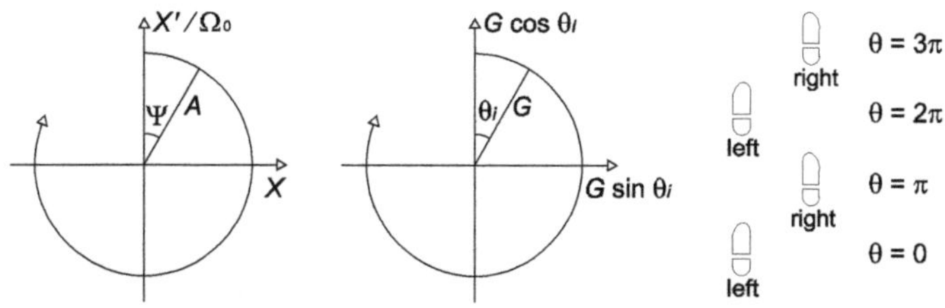

Fig. 3 Definitions of X, A, Ψ and Θ_i

The constant of proportionality, which appears in f, is made of two terms which take into account two different aspects: the effect of the oscillations amplitude and the intrinsic capability of pedestrians to be affected by that amplitude. As A increases, its influence on the pedestrian becomes stronger, according to the linear relationship between f and A. The parameter C_i, on the other hand, quantifies the effect on the pedestrian of the stimulus of amplitude A and phase Ψ , and in this sense it acts like sensitivity to bridge motion.

In the absence of bridge-pedestrians interaction ($C_i = 0$) we have that $\dot{\Theta}_i = \Omega_i$ is the governing equation for the walker dynamics [46], so that each pedestrian walks unconditionally at his own natural constant frequency Ω_i, which has a certain statistical distribution across the population. Then, $\sum_i \sin\Theta_i = \sum_i \sin(\Omega_i t)$ is a distribution with zero mean value, and the bridge is practically still.

In general, from a biological point of view, it is realistic to consider a variation of sensitivity among individuals in the population, and therefore a random distribution of values C_i depending on a person's age, size, health and so on; for sake of simplicity, lacking a specific study in this direction in the literature, in the following it will be used a single constant value for all walkers: $C_i = C$.

Finally it is worthy to note that the model is able to describe the real scenario in which the number of people walking on the bridge deck varies with time: as a new pedestrian enters the bridge, a new equation is added to the system; therefore the number of equations in the model varies with time depending on the number of pedestrians who are entering or leaving the bridge, so that the model can be classified as a time varying system.

2.1 Parametric Investigations: Model Implementation and Computational Aspects

The SAMEO model is governed by highly nonlinear ordinary differential equations; therefore we have to solve them numerically. We use a self-made code which joins

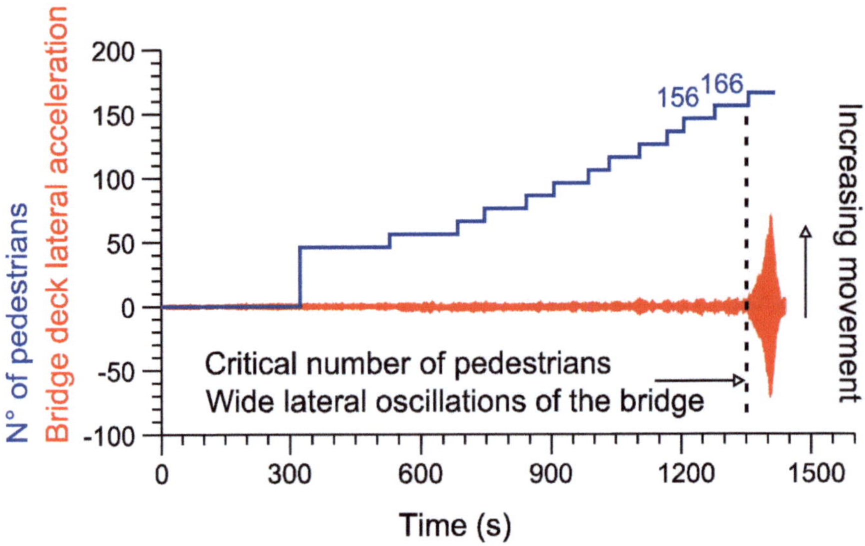

Fig. 4 Results of *in situ* tests on the north span of the Millennium Bridge (Arup figure from [1], [35]): time histories of the number of walkers (staircase-like trace) and of the bridge deck lateral acceleration

traditional numerical methods (the classical routines of Matlab© are used to numerically integrate the system of ODEs (1), (2) and (3) governing the phenomenon) with self-developed algorithms aimed at capturing the main dynamical aspects. Despite the generality of the model, we need data from a real case study and therefore we refer to the Millennium Bridge, as it is the most famous case of wobbling footbridge and also one of the most well-documented and studied in the literature. In fact, in 2000, during its temporary closure, researches were undertaken both through laboratory tests on moving platforms (Imperial College and University of Southampton) and *in situ* tests on the bridge itself (Arup), [12]–[16]. In particular, the latter allowed determining the critical number of pedestrians necessary to destabilize a given span of the bridge, which was found to be about 160 on the north span (Fig. 4).

Because the majority of published experimental data pertains to the fundamental lateral mode of the north span, those numerical values are used as benchmark and reference for our analysis: $M = 113000\,\text{kg}$, $B = 11000\,\text{kg/s}$, $K = 4730000\,\text{kg/s}^2$, which imply $\xi = B/2\sqrt{MK} = 0.0075$ and $\Omega_0 = \sqrt{K/M} = 6.47\,\text{rad/s}$, in turn corresponding to 1.03 Hz. Furthermore, according to [47], we assume $G = 30\,\text{N}$, $C_i = C = 16\,\text{m}^{-1}\,\text{s}^{-1}$ and, as said, $\alpha = \pi/2$.

We adopt for the native frequencies Ω_i a Gaussian distribution (Fig. 5a) with mean value 1.03 Hz (6.47 rad/s) and standard deviation 0.1 Hz (0.63 rad/s) [54]. The initial phases (Fig. 5b) are supposed uniformly distributed in the interval $[0, 2\pi]$. We choose these phases randomly with the idea to reproduce the different moments when each new individual enters the footbridge; these random constants are used

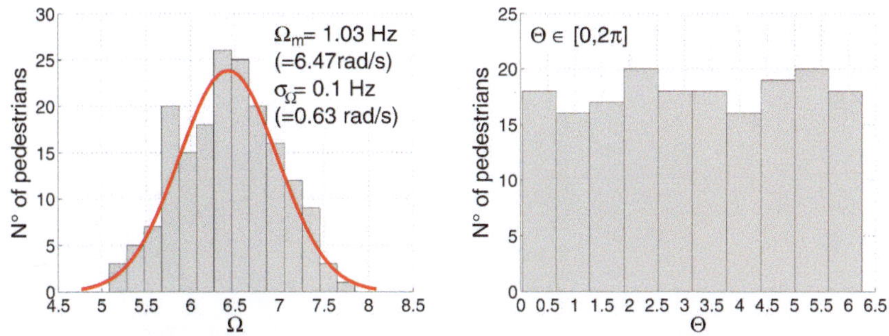

Fig. 5 Distribution of **a** native frequencies Ω_i and **b** initial phases Θ_i for a generic group of 180 pedestrians considered in our simulations

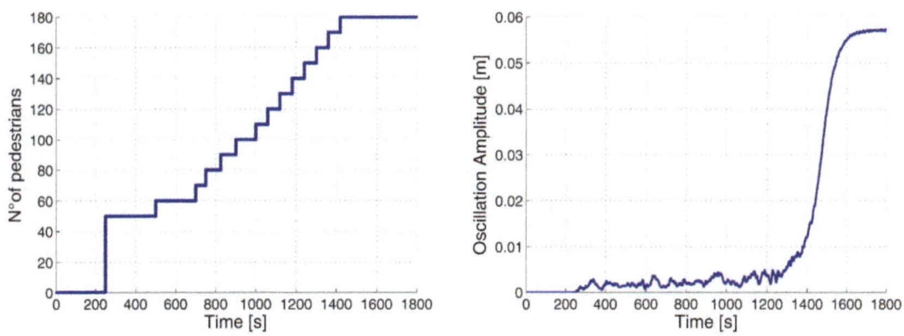

Fig. 6 Number of walkers on the bridge **a** and amplitude of vibration **b** versus time

exclusively as initial conditions for numerically integrate the system of ordinary nonlinear differential equations in the Monte Carlo simulations.

The initial position and velocity of the bridge are zero. Fig. 6 shows the pedestrians ramp and a typical plot of the oscillation amplitude: we consider a staircase loading path in order to reproduce Arup test conditions on the Millennium Bridge (see the upper staircase-like trace in Fig. 4) and the relevant results (see the lower trace in Fig. 4).

For small crowds, the oscillation amplitude (Fig. 6b) as well as the bridge deck lateral acceleration (Fig. 4) is near zero, with a fluctuating trend, as walkers are still desynchronized and randomly phased. Then, as more and more people walk on the deck, there is no hint of instability until the crowd reaches a critical size N_c, after that wobbling and synchrony suddenly emerge simultaneously, as dual aspects of a single instability mechanism.

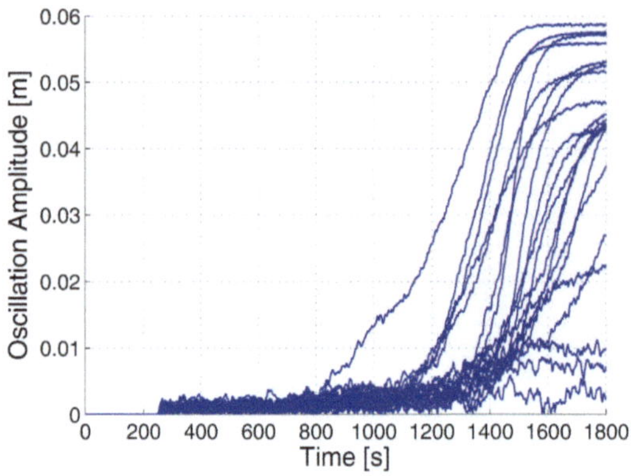

Fig. 7 Amplitude of vibration versus time for a first group of 20 simulations

The practical estimation of N_c from our simulations is made by an *ad hoc* numerical procedure based on the identification of the onset time for the exponential growth of the instability t_c (see the following subsection 2.1.1): from the generic amplitude versus time curve (Fig. 6b), we determine this critical threshold t_c, then with this value we enter the staircase loading path graph and we read the corresponding critical number of pedestrians N_c triggering the synchronization.

For the reference case we perform a Monte Carlo analysis with 200 simulations; the results, in terms of crowd's critical size N_c, are randomly distributed: we evaluate its mean value $N_a = 155$ and its standard deviation $\sigma_N = 27.7$. Our critical number is practically coincident with the Arup's results, and the predicted final amplitude of the bridge motion is very close to the observed values of about 5–7 cm on the opening day. For a reason of graphic readability we plot in Fig. 7 only a sample of 20 simulations, which is in any case visually representative of the global outcome of the whole group of 200.

From Fig. 7 it clearly emerges that, despite the loading path is the same for all the simulations, the response in terms of amplitude changes significantly: even if in most simulations pedestrians synchronize, there are also some cases in which they synchronize a little and others (only 3 in the sample of 20 simulations reported) in which pedestrians do not synchronize at all in the considered time interval. The variations from one run to the next depend on the initial values of Θ_i and on Ω_i, randomly assigned (Fig. 5) to pedestrians at each simulation; the consequence is a dispersion of the critical values N_c, as confirmed by the quite large standard deviation ($\sigma_N = 27.7$).

The results of Fig. 7 constitute the reference case with respect to which we develop all the subsequent analysis.

In the following we perform 200 simulations for each tested case, in order to obtain statistically reliable considerations, even if, as before, for readability reasons we always plot only a sample of 20. For each group of simulations, the average N_a and the standard deviation σ_N of the critical numbers N_c are computed: they summarize the outcome of the performed investigations.

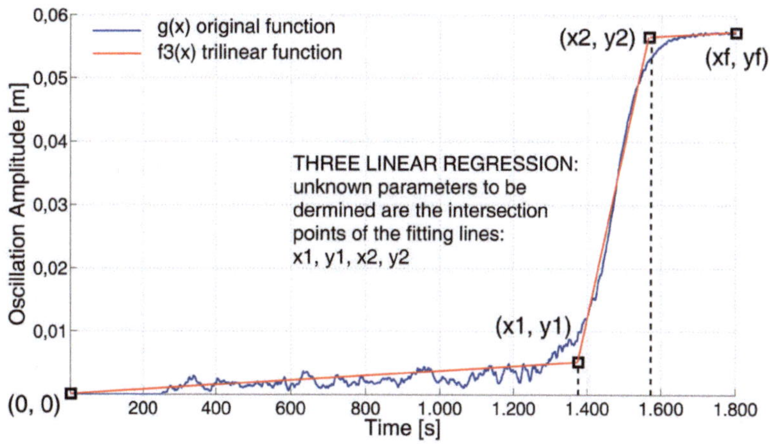

Fig. 8 Multi-phase linear regression: data points are divided into 3 segments on the x axis and then a different straight line is fitted to each segment; the intersection points are not known *a priori*

In order to perform an extensive parametric analysis, which is one of the aims of this section, it is necessary to automate as much as possible the computation of N_a and σ_N. This is done in the next subsection.

2.1.1 Automatic Detection of N_c

To automatically detect the time t_c for the onset of the instability, and consequently the critical number N_c of pedestrians triggering the unwanted dynamical phenomena, we develop a numerical technique to fit the oscillations amplitude versus time curve, $g(x)$, by a piecewise linear averaging curve. Actually, we use three straight segments (Fig. 8), which are sufficient for our purposes. In fact, by observing the SAMEO model results (Fig. 6 and Fig. 7), we can clearly identify three different ranges, in each of them the data points having an approximately linear trend. This three-linear regression is not so trivial, because it is, in fact, a nonlinear regression problem due to the two unknown intersection points of the fitting segments; the problem is continuous but not differentiable at those points and this messes up the local linearization approach often used for weakly nonlinear problems. In particular, here we are interested in finding the intersection point between the first and the second segment, as it corresponds to a reliable and automatic estimate of t_c and therefore, from the load path, of N_c.

The goal is to approximate our real rippled function $g(x)$ with a *tri-linear* function $f^{(3)}(x)$ defined as follows (Fig. 8):

$$f^{(3)}(x)=\begin{cases}\dfrac{y_1}{x_1}x, & \text{for } x<x_1,\\[2mm] y_1+\dfrac{y_2-y_1}{x_2-x_1}(x-x_1), & \text{for } x_1<x<x_2,\\[2mm] y_2+\dfrac{y_f-y_2}{x_f-x_2}(x-x_2), & \text{for } x_2<x.\end{cases} \tag{5}$$

We fix the first and the last point, respectively $(0,0)$ and (x_f,y_f), as those of $g(x)$, so in Eq.(5) the unknown parameters are the four coordinates of the two intersection points of the fitting lines: x_1, y_1, x_2, y_2. We determine them by minimizing the total quadratic error between the original curve and our piecewise linear approximation:

$$\varepsilon(x_1,y_1,x_2,y_2)=\int_0^{x_f}\left(g(x)-f^{(3)}(x)\right)^2dx \propto \sum_{i=1}^{M}\left(g\left(x^{(i)}\right)-f^{(3)}\left(x^{(i)};x_1,y_1,x_2,y_2\right)\right)^2. \tag{6}$$

We underline that in the previous Eq. (6) we pass from the rigorous integral definition of total quadratic error to a discrete formulation, by simply integrating with the trapeziums rule: M is the number of points, equally spaced in time, in which we discretize the time-history. In the right hand side of (6) we omit to multiply for the constant integration interval Δx because, when we minimize, it does not affect the minimum *point* we are looking for.

To minimize the function ε with respect to the parameters x_1, y_1, x_2 and y_2 we use the simplex method of Nelder and Mead [26], [18], which is a direct method that does not use numerical or analytic gradients. From any 'initial' guess of x_1, y_1, x_2 and y_2 the algorithm runs and provides a 'minimum' of the function. In our case the parameters to initialize are only x_1 and x_2 since for sake of simplicity we choose as initial guess $y_1 = g(x_1)$ and $y_2 = g(x_2)$. For particular situations the algorithm can fail to converge or converge to a *local* minimum. When the *global* minimum is not achieved, we solve the impasse by automatically (and randomly) changing the initial guess. To improve the reliability of the results, we consider in any case different initial guesses even when the solution does not show drawbacks.

This procedure, despite the thousands of function evaluations to determine the optimum, is computationally efficient, and it takes only few seconds to give the result. This aspect can be further improved if we filter the input data with a lowpass filter and then resample the resulting smoothed signal at a lower rate; with this trick we are also able to reduce the problem connected with multiple local minima. In any case, for each group of simulations, a global visual supervision of the plotted results is required in order to be sure of the correct prediction of N_c. In this sense our code does not permit completely automated results.

Despite this limit, the method we use is able to detect in a sufficiently automatic manner the number of pedestrians which trigger the synchronization, and it allows us to perform a wide set of simulations with an acceptable CPU time, thus deriving statistically reliable considerations.

2.2 Numerical Simulations

We perform extensive parametric investigations with the following practical motivations:

1. to test the robustness of the model;
2. to improve the understanding of some aspects of the underlying mechanical phenomena (e.g. to give a deeper insight into the synchronization phenomena);
3. to increase the agreement of the model results to the effective behaviour of the walkers, as observed in real conditions;
4. to highlight the role of the main parameters involved on the system dynamics.

We choose as parameters to be varied those which mainly affect the model behaviour and which are sensitive with respect to the real situation to be modelled:

1. the initial phases and native frequencies for the walkers;
2. the constant phase-lag parameter α;
3. the amplitude of the pedestrian lateral forcing during walking;
4. the shape of the pedestrian loading wave;
5. the pedestrians' sensitivity to the bridge motion, C;
6. the coherence of the model results with respect to an 'inverse' approach analysis;
7. the synchronization between pedestrians, D (to be introduced later);
8. the interaction between the two types of synchronization, through the tuning of the respective parametric indicators C and D;
9. the effects of different loading paths, i.e. different modality and number of pedestrians introduced on the bridge deck per unit time.

We remark that we also introduce and test some modifications to the original SAMEO model, maintaining unaltered its essence and plainness of description of the physical event but improving its effectiveness. Our enhancements concern:

1. the addition of a further level of synchronization, between pedestrians;
2. a different, and more conservative, relationship between the amplitude of the pedestrian lateral forcing and the amplitude of the bridge lateral vibrations, which we determine according to experimental tests [12];
3. a more realistic square-type shape of the pedestrian loads on the bridge, according to experimental evidences on treadmill [5].

2.2.1 Influence of Pedestrians Initial Phases and Native Frequencies

What happens if the pedestrians enter the bridge with some initial level of synchrony, as it may occur in overcrowding conditions, or if they are a typologically homogeneous group with native frequencies slightly spread around the mean value?

In Fig. 9 we analyze the influence of the initial conditions of the walkers on the temporal course of the phenomenon: we assign random phases Θ_i uniformly distributed in an interval $[0,1]$ instead of $[0,2\pi]$ (see for comparison Fig. 5b), and

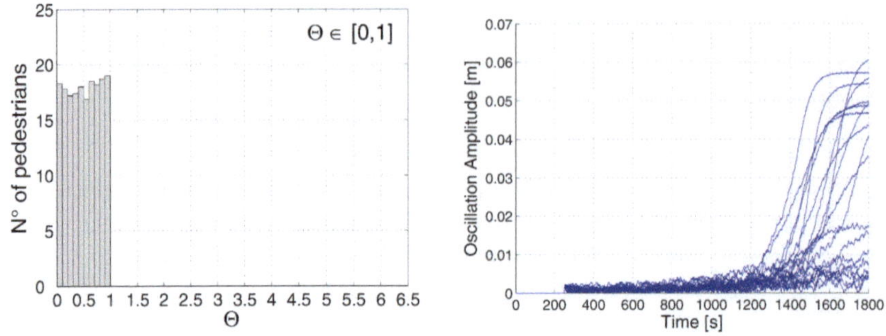

Fig. 9 Effect of the initial level of synchrony among pedestrians: **a** phases distribution and **b** amplitude of vibration versus time curves

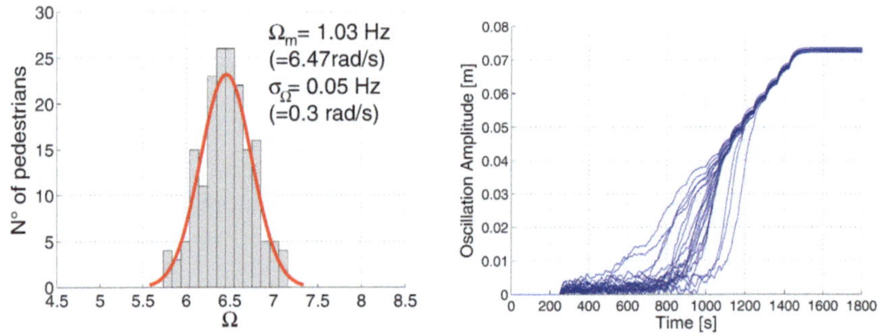

Fig. 10 Effect of the typological homogeneity level of the pedestrians: **a** native frequencies distribution and **b** amplitude of vibration versus time curves

we keep all the other parameters unchanged with respect to the reference case. This entails to admit an initial synchrony among pedestrians.

With these initial conditions, we obtain wobbling for an almost identical critical value, $N_a = 160$, which is affected by a similar dispersion, $\sigma_N = 24.32$. On the other hand, if we compare the amplitude of vibration versus time curves reported in Fig. 9b with those of the reference case (Fig. 7), we observe that substantial differences do not exist. Here pedestrians enter the bridge next to the synchrony, so their initial behaviour is less random than the previous one (Fig. 5b); but the point is that they are synchronized among them, not with the bridge. Therefore, as they have different native frequencies, after one step they will not be anymore synchronized, and the phenomenon will proceed almost as in the reference case.

In Fig. 10 we analyze the influence of the intrinsic properties of the pedestrians: we assign random native frequencies according to a Gaussian distribution with the same mean value (6.47 rad/s), but with a smaller standard deviation ($\sigma_\Omega = 0.3$ rad/s instead of 0.63 rad/s) and we keep all the other parameters unchanged with respect to the reference case.

In general walkers tend to have a natural frequency of pacing that varies depending on the height and weight of the individual, on his/her age, on the travel purpose, on physical, social and psychological factors. Here we suppose that an almost homogeneous group of pedestrians enter the bridge; this means they have natural frequencies near the mean value for the population (1.03 Hz), which is near the bridge resonant frequency. It is obvious that, in such situation, the synchrony with the bridge lateral movements is fostered and sped up. In fact, as expected, we obtain wobbling for a quite smaller critical value, $N_a = 84$, which is affected by a smaller dispersion, $\sigma_N = 15.69$. By observing the amplitude of vibration versus time curves (Fig. 10b), we highlight that, unlike the other cases (Fig. 7 and Fig. 9), in all the simulations pedestrians synchronize in a quite restricted time scale, inside the considered time interval.

Since the reduction from $N_a = 160$ (Fig. 9b) to $N_a = 84$ (Fig. 10b) is very marked, we conclude that the typological homogeneity degree of the crowd is a determining factor both for the trigger point of instability and for the temporal probability of the event to occur, while the initial level of synchrony among pedestrians does not affect the onset of the phenomenon.

2.2.2 The Constant Phase-Lag Parameter α

The constant phase-lag parameter α is determined by a pedestrian's desired phase relationship with the moving surface. It is recognized that on a laterally moving surface, the generic walking pedestrian tends to counterbalance the oscillations with a snaking gait: this instinctive behaviour fosters the bridge lateral vibrations.

If we observe the typical variation of the dynamic amplification factor (DAF) and of the phase angle (ϕ) as a function of the frequency ratio $\beta = \dot{\Theta}_i/\Omega_0$ (see Fig. 11), we note that $\alpha = \pi/2$ gives the worst-case scenario in which the bridge is maximally destabilized because pedestrians drive it most 'efficiently', so that the resulting prediction of the critical number of walkers is conservative. On the contrary $\alpha = 0$ corresponds practically to the 'static' case ($DAF = 1$) in which the synchronized pedestrians are not effective in applying their force and in amplifying bridge oscillations. Therefore $\alpha = \pi/2$ is the correct value for human response to lateral vibrations.

To confirm the previous interpretation we have reported in Fig. 12 the results obtained for $\alpha = 0$. We can observe that the bridge lateral vibrations remain in the field of the little oscillations (in the order of mm): pedestrians are not able to trigger wide lateral oscillations of the bridge, not even if we consider a longer time of observation under the same maximum final value of pedestrian loading (Fig. 13).

2.2.3 Pedestrian Forcing Amplitude

In the previous simulations we have assumed G to be a constant independent of A, i.e. independent of how much the bridge is wobbling. Its value (30 N) corresponds to the mean amplitude of the lateral force by an average pedestrian during normal walking,

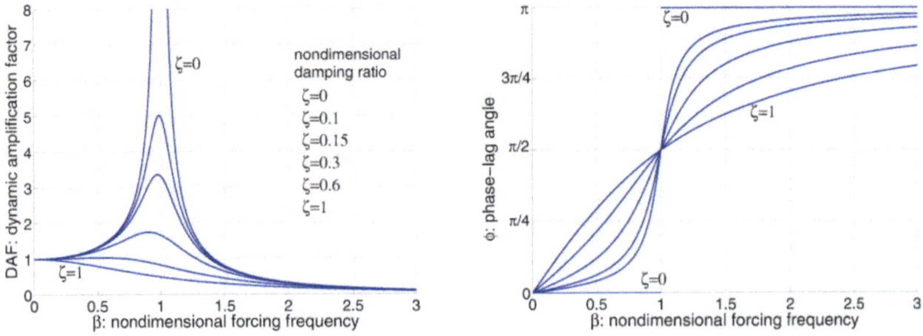

Fig. 11 Variation of dynamic amplification factor ($DAF = 1/\sqrt{(1-\beta^2)+4\zeta^2\beta^2}$) **a** and of phase angle ($\phi = \arctan(2\zeta\beta/(1-\beta^2))$) **b** with damping ($\zeta$=0, 0.1, 0.15, 0.3, 0.6, 1) and frequency (β), for a single d.o.f. damped oscillator harmonically excited

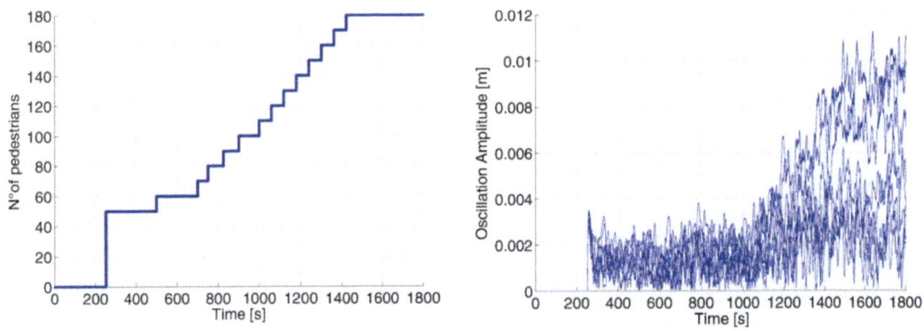

Fig. 12 Number of walkers on the bridge **a** and amplitude of vibration **b** versus time for $\alpha = 0$

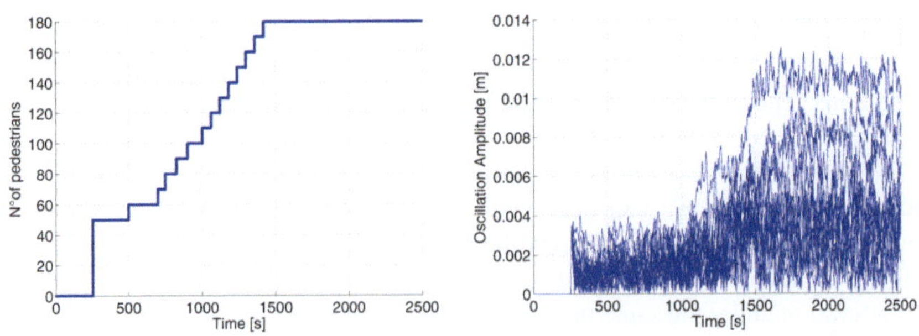

Fig. 13 Number of walkers on the bridge **a** and amplitude of vibration **b** versus time for $\alpha = 0$; a longer time of observation under the same maximum final number of pedestrians is considered here

Fig. 14 Lateral forcing amplitude versus bridge vibration amplitude (gait function): experimental data by Arup obtained at the London Imperial College tests (dots), Abrams' model (7) and our model (8)

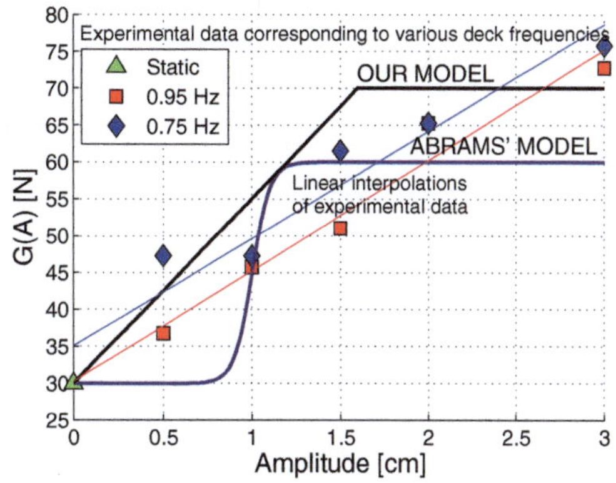

as experiments on a treadmill confirm [5]. It is known that the magnitude of this force can increase when the pedestrian is on a laterally moving surface, because he widens his stance and adopts a different gait in order to balance himself [12], [31], [4].

We consider this effect of changing gait by assuming a model of pedestrian force proposed by Abrams [1] and motivated by the experiments of McRobie et al. [31] (Fig. 14):

$$G(A) = (1/2)\left(G_{low} + G_{high}\right) + (1/2)(G_{low} - G_{high})\tanh\left[C_2\,(A - C_1)\right]. \quad (7)$$

Here $G_{low} = 30\,\mathrm{N}$ and $G_{high} = 60\,\mathrm{N}$ are the minimum and the maximum forcing amplitude; $C_1 = 1$ cm and $C_2 = 10\,\mathrm{cm}^{-1}$ are the amplitude at which the force increases, and the rate at which the force increases with the oscillation amplitude, respectively [1].

Numerical results show that in this case the amplitude curves lift off for a number of pedestrians, $N_a = 150$, slightly smaller than that corresponding to constant G, but the standard deviation is strongly reduced, $\sigma_N = 14.65$. Moreover, it is interesting to note that the maximum final value of A is higher (14 cm against about 7 cm in the reference case), according to the fact that the model is linear in the mechanical part. We can then affirm that doubling G when oscillations reach only 1.3 cm, as it happens with the expression (7), see Fig. 14, has the effect of doubling the bridge final oscillation amplitude and of reducing the dispersion, while it does not affect significantly the instability critical threshold (Fig. 15a).

We also implement a different and more conservative bilinear function $G(A)$:

$$G(A) = \begin{cases} 30 + 2500A, & \text{for } A \leq 0.016\,\mathrm{m}, \\ 70, & \text{for } A > 0.016\,\mathrm{m}. \end{cases} \quad (8)$$

This expression is chosen in order to overlay, for sake of security, the greatest possible number of laboratory tests' data [12], as shown in Fig. 14. The simulations provide a fairly lower critical number of pedestrians, $N_a = 130$, and a larger standard

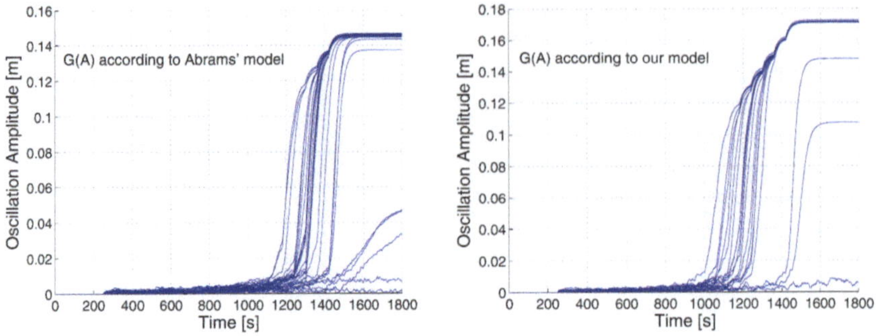

Fig. 15 Amplitude of vibration versus time: simulations performed with $G(A)$ according to **a** Abrams' model and **b** our model

deviation, $\sigma_N = 24.28$, according to the fact that the force applied to the bridge is higher with respect to Abrams' gait function (7). The same considerations of Eq. (7) apply to the maximum final value of A (Fig. 15b).

Both the models assume $G(A)$ increasing, more or less rapidly, until a certain saturation threshold and then constant (Fig. 14); this trend is motivated by the observation that both the magnitude of the pedestrian lateral forcing and the phenomenon itself are naturally self-limiting: in fact a human sensitivity limit to lateral vibrations exists, after which pedestrians begin to have difficulty in walking and finally stop.

From Fig. 15 we also observe that incorporating in the model the dependence $G(A)$, with both the Eqs. (7) and (8), speeds up the time scale for the growth of bridge oscillations towards the steady state, after the critical crowd size has been exceeded.

2.2.4 Pedestrian Forcing Shape

In the original SAMEO model the pedestrian force is idealized as sinusoidal (2), even though experiments on treadmill reveal a periodic trend more similar to a square wave [5] (Fig. 16):

$$F_{ped,i} = G \, \mathrm{sign}\left(\sin \Theta_i\right) . \tag{9}$$

Our simulations show that in the case of square-type force (9), the phenomenon triggers for a number of pedestrians $N_a = 125$, smaller than that of the sinusoidal wave case, while the standard deviation practically does not vary, $\sigma_N = 24.55$. The maximum final value of A is fairly higher: about 8 cm (Fig. 17).

This behaviour was expected as for the square wave a higher force is applied to the bridge (on the average the force is 1.41 times larger, since $\int_0^{2\pi} [\sin(\Theta_i)]^2 d\Theta_i = \pi$ while $\int_0^{2\pi} [\mathrm{sign}(\sin(\Theta_i))]^2 d\Theta_i = 2\pi$), thus strengthening the 'positive' feedback loop between synchrony and wobbling.

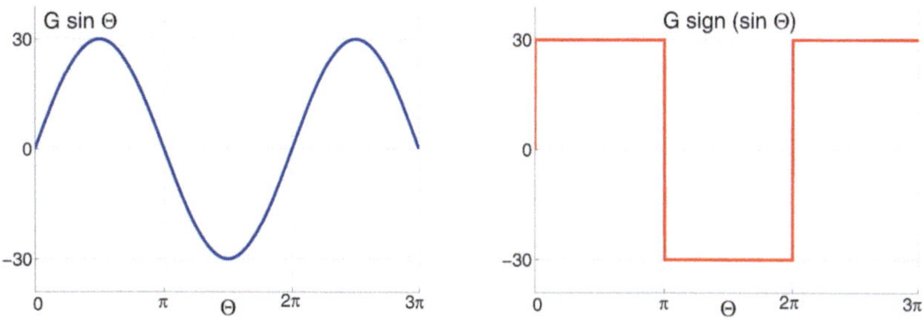

Fig. 16 Pedestrian lateral forcing: **a** sinusoidal and **b** square-wave

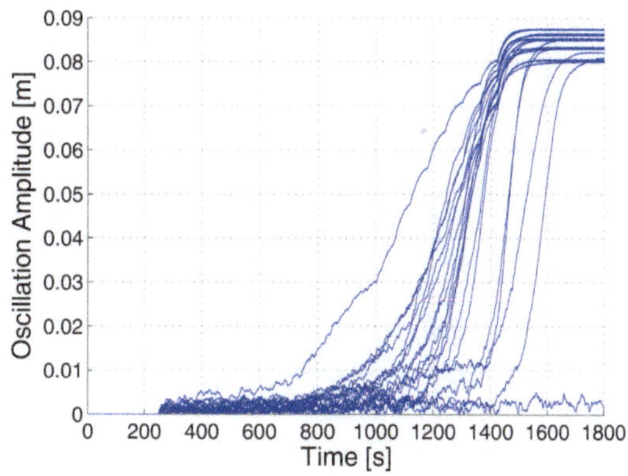

Fig. 17 Amplitude curves obtained in the case of square-type force

We can then conclude that the temporal shape of the pedestrian force plays a more important role than the variation of G with A, being the decrement in the critical crowd size more marked. Therefore, the square wave force shape is both more realistic and more safe.

2.2.5 Pedestrian Sensitivity to Bridge Motion

The parameter C controls how fast a pedestrian, unconsciously, shifts the phase of his walking cycle in response to the sideways oscillations of the platform on which he is walking. There are both physical (related to its meaning of pedestrian sensitivity) and mathematical reasons (related to its definition in (3)) to assume that C is positive; in fact, since A is positive, the sine function in (3) is already able to alter the instantaneous pedestrian frequency in the desired way, i.e. slowing down the pedestrian if he is walking too early with respect to the bridge or speeding up him otherwise.

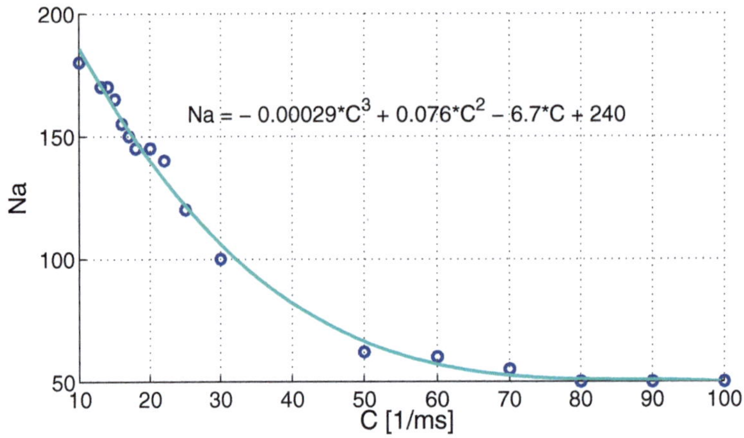

Fig. 18 Average critical number of pedestrians versus C: data points and fitted cubic curve

Apart from the previous one, we have no other overall information on C. In fact, being the parameter which, in some sense, links the mechanical behavior of the bridge, governed by the laws of the physics, with the human behavior of pedestrians, which is not subjected to well-known and mathematically established governing law, C is the most delicate parameter of the model, and actually the most difficult to be determined; practically, it cannot be determined directly, but only indirectly by comparing experimental and model outcomes. For this reason it is very important to perform a parametric analysis with respect to C or, more precisely, to determine how N_a depends on C, beyond the value $C = 16\,\mathrm{m}^{-1}\mathrm{s}^{-1}$ suggested by [47].

In Fig. 18 we report the average critical number of pedestrians corresponding to values of C in the range $10 - 100\,\mathrm{m}^{-1}\mathrm{s}^{-1}$: each point in the graph is the result, as usual, of 200 simulations performed considering Arup loading path. The best (in the least squares sense) cubic curve fitting the data is:

$$N_a = -2.9 \times 10^{-4} \times C^3 + 7.6 \times 10^{-2} \times C^2 - 6.7 \times C + 240\,. \tag{10}$$

Values of $C < 10\,\mathrm{m}^{-1}\mathrm{s}^{-1}$ are investigated but they are not plotted in Fig. 18, because in these cases the pedestrians are never able to synchronize and only small oscillations are observed. In Fig. 19a we show a sample of 20 of the 200 simulations performed with $C = 10\,\mathrm{m}^{-1}\mathrm{s}^{-1}$: in only one case the system seems to be able to synchronize in the considered time interval. In general only carrying forward the simulations for unrealistically high values of time and number of pedestrians, synchronization is a possible outcome. Thus, $C = 10\,\mathrm{m}^{-1}\mathrm{s}^{-1}$ can be assumed as a lower bound for the human sensitivity parameter to bridge lateral vibrations.

Going forward, values of C up to $14\,\mathrm{m}^{-1}\mathrm{s}^{-1}$ are of little practical interest as the trigger of the phenomenon is still characterized by great uncertainty and only some samples synchronize.

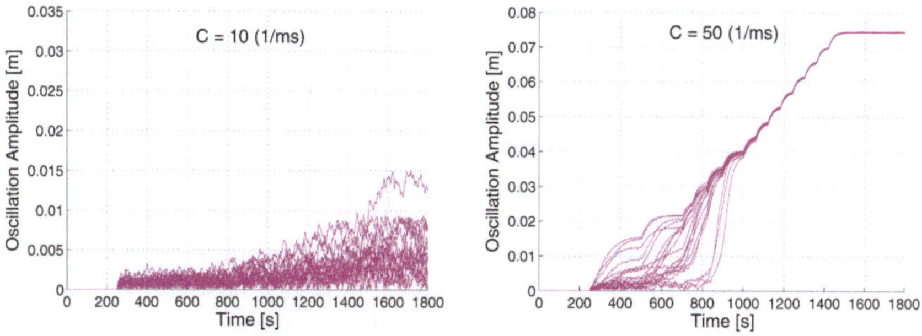

Fig. 19 Amplitude of vibration versus time: **a** curves obtained with $C = 10\,\mathrm{m}^{-1}\mathrm{s}^{-1}$ and **b** with $C = 50\,\mathrm{m}^{-1}\mathrm{s}^{-1}$

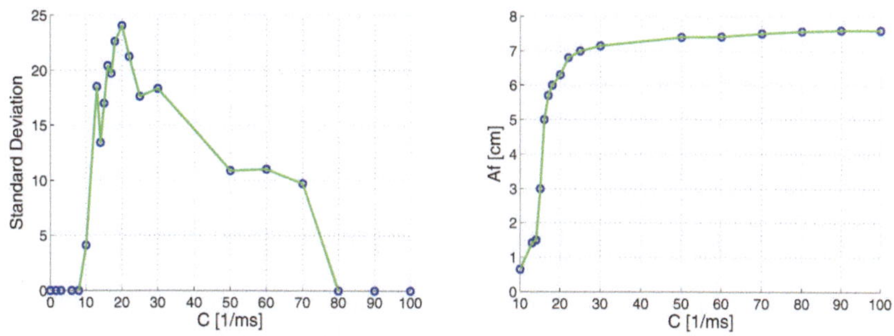

Fig. 20 Standard deviations **a** and maximum final amplitude of bridge vibrations **b** associated to the mean values of the pedestrians critical number triggering the synchronization: simulations performed by varying C

Overall as C increases, the critical crowd size decreases (Fig. 18) and the maximum final amplitude of the bridge oscillations increases (Fig. 20b). Then, for values of $C > 50\,\mathrm{m}^{-1}\mathrm{s}^{-1}$, the critical threshold keeps almost identical: synchronization instantly occurs, when only the first loading step of 50 pedestrians is applied on the bridge. This means that $C = 50\,\mathrm{m}^{-1}\mathrm{s}^{-1}$ can be assumed as an upper bound for the human sensitivity parameter to bridge lateral vibrations. In Fig. 19b we show the result of simulations performed with $C = 50\,\mathrm{m}^{-1}\mathrm{s}^{-1}$ ($N_a = 60$ and $\sigma_N = 10.89$).

Thus we clearly define the limits for the pedestrians-bridge synchronization parameter C: values lower than $10\,\mathrm{m}^{-1}\mathrm{s}^{-1}$ and higher than $50\,\mathrm{m}^{-1}\mathrm{s}^{-1}$ are not meaningful; we highlight that the range of most practical interest is for $C \cong 14 - 25\,\mathrm{m}^{-1}\mathrm{s}^{-1}$, as these are the most realistic values of pedestrian sensitivity (this range includes $C = 16\,\mathrm{m}^{-1}\mathrm{s}^{-1}$, value for which the model results match the experimental data for the north span of the Millennium Bridge) and also the most delicate (in this range the average critical number N_a decreases more quickly, see Fig. 18).

For completeness of investigation, in Fig. 20a we report also the standard deviation σ_N associated to the mean value N_a of the critical number of pedestrians, for each group of simulations performed by varying C. If we neglect the irregularities of statistical nature, we can observe that σ_N starts from zero, follows an increasing trend until a peak threshold for $C \cong 14-25\,\mathrm{m}^{-1}\mathrm{s}^{-1}$ and then decreases again to zero. A possible interpretation is that the more the pedestrians are sensitive to bridge lateral vibrations, the more the initial values of phases and native frequencies, randomly assigned at each simulation, influence the temporal evolution of the phenomenon in terms of results dispersion; as a consequence the critical threshold N_a is affected by an increasing standard deviation, and this trend is maintained until the peak threshold. Then when the sensitivity increases beyond that range, a trend reversal occurs: pedestrians are so sensitive to bridge lateral vibrations that they are progressively less and less influenced by the initial conditions, and the critical threshold N_a is affected by a standard deviation which drops to zero. Therefore $C \cong 14-25\,\mathrm{m}^{-1}\mathrm{s}^{-1}$ is also the range in which the initial conditions are mainly able to influence the phenomenon, with a resulting higher dispersion in our prediction of N_a. In addition, we observe that the maximum final amplitude of the bridge oscillations increases rapidly in this range before reaching a plateau (Fig. 20b).

These considerations confirm that $C \cong 14-25\,\mathrm{m}^{-1}\mathrm{s}^{-1}$ is the most critical range of values for C, and therefore our interest will be concentrated on it (see the following subsection 2.2.8).

2.2.6 Inverse Approach Analysis: Coherence of the Model Results

With reference to the specific example of the Millennium Bridge, it may be interesting to analyse also an 'inverse' approach in the evaluation of the model behaviour. Instead of investigating what happens to the crowd critical size when we vary the model parameters or the description of the pedestrian load, we can study which is the value of C able to give a model critical threshold coinciding with the experimental one. Actually, this is the way with which C can be determined for each 'variant' of the SAMEO model.

With this aim we perform simulations by looking for the value of C such that $N_a = N_{c,exp}$, when:

1. G is assumed variable with the bridge oscillation amplitude according to Eq. (7) (see the following Fig. 21);
2. the pedestrian force is assumed of 'square-type' according to Eq. (9), instead of sinusoidal (see Fig. 22 hereunder).

In this first case, if we compare our simulations results with data obtained from crowd tests on the Millennium Bridge [12], we confirm the value $C = 16\,\mathrm{m}^{-1}\mathrm{s}^{-1}$ obtained with $G = \text{constant} = 30\,\mathrm{N}$. This is not surprisingly, since in the subsection 2.2.3 we have shown that with (7) N_a is only slightly smaller than that corresponding to $G = 30\,\mathrm{N}$ (150 vs 155), the difference being hidden by the statistical nature of the problem.

In this second case, in order to match the model predictions to the Arup experiment, the correct value for C is slightly smaller: $15\,\mathrm{m}^{-1}\mathrm{s}^{-1}$. This is a reasonable result as in this case we put suddenly more energy into the system and therefore, in order to obtain the same response of the reference case, we have to consider a

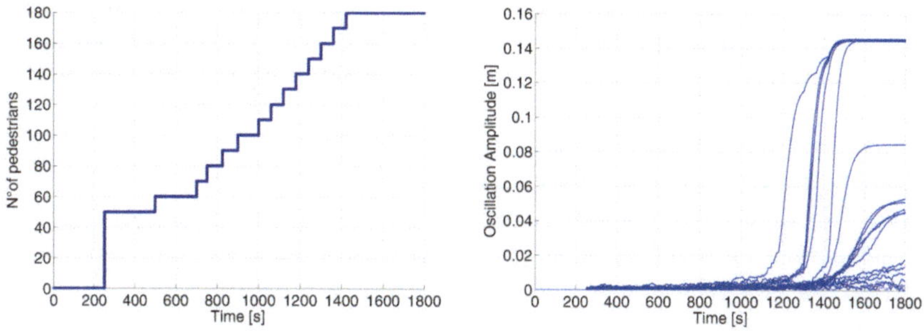

Fig. 21 Number of walkers on the bridge **a** and amplitude of vibration **b** versus time when G is variable with A according to Abrams' model (7). Simulations performed with $C = 16\,\mathrm{m}^{-1}\mathrm{s}^{-1}$

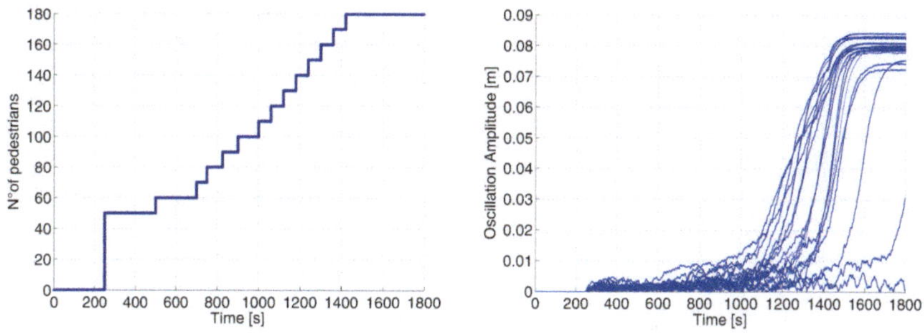

Fig. 22 Number of walkers on the bridge **a** and amplitude of vibration **b** versus time when the pedestrian force is assumed of 'square-type' instead of sinusoidal (9). Simulations performed with $C = 15\,\mathrm{m}^{-1}\mathrm{s}^{-1}$

lower value of the sensitivity parameter. These results are in full agreement with those obtained in subsection 2.2.4, where for the case $C = 16\,\mathrm{m}^{-1}\mathrm{s}^{-1}$ and with a 'square-type' pedestrian force, we had obtained a fairly lower critical number N_a.

Overall the SAMEO model seems to provide coherent results with respect to an inverse approach analysis, in spite of the statistical data involved in its predictions.

2.2.7 Synchronization between Pedestrians

The original formulation of the SAMEO model considers only the bridge-pedestrians interaction, and not an interaction between pedestrians. Since it was found that a certain level of synchronization among people within the crowd exists [54], [20], it should be considered also a correlation due to people falling into step with each other, simply responding to visual clues such as the movement of people in front of them [25]. There is experimental evidence that the brain's control center syncs up to its visual center with high-frequency brain waves, directing attention to select features of the visual world [19]. In a crowded bridge, lacking other external stimuli, human attention is caught by the people who are walking ahead, with a tendency to synchronize in order to minimize the reciprocal interferences and to achieve a more comfortable and fluent gait [32]–[24].

To describe this phenomenon we add to the SAMEO relation for the bridge-pedestrians interaction (3) a term which takes into account the pedestrian- to- pedestrian interaction. However, without complicating too much the model, we consider that a pedestrian interacts only with the one who is in front of him (Fig. 23), who is the one who mainly influences his motion. From a practical point of view, this means to consider pedestrians walking 'in single file' (i.e. lined up one behind another): this is obviously a simplification of the real crowd interaction, although it exactly corresponds to the Arup experiments on the Millennium Bridge [2]. It can be considered as a rough preliminary proposal and additional features could be dealt with after a fundamental understanding has been established. Indeed our modified SAMEO model is very flexible in its present form, and the real effect on the single individual of all the pedestrians who are ahead in his/her visual cone could be easily introduced through an 'average effect coefficient', determinable with proper experimental analyses of human behavior.

Instead of (3) we then assume:

$$\dot{\Theta}_i = \Omega_i + C_i A \sin(\Psi - \Theta_i + \alpha) + D_i \sin(\Theta_{i-1} - \Theta_i)\,. \tag{11}$$

The new term $g = D_i \sin(\Theta_{i-1} - \Theta_i)$ is chosen in analogy with the bridge-pedestrians interaction term and on the basis of the following considerations:

1. it is a function of the phase difference between pedestrians 'in single file', being Θ_{i-1} the phase of the generic leading pedestrian $i-1$ which acts as stimulus signal for the following pedestrian i walking just behind;
2. it has the effect of shifting each walker to a phase closer to that of the previous one. Therefore, when the phase difference $(\Theta_{i-1} - \Theta_i)$ is positive, i.e. Θ_i lags Θ_{i-1}, g must be globally positive, in order to increase the frequency of pedestrian i, thus fostering synchrony with pedestrian $i-1$; similarly when $(\Theta_{i-1} - \Theta_i)$ is negative, i.e. Θ_i leads Θ_{i-1}, the term in question must be globally negative. Of course g must be periodic in $(\Theta_{i-1} - \Theta_i)$, and the simplest periodic function that satisfies these requirements is the sine function;
3. the constant of proportionality D_i measures the effect of the pedestrian $i-1$ on the following pedestrian i. Since D_i is the amplitude of the maximum phase shift corrections between walkers, it can be considered as a sort of 'visual sensitivity' of pedestrians to the crowd self-synchronization. Thus we will have $g = 0$ when $D_i = 0$, i.e. the pedestrians are visually insensitive. It is reasonable to consider a certain variation of visual perception among individuals in the population. We should in general use a random distribution D_i for these sensitivities, but lacking specific studies in this direction, we will later make the simplifying assumption that $D_i = D$, a single constant value for all walkers (similarly to what have been done for the C_i).

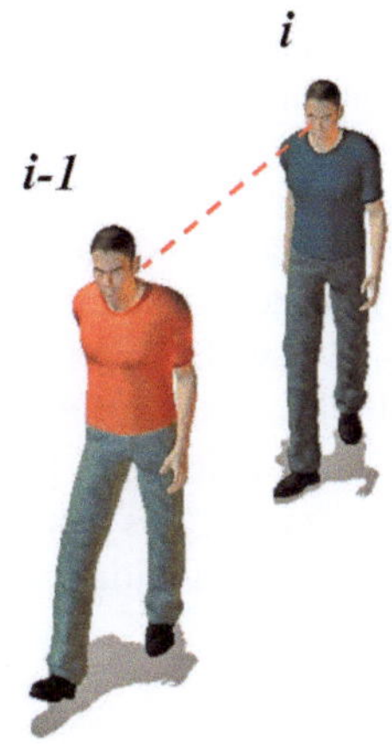

Fig. 23 Scheme of the visual pedestrian-to-pedestrian interaction

Hence with this model, when pedestrian i sees pedestrian $i-1$ walking near in front of him, he is visually influenced and he slows down or speeds up so as to walk more nearly in phase on the next step. This effect certainly contributes to speed up the trigger of the bridge first lateral movements. It is obvious that if the visual stimulus of the previous pedestrian $i-1$ is too fast or too slow, the pedestrian i cannot keep up and entrainment is lost. Therefore, this type of synchronization is certainly fostered in the case of a typologically homogeneous crowd, i.e. with a Gaussian distribution of native frequencies characterized by a small standard deviation (Fig. 10).

Moreover the synchronization between pedestrians is clearly possible only if the number of persons on the bridge is sufficiently large so that they are able to influence each other. For this reason our modification to SAMEO model applies only in case of yet crowded bridge, with a density over $0.6\,\text{pers}/\text{m}^2$ [15] (corresponding to a relative distance between pedestrians of about 1 m or even less): over this density value the single pedestrian is no longer able to walk with his individual undisturbed step frequency and walking velocity. Within this limit, we have to detect the admissible scale range for D_i (i.e. to fix its lower and upper bounds) and to predict the effects of D_i on the model.

Some preliminary qualitative physical considerations allow affirming that:

1. D_i cannot be negative both for a physical reason related to its meaning, and because the sine function already controls, in the mathematically right way, the increase/decrease of the walking frequency due to 'self'-synchronization between pedestrians;
2. as the visual sensitivity D_i increases, we expect a decrease of the critical number N_c, because the 'self'-synchronization between pedestrians tends to facilitate the 'global' synchronization with the bridge;
3. large values of D_i could mean that each pedestrian is so much influenced by the previous one because of their minimal distance (crowd close to the densest possible packing) that he doesn't mind the bridge; namely the pedestrians could synchronize each other on a frequency different from the bridge native one, thus

the critical number N_c triggering the phenomenon can increase instead of decreasing. This is obviously a limit case.

On the basis of these qualitative physical considerations, $D_i = 0\,\mathrm{s}^{-1}$ can be assumed as a lower bound for the visual sensitivity parameter to the 'self'- synchronization.

In order to determine also an upper bound for D_i, hereafter we consider the Eq. (11) with $C = 0\ \mathrm{m}^{-1}\mathrm{s}^{-1}$, in order to deal with the effects of D on the model only. The system, for N pedestrians, is made of $N-1$ equations (note that the first pedestrian has not a person ahead him and so it is not involved in the 'self'- synchronization. It is just a leader of the crowd and its native frequency, $\dot{\Theta}_1 = \Omega_1 \neq 0$, constitutes an initial condition for the problem):

$$\begin{cases} \dot{\Theta}_2 = \Omega_2 + D_2 \sin(\Theta_1 - \Theta_2)\,, \\ \vdots \\ \dot{\Theta}_i = \Omega_i + D_i \sin(\Theta_{i-1} - \Theta_i)\,, \\ \vdots \\ \dot{\Theta}_N = \Omega_N + D_N \sin(\Theta_{N-1} - \Theta_N). \end{cases} \tag{12}$$

The fixed points, corresponding to steady-states or equilibriums of the system, are given by $\dot{\Theta}_i = 0$, $\forall i = 2,\ldots,N$. From a qualitative point of view this is the limit situation, when overcrowding is such that pedestrians are packed and therefore they are forced to stop: $\dot{\Theta}_i$ is nil and the maximum value of D_i is achieved (it is the upper bound we are looking for).

For the generic pedestrian i the solution for $\dot{\Theta}_i = 0$ provides:

$$-\frac{\Omega_i}{D_i} = \sin(\Theta_{i-1} - \Theta_i)\,. \tag{13}$$

We highlight that, in order to have the equilibrium position, the condition $\left|-\Omega_i/D_i\right| < 1$, i.e. $D_i > \Omega_i$, must be satisfied.

From (13) we have

$$\Theta_i = \Theta_{i-1} + \arcsin\frac{\Omega_i}{D_i}\,, \quad i = 2,\ldots,N\,. \tag{14}$$

The first term Θ_1^* in this chain of equalities (14) is known, as the first pedestrian is not influenced by anyone else on the bridge, while the second is visually influenced by the first and so on. Thus we determine in cascade the solution of the system (12): $(\Theta_2^*, \Theta_3^*, \ldots, \Theta_N^*)$; the Jacobian matrix $\mathbf{J}$ evaluated at that point is:

$$\mathbf{J} = \begin{bmatrix} -\mu_2 & 0 & 0 & \cdots & 0 \\ \mu_3 & -\mu_3 & 0 & & 0 \\ 0 & \ddots & \ddots & \ddots & \vdots \\ \vdots & \ddots & \mu_{N-1} & -\mu_{N-1} & 0 \\ 0 & \cdots & 0 & \mu_N & -\mu_N \end{bmatrix}, \quad \mu_i = D_i\sqrt{1-(\Omega_i/D_i)^2}\,. \tag{15}$$

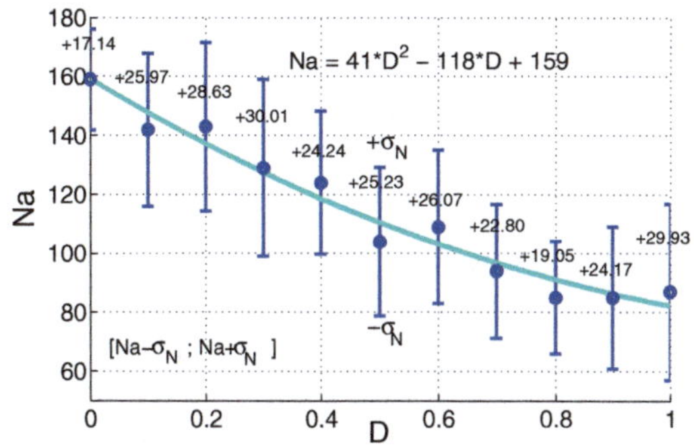

Fig. 24 Critical number of pedestrians versus D: data points, fitted quadratic curve and graphical display of the standard deviations associated to each mean value N_a

This matrix has negative eigenvalues $\lambda_i = -D_i\sqrt{1-(\Omega_i/D_i)^2}$, as it can be seen by inspection (remember that $D_i > \Omega_i$ otherwise the equilibrium solution does not exist); hence the fixed point $(\Theta_2^*, \Theta_3^*, ..., \Theta_N^*)$ is a stable node. In other words, as D_i increases, the phases tend towards those values $(\Theta_2^*, \Theta_3^*, \ldots, \Theta_N^*)$, equilibrium solutions of system (12). Therefore we infer that to not have the phenomenon of stable equilibrium, corresponding to overcrowding condition, we have to assume $D_i < \Omega_i$, and this constitutes an upper bound for D_i.

It is important to remark that from a physical point of view, for equilibrium of the system we mean the limit situation such that overcrowding is close to the densest possible packing and therefore it prevents any further form of motion: pedestrians are obliged to stop and their relative phases are blocked; the fixed point $(\Theta_2^*, \Theta_3^*, ..., \Theta_N^*)$ provides the pedestrian phases an instant before the packed configuration. This equilibrium is also stable, if nobody leaves the bridge. We impose that this equilibrium must be avoid because it is the upper possible limit for real situations; in this sense it allows an estimation of the maximum value for D_i: the maximum visual sensitivity threshold occurs when overcrowding is maximum.

For sake of simplicity, as already mentioned, D_i will be assumed constant, $D_i = D$, as it has been done with the other sensitivity parameter C (Sect. 2). Therefore, the bounds we have just found can be rewritten as $0 \le D < \Omega_a = 1.03\,\mathrm{s}^{-1}$ (note that we refer to the mean of the Gaussian distribution of pedestrian lateral frequencies).

Afterwards we reintroduce $C = 16\,\mathrm{m}^{-1}\mathrm{s}^{-1}$ (original value from SAMEO model) in the governing equation (11), and we investigate values of D in the interval [0,1] s^{-1} in order to test the effects on the model. We obtain, as expected, a monotonically decreasing trend of the critical number of pedestrians triggering the synchronization versus D (Fig. 24).

We use once again the method of least squares to characterize data using a global fit; we obtain the following quadratic correlation function:

$$N_a = 41 \times D^2 - 118 \times D + 159\,. \tag{16}$$

In Fig. 24, for each group of simulations, we also report the standard deviation σ_N associated to the mean value N_a of the crowd critical number. We remark that for

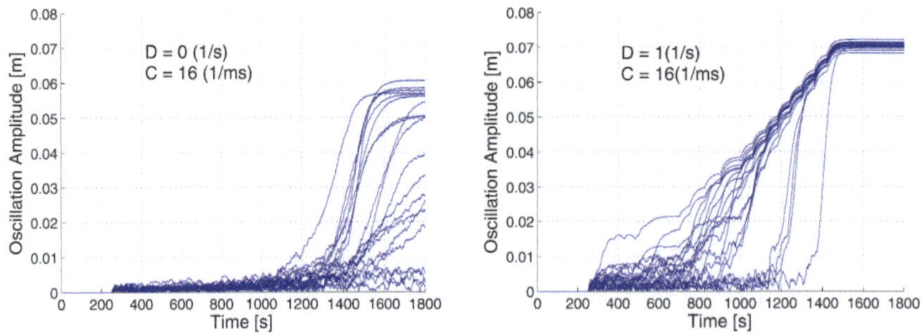

Fig. 25 Amplitude of vibration versus time: curves obtained with **a** $D = 0\,\mathrm{s}^{-1}$, $C = 16\,\mathrm{m}^{-1}\mathrm{s}^{-1}$ and with **b** $D = 1\,\mathrm{s}^{-1}$ and $C = 16\,\mathrm{m}^{-1}\mathrm{s}^{-1}$

$D = 0$ we are not actually considering pedestrian-to-pedestrian interaction and we fall back in the reference case, thus obtaining $N_a = 155$–160 and $\sigma_N = 27.7$–25.4; for $D = 1\,\mathrm{s}^{-1}$, we obtain a strongly reduced critical number $N_a = 85$ and $\sigma_N = 29.93$, being increased the pedestrian sensitivity to visual clues. The amplitude versus time curves for these two border cases are shown hereunder: Fig. 25a is the equivalent of Fig. 7, re-proposed to facilitate the comparison with Fig. 25b.

We note an increase of the final predicted amplitude of the bridge motion, which is, in any case, very close to the observed values on the Millennium Bridge.

It is evident, as expected, that this additional synchronization mechanism self-excites the phenomenon, as the critical threshold almost halves (from 155–160 to 85) with respect to the reference model.

2.2.8 Effects of the Two Synchronization Parameters C and D

When does the pedestrian-pedestrian visual interaction become irrelevant with respect to the interaction with the bridge lateral vibrations? How much N_a is influenced by C and D? Unfortunately these remain partly open questions, lacking a proper complementary experimental investigation.

The only thing we can do here is to perform a systematic numerical simulation aimed at determining the joint effect of C and D on the model critical threshold N_a. We report in Fig. 26 the contour plot of N_a as a function of C and D, which just permits understanding the joint effect of the synchronization parameters on the model behaviour.

This map is obtained by considering a grid of 12×11 points: each of them is the result of 200 simulations. The plot window is performed for D ranging in its definition interval $[0,1]\,\mathrm{s}^{-1}$ (see subsection 2.2.7), and for C varying in a restricted range of values, from $14\,\mathrm{m}^{-1}\mathrm{s}^{-1}$ up to $25\,\mathrm{m}^{-1}\mathrm{s}^{-1}$; this last choice is admissible in the light of the observations drawn in subsection 2.2.5.

Figure 26 highlights the complementarity of C and D in describing the phenomena. The simulations point out that as C and D increase, the critical number decreases, according to common sense; moreover the same number N_a is obtained for smaller D when C is larger and vice versa. This trend has an intuitive physical confirmation; it is justified by the real difficulties pedestrians encounter when they walk on a laterally moving surface: as C increases, their sensory system, both neural and musculoskeletal, is so much involved in balance control that it does not care of the visual stimuli due to the presence of other people and the natural consequence is a lower D.

It is worthy of note that the two synchronization parameters, C and D, have different sensitivities, as we can see from their different scales of values (Fig. 26). By fixing one parameter and varying the other, if we increase D of $0.1\,\mathrm{s}^{-1}$ we obtain a decrease of N_a equal 10 pedestrians, but if we increase C of $0.1\,\mathrm{m}^{-1}\mathrm{s}^{-1}$ we obtain a decrease of N_a equal 1 (i.e. fixed the decrement of N_a, there is a ratio $10:1$ in the increase of C with respect to D). In mathematical terms, we can affirm that the derivative of N_a with respect to C is smaller than the derivative of N_a with respect to D, although care must be used in this comparison because these are not dimensionless quantities. The direct consequence is that we have to pay more attention on the correct evaluation of the visual sensitivity parameter D.

Another aspect to take into account is that, by definition, the parameter C depends only on the pedestrians' sensitivity to the bridge lateral movements; however in the way it is used in the SAMEO model, actually it depends on the mode shape, even if, for sake of simplicity, only one value of C is considered for all the pedestrians on the bridge, independently of their position and distribution over the whole span. From a practical point of view its numerical value, as defined in the model, is acceptable only in the case of bridges having the same mode shape (sinusoidal) as the north span of the Millennium Bridge: for other bridges, C should be different. On the contrary, the parameter D, again by definition, depends only on the pedestrians' visual sensitivity to the motion of the people who are walking ahead, within a certain visual-psychological influence distance; therefore it is related to the crowd density and it can be estimated also through experiments with pedestrians walking on a fixed floor. Thus the numerical value of D is independent on the mode shape and it is valid for every bridge.

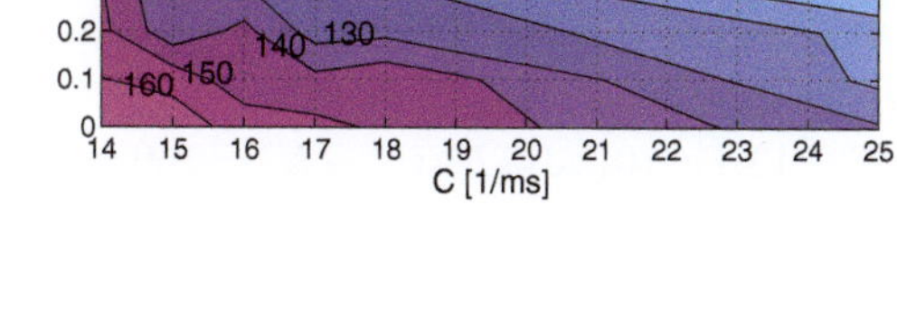

Fig. 26 Contour map for N_a obtained by varying C and D

In any case it is confirmed the need of tuning C and D through experimental tests, taking into account the simultaneous existence of the two synchronization degrees in relation to the crowd density and to the physical characteristics of the pedestrians involved (e.g., we expect that senior people or pedestrians with slight difficulties of locomotion are less prone to synchronize with the other persons, but are much more sensitive to the synchronization with the bridge). Once performed such tests, the contour plot in Fig. 26 could be a useful design map to evaluate N_a, depending on the expected predominant typology of pedestrian traffic.

2.2.9 Linear and Random Loading Paths

The reference (Arup's) loading path (Fig. 6a) is useful to detect the onset of the instabilities and to assess the model, but it is not general enough, and thus, for example, it is useless to design a control device. Here we study different, in particular *linear*, paths with the aim of investigating the variations of the critical threshold, and the effects of a larger number of pedestrians walking on the bridge.

It was estimated that between 80000 and 100000 people crossed the Millennium Bridge during the opening day, with a maximum of 2000 people on the deck at any time (approximately 450 only on the north span), resulting in a density of 1.3–1.5 $\text{persons}/\text{m}^2$ [12]. Observations on crowd indicate that the upper limit density for unconditioned free motion is only about 0.3 $\text{persons}/\text{m}^2$, while normal walking becomes practically impossible for densities above 1.7 $\text{persons}/\text{m}^2$ [54]. Therefore, in the following simulations, we arrive at a maximum of 550 ($= 1.7 \times 324$ square meters [12]) pedestrians walking simultaneously on the bridge (Fig. 27a). However it is important to mention that the number of pedestrians exciting the bridge is related to their distribution over the whole span, so that the local amplitude of the oscillation is related to the mode shape amplitude at that point, and thus it is a fraction of the amplitude of the excitation; moreover the theoretical limit numbers are for people who are walking, while usually some people will stop on the bridge, possibly contributing extra damping. This increases the practical critical threshold. Thus, we have competing phenomena practically reducing/increasing the theoretical critical threshold, and therefore our simulations have to be considered as a first, non systematic, study in this direction.

We consider five linear loading paths (Fig. 27) which differ in the number of pedestrians introduced on the bridge deck per unit time (we refer to it as 'loading velocity'); we identify them with a ramp factor, RF, defined as the inverse of their average slope. We obtain:

1. $RF = 0.5$, i.e. $10\,\text{ped}/5\,\text{sec}, \to N_a = 240, \sigma_N = 34.98$;
2. $RF = 1$, i.e. $10\,\text{ped}/10\,\text{sec}, \to N_a = 225, \sigma_N = 35.76$;
3. $RF = 3$, i.e. $10\,\text{ped}/30\,\text{sec}, \to N_a = 190, \sigma_N = 36.21$;
4. $RF = 6$, i.e. $10\,\text{ped}/60\,\text{sec}, \to N_a = 165, \sigma_N = 38.67$;
5. $RF = 12$, i.e. $10\,\text{ped}/120\,\text{sec}, \to N_a = 160, \sigma_N = 39.00$.

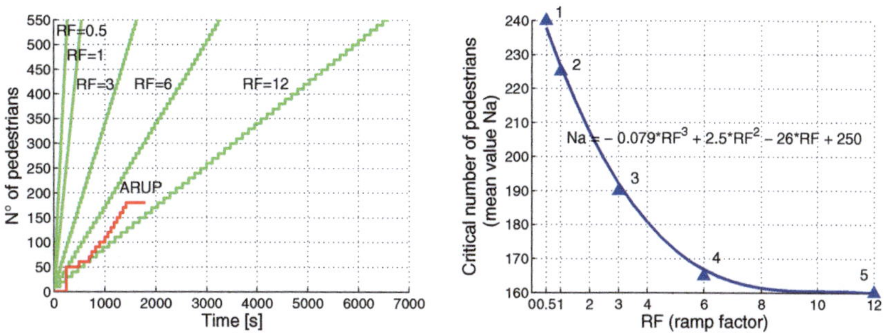

Fig. 27 Linear loading paths **a** and the corresponding crowd critical size **b**

For each of these RFs, the reported values of N_a and σ_N are, as usual, the results of 200 simulations. The best (in the least squares sense) cubic curve fitting the data is (Fig. 27b):

$$N_a = -0.079 \times RF^3 + 2.5 \cdot RF^2 - 26 \times RF + 250. \tag{17}$$

From Fig. 27 and from Eq. 17 we can understand and quantify the influence of the loading velocity on the instability threshold. For high loading velocities, i.e. low RF, the system has not enough time to develop synchronization. Pedestrians who enter the bridge will not be driven to adapt soon their footsteps to the slight motion of the structure: the crowd critical size rises. Instead, if we decrease the loading velocity, i.e. we consider high RF, when new pedestrians are introduced on the bridge, they will find a situation in which probably the bridge is already unstable and its wobbling is enough to force them to synchronize before the next load package arrives. So the critical threshold will be lower.

We also observe that the loading paths 4 (which, after the initial transient, has the same slope of the Arup's ramp) and 5 are characterized by a critical crowd size essentially identical; we can infer that by further decreasing the loading velocity likely the critical threshold no longer decreases (Fig. 27 b): Arup's critical threshold can be considered a lower bound of critical thresholds for other deterministic paths, thus being, in this respect, a conservative estimation.

We have also considered some *random* loading paths in order to test the general validity of the model. For each of the 4 groups of simulations summarized in Fig. 28, we have varied stochastically (inside fixed intervals) both the pedestrian increment and the time step. These intervals have been chosen in order to obtain, on average, loading paths comparable with those we have studied above. The numerical simulations give:

1. $RF_a = 1$, i.e. 5–15 pedestrians/5–15 sec (on average 10 ped/10 sec) $\rightarrow N_a = 184, \sigma_N = 22.29$;
2. $RF_a = 3$, i.e. 5–15 pedestrians/10–50 sec (on average 10 ped/30 sec) $\rightarrow N_a = 165, \sigma_N = 18.72$;

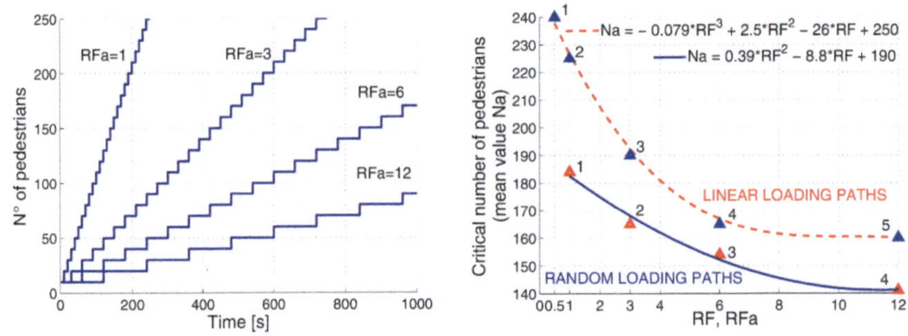

Fig. 28 Random loading paths (average trends) **a** and the corresponding crowd critical size **b**: comparison with the results from the linear loading paths

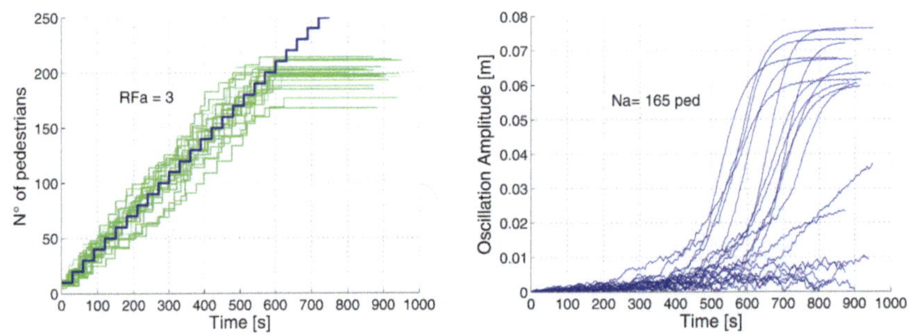

Fig. 29 Random loading paths with $RF_a = 3$ **a** and corresponding amplitude curves **b**

3. $RF_a = 6$, i.e. 5–15 pedestrians/40–80 sec (on average 10 ped/60 sec) $\rightarrow N_a = 154, \sigma_N = 21.49$;
4. $RF_a = 12$, i.e. 5–15 pedestrians/70–170 sec (on average 10 ped/120 sec) $\rightarrow N_a = 141, \sigma_N = 29.19$.

We use once again the method of least squares to characterize data using a global fit; we obtain the following quadratic correlation function (see Fig. 28b):

$$N_a = 0.39 \times RF_a^2 - 8.8 \times RF_a + 190\,. \qquad (18)$$

We can observe that the decrease of N_a with RF, previously observed in deterministic ramps, is maintained, with the same qualitative behaviour. Passing from deterministic to stochastic ramps we observe a systematic decrease of the critical threshold of about 13% (Fig. 28b) which should be carefully considered by the designer. In any case the values obtained are in quite reasonable agreement both with the experimental and with the numerical results of the previous simulations.

It is interesting to show at least a sample of the output from our simulations, e.g. for $RF_a = 3$ (Fig. 29).

Despite the dispersion due to the randomness of the loading paths (Fig. 29a), the results are encouraging as the qualitative trend of the amplitude curves is maintained on average (Fig. 29b): the dynamic response of the bridge continues to be stable until a critical threshold and then increases rapidly towards a final landing. Furthermore, the predicted amplitude of the bridge motion is very close, on average, to the observed value of about 5–7 cm (see Fig. 29b). This is a remarkable robustness property of the SAMEO model, especially if we think to the many uncertainties which affect this type of problems.

3 A Discrete-Time Model

The critical overview of the existing literature, presented in Sect. 1, highlights many continuous-time models explaining the excessive lateral sway motion induced by a crowd crossing a footbridge. They are governed by partial or ordinary nonlinear differential equations (ODEs), and commonly cannot be solved in closed form but require extensive numerical simulations to be utilized in practice [52], as shown in the previous section. To overcome this drawback, in this second part of the chapter we present a nonlinear discrete-time model able to describe the synchronous lateral excitation without numerical simulations [27].

The basic idea is to work in the context of discrete dynamics, by an appropriate choice of a Poincaré section, thus turning a continuous dynamical system into a discrete one. If the Poincaré section is carefully chosen, no information is lost concerning the qualitative behavior of the dynamics [44].

Some approximations and simplifications are assumed in order to obtain a model which is as simple as possible and with the least possible number of parameters, while keeping the description of the underlying mechanical event. Some of these approximations can then be removed in order to have a model of more general validity. In spite of the approximations, however, the model is able to provide a reliable value of the number of pedestrians which trigger the synchronization, thus predicting the onset of instability which is also the onset of crowd synchronization.

From a dynamical system point of view, the main result is that the model highlights how the phenomenon can be seen as a perturbation of a classical pitchfork bifurcation, which is then shown to be the underlying dynamical event. It is worth to note that the proposed model is independent of the specific case of the Millennium Bridge, which is considered as a reference (Sect. 3.4), so it is applicable to any bridge where a similar problem is observed or expected to occur.

Besides improving the understanding of the physical phenomenon, our model proves simple and reliable in its previsions; therefore, it may be useful for estimating the damping needed to stabilize other exceptionally crowded footbridges against synchronous lateral excitations by pedestrians, or for designing other technologies aimed at eliminating the phenomenon in real structures.

3.1 Single Degree of Freedom Oscillator and Discrete Dynamic Model

According to the considerations made in Sect. 1, the dynamics of the bridge are governed by the Eq. (1). Introducing the quantities

$$t=\sqrt{\frac{M}{K}}\bar{t}=\frac{\bar{t}}{\Omega},\ X(t)=x(\bar{t}),\ B=2\sqrt{MK}\,\xi,\ F_{ped}(t)=Kf(\bar{t}), \tag{19}$$

the Eq. (1) can be re-written in the dimensionless form:

$$\ddot{x}+2\xi\dot{x}+x=f(\bar{t}), \tag{20}$$

where dots are derivatives with respect to the dimensionless time $\bar{t}$. In the following the hat is neglected for simplicity. Although equivalent, Eq. (20) is easier to be analyzed than (1) (it has less parameters), and thus it will be considered in the following.

3.1.1 Free Dynamics

In the unforced case $f(t)=0$, the general solution of (20) is:

$$x(t)=c_1e^{-\xi t}\sin\left(\sqrt{1-\xi^2}t\right)+c_2e^{-\xi t}\cos\left(\sqrt{1-\xi^2}t\right), \tag{21}$$

so that $\omega=\sqrt{1-\xi^2}$ is the natural circular frequency and $p=2\pi/\omega$ is the period, i.e. the time distance between two successive relative maximum points of the system motion.

We call x_n the amplitude of the generic relative maximum of the motion (see Fig. 30). We want to find an analytical relation, $x_{n+1}=l(x_n)$, between one peak and the subsequent in time. This map describes the so-called peak-to-peak dynamics (PPD) introduced by Lorenz in the study of chaos [10]. With this aim we assume that the starting point is a peak: $x(0)=x_n$ and $\dot{x}(0)=0$. With these initial conditions we obtain:

$$x(t)=x_ne^{-\xi t}\left[\frac{\xi}{\sqrt{1-\xi^2}}\sin\left(\sqrt{1-\xi^2}t\right)+\cos\left(\sqrt{1-\xi^2}t\right)\right], \tag{22}$$

which is depicted in Fig. 30 for $\xi=0.05$.

From Fig. 30 we can see that the successive peak occurs after one period. Thus, if we consider $x_{n+1}=l(x_n)$ we are actually considering a stroboscopic Poincaré map sampled at each period. We then have:

$$x_{n+1}=x(p)=x_ne^{\frac{-2\pi\xi}{\sqrt{1-\xi^2}}}=\alpha x_n, \tag{23}$$

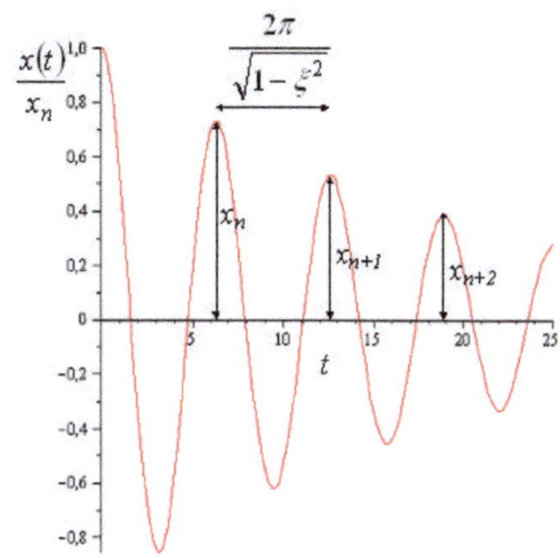

Fig. 30 The solution $x(t)/x_n$ of the damped harmonic oscillator with $\xi = 0.05$

where the parameter $\alpha \leq 1$ is defined by

$$\alpha = e^{-\frac{2\pi\xi}{\sqrt{1-\xi^2}}} \cong 1 - 2\pi\xi + \ldots . \tag{24}$$

Equation (23) is the simplest (it is linear) discrete-time dynamical system we are considering. We exclude the value $\alpha = 1$ because it corresponds to the unrealistic case of absence of damping. Its dynamic behavior is trivial: by iterating the map we obtain $x_n = \alpha^n x_0$, x_0 being the initial condition. Thus, all initial conditions tend, as expected, to the rest position $x = 0$.

3.1.2 Forced Dynamics: Single Pedestrian

We assume that a walking pedestrian acts as a periodic forcing on the oscillator. Here we idealize the pedestrian force as sinusoidal, even though experiments on a treadmill reveal a periodic trend more similar to a square wave (see Sect. 2.2.4). Practically, we consider the first term in the Fourier series expansion of the generic periodic function which approximates the experimental data concerning lateral walking forces on a still surface [41], [42], [8]:

$$f(t) = g\sin(\omega_p t + \phi) , \tag{25}$$

where:

1. $g > 0$ is the dimensionless force amplitude, such that $G = gK = 30\,\mathrm{N}$ on average [1]. G is the maximum lateral force exerted by a pedestrian. We remember that we are focusing on the lateral component of pedestrian forcing, the only one responsible for synchronous lateral excitation;
2. ω_p is the pedestrian footstep native frequency and $p_p = 2\pi/\omega_p$ his/her period;
3. $\phi \in [0, 2\pi]$ is the pedestrian temporal phase.

Substituting Eq. (25) into Eq. (20) we obtain the equation of motion for the case of single pedestrian:

$$\ddot{x}+2\xi\dot{x}+x=g\,\sin(\omega_p t+\phi)\,. \tag{26}$$

The exact solution of Eq. (26), obtained with $x(0)=x_n$ and $\dot{x}(0)=0$, calculated at $t=p$ (i.e. after one period) and then named x_{n+1} gives:

$$x_{n+1}=\alpha x_n+g\left[k_1\sin(\phi)+k_2\cos(\phi)\right], \tag{27}$$

where

$$k_1=\frac{(\omega_p^2-1)e^{\frac{-2\pi\xi}{\sqrt{1-\xi^2}}}+(1-\omega_p^2)\cos\left(\frac{2\pi\omega_p}{\sqrt{1-\xi^2}}\right)+2\xi\,\omega_p\sin\left(\frac{2\pi\omega_p}{\sqrt{1-\xi^2}}\right)}{(1-\omega_p^2)^2+4\xi^2\omega_p^2},$$

$$k_2=\frac{2\xi\,\omega_p e^{\frac{-2\pi\xi}{\sqrt{1-\xi^2}}}+(1-\omega_p^2)\sin\left(\frac{2\pi\omega_p}{\sqrt{1-\xi^2}}\right)-2\xi\,\omega_p\cos\left(\frac{2\pi\omega_p}{\sqrt{1-\xi^2}}\right)}{(1-\omega_p^2)^2+4\xi^2\omega_p^2}. \tag{28}$$

Defining:

$$\beta=g\sqrt{k_1^2+k_2^2},\ \ \sin(\gamma)=\frac{k_1}{\sqrt{k_1^2+k_2^2}},\ \ \cos(\gamma)=\frac{k_2}{\sqrt{k_1^2+k_2^2}}, \tag{29}$$

we can re-write Eq. (27) in the simpler form:

$$x_{n+1}=l(x_n)=\alpha x_n+\beta\cos(\phi-\gamma)\,. \tag{30}$$

Comparing Eq. (30) with Eq. (23), we see that the influence of the single pedestrian's motion on the map is described by the additional term on the right hand side.

We observe that if the pedestrian period is sufficiently close to the structure's natural period (i.e. $\omega_p\cong\omega$) it is possible a bridge-pedestrian interaction. In the worst-case scenario of perfect resonance, $\omega_p=\omega=\sqrt{1-\xi^2}$, it follows that:

$$k_1=\frac{1-e^{\frac{-2\pi\xi}{\sqrt{1-\xi^2}}}}{4-3\xi^2}=\frac{1-\alpha}{4-3\xi^2},$$

$$k_2=-2\frac{\sqrt{1-\xi^2}}{\xi}\frac{1-e^{\frac{-2\pi\xi}{\sqrt{1-\xi^2}}}}{4-3\xi^2}=-2\frac{\sqrt{1-\xi^2}}{\xi}\frac{1-\alpha}{4-3\xi^2},$$

$$\beta=g\frac{1-e^{\frac{-2\pi\xi}{\sqrt{1-\xi^2}}}}{\xi\sqrt{4-3\xi^2}}=g\frac{1-\alpha}{\xi\sqrt{4-3\xi^2}},\quad \tan(\gamma)=-\frac{1}{2}\frac{\xi}{\sqrt{1-\xi^2}}. \tag{31}$$

The problem becomes simpler, since being α and γ functions of ξ, the only parameters are ϕ, ξ and β (i.e. g). Furthermore, in this case, the peak-to-peak map corresponds to the stroboscopic Poincaré map.

Also in this case the dynamics of the (discrete) Eq. (30) are trivial. All initial conditions tend to the map fixed point

$$x = \frac{\beta \cos(\phi - \gamma)}{1 - \alpha}, \tag{32}$$

corresponding to a periodic oscillation of the real system. We have that $1 - \alpha = 2\pi\xi + \ldots$ is a small quantity, so that x is large, according to the fact that we are in (perfect) resonance.

3.1.3 Forced Dynamics: Crowd of Pedestrians

When a crowd of N pedestrians is walking on the bridge, the force is:

$$f(t) = \sum_{i=1}^{N} g_i \sin(\omega_{p,i} t + \phi_i), \tag{33}$$

where g_i is the lateral force exerted by the ith pedestrian and $p_{p,i} = 2\pi/\omega_{p,i}$ is the natural period of his/her footstep; both are stochastic variables depending on the characteristics of the pedestrian himself. The pedestrian phase ϕ_i, still a stochastic variable, depends instead on the instant of time in which the pedestrian enters the bridge.

We consider a randomly walking crowd and we suppose a free entrance of pedestrians on the bridge. Therefore, ϕ_i is a stochastic variable uniformly distributed over its interval of existence $[0, 2\pi]$.

We instead introduce a simplification by assuming that all the pedestrians have the same native frequency, $\omega_{p,i} = \omega_p$. This assumption corresponds to the worst-case scenario, and actually occurs during synchronization because all the pedestrians tend to walk with the same period by feedback modifications to their native period, as shown by the SAMEO model in Sect. 2.

With the hypothesis $\omega_{p,i} = \omega_p$ the excitation Eq. (33) is a $(2\pi/\omega_p)$-periodic function, which can be re-written in the form:

$$f(t) = \bar{g} \sin\left(\omega_p t + \bar{\phi}\right), \tag{34}$$

with

$$\bar{g} = \sqrt{\left[\sum_{i=1}^{N} g_i \cos(\phi_i)\right]^2 + \left[\sum_{i=1}^{N} g_i \sin(\phi_i)\right]^2}, \quad \tan\left(\bar{\phi}\right) = \frac{\sum_{i=1}^{N} g_i \sin(\phi_i)}{\sum_{i=1}^{N} g_i \cos(\phi_i)}. \tag{35}$$

The expression (34) is formally similar to Eq. (25), so that we bring back the crowd's forcing to an *equivalent pedestrian's* force:

1. $\bar{g}$ is the overall lateral force amplitude due to crowd's motion;
2. $\bar{\phi}$ is the mean phase.

In the case of *perfectly synchronous* pedestrians, i.e. $\phi_i = \phi \pm 2n\pi$, the expressions (35) become:

$$\bar{g} = \sum_{i=1}^{N} g_i = N g_{average}\,, \quad \bar{\phi} = \phi\,. \tag{36}$$

In case of *perfectly asynchronous* pedestrians, i.e. for each pedestrian there is, on average, another with opposite phase, we have $\bar{g} = 0$.

Therefore $\bar{g}$ depends on the pedestrians' degree of synchronization, an observation which is crucial for the following developments.

Comparing Eq. (34) with Eq. (25), the map (30) becomes:

$$x_{n+1} = l(x_n) = \alpha x_n + \bar{\beta}\cos\left(\bar{\phi} - \gamma\right), \quad \bar{\beta} = \bar{g}\sqrt{k_1^2 + k_2^2}\,, \tag{37}$$

where k_1, k_2 and γ are unmodified.

The map (37) is formally identical to the previous one for single pedestrian.

Hereafter we will describe the crowd's force only by the parameters $\bar{g}$ (or $\bar{\beta}$), $\bar{\phi}$ and ω_p. If we further assume, as done for the single pedestrian, the worst-case scenario of perfect resonance $\omega_p = \omega = \sqrt{1-\xi^2}$, we then have only the two parameters $\bar{g}$ (or $\bar{\beta}$) and $\bar{\phi}$ describing the excitation acting on the bridge.

3.2 *Interaction Oscillator-Pedestrians*

In order to model the dynamical bridge-pedestrians interaction and to describe the natural tendency of the systems to synchronize, we must assume that they influence each other. Therefore, not only the motion amplitude x_n, but also the forcing characteristics $\bar{\beta}$ and $\bar{\phi}$ are assumed to vary on (discrete) time. Our state variables are then x_n, $\bar{\beta}_n$, $\bar{\phi}_n$, and Eq. (37) becomes:

$$x_{n+1} = \alpha x_n + \bar{\beta}_n \cos\left(\bar{\phi}_n - \gamma\right)\,. \tag{38}$$

We note that the argument of the cosine function, $\sigma_n = \bar{\phi}_n - \gamma$, has a precise meaning in terms of synchronization:

1. when $\sigma_n = 0$, i.e. $\bar{\phi}_n = \gamma$, the crowd is perfectly synchronized with the bridge motion and the force exerted on the bridge is maximum;
2. when $\sigma_n = \pm\pi/2$, i.e. $\bar{\phi}_n = \gamma \pm \pi/2$, the crowd is perfectly asynchronous and the net force on the bridge is zero; this means that the crowd does not alter the bridge equilibrium state ($x = 0$) in this case.

On the basis of the previous observation we define σ_n to be the 'synchronization parameter'. Equation (38) takes the form:

$$x_{n+1} = \alpha x_n + \bar{\beta}_n \cos\left(\sigma_n\right)\,, \tag{39}$$

and the system state variables are x_n, $\bar{\beta}_n$, σ_n.

The next step consists in finding the evolution laws for $\bar{\beta}_n$ and σ_n.

By noting that pedestrians tend to synchronize with the bridge lateral motion, we propose for σ_n the following evolution law:

$$\sigma_{n+1} = \frac{a}{a+x_n}\sigma_n\,, \tag{40}$$

where $a > 0$ is a parameter measuring the pedestrians' sensitivity to bridge lateral vibrations; it has the dimension of a length, and its value can be determined experimentally.

Equation (40) fulfils the following physical requirements which suggest its use:

1. the bridge-pedestrians system tends naturally to the maximally synchronous state. In fact $\lim_{n\to\infty}\sigma_n = 0$;
2. for small vibration amplitudes x_n, the pedestrians are not influenced by the bridge. In fact $\lim_{x_n\to 0}\sigma_{n+1} = \sigma_n$;
3. for large values of x_n, the pedestrians quickly synchronize, in fact $\lim_{x_n\to\infty}\sigma_n = 0$.

It is worth to note that Eq. (40) is susceptible of extension in order to obtain a more general theory. In fact, it is realistic to think that the synchronization is not asymptotically perfect; in mathematical terms, this means that $\lim_{n\to\infty}\sigma_n = \eta$ (η close but different from 0), or better, $\lim_{n\to\infty}\sigma_n = \Upsilon$, with Υ near but different from γ. We can introduce this aspect by simply substituting γ with Υ in the definition of σ_n.

As regards $\bar{\beta}_n$, its evolution law is known once we know the law for $\bar{g}_n$, being $\bar{\beta}_n = \bar{g}_n\sqrt{k_1^2+k_2^2}$. Considering that $\bar{g}_n$ depends on the bridge-pedestrians synchronization level, we could relate it to the synchronization parameter σ_n and write $\bar{g}_n = \bar{g}_n(\sigma_n)$ such that:

1. if $\sigma_n = 0$, we should have $\bar{g}_n(0) = N g_{average}$ (perfectly synchronous crowd);
2. if $\sigma_n = \pm\pi/2$, we should have $\bar{g}_n(\pm\pi/2) = 0$ (perfectly asynchronous crowd).

At the same time, it is also true that synchronization depends on the oscillation amplitude x_n. Therefore we can assert that $\bar{g}_n = \bar{g}_n(x_n)$, and in particular:

1. if the oscillation amplitude is nil, i.e. $x_n = 0$, we expect $\bar{g}_n(0) = 0$ (perfectly asynchronous crowd);
2. if the oscillation amplitude becomes large, $x_n \to \infty$, we expect $\bar{g}_n(x_n = \infty) = N g_{average}$ (perfectly synchronous crowd). This is of course only a mathematical limit: from an engineering point of view we expect $\bar{g}_n \cong N g_{average}$ even for finite, although 'large', values of x_n.

We observe that $\bar{g}_n(x_n)$ is a monotonically increasing function and it is easier to be 'invented' than $\bar{g}_n(\sigma_n)$. Therefore, we propose for $\bar{g}_n$ the following law which satisfies the aforementioned requirements:

$$\bar{g}_n = N g_{average}\tanh\left(\frac{x_n}{\delta}+\varepsilon\right). \tag{41}$$

$\delta > 0$ is a parameter with the dimensions of a length. It can be determined experimentally and it measures how fast the asymptotic value $Ng_{average}$ becomes saturated while the oscillations amplitude is growing up. δ can also represent the bridge displacement for which synchronization is almost completed, being $\tanh(1+\varepsilon) \cong 0.76-0.80$. As a consequence, its value can be estimated slightly lower than the bridge maximum lateral displacement.

$\varepsilon > 0$ is a perturbation/imperfection dimensionless parameter with respect to the limit ideal case $\varepsilon = 0$ (non-perturbed/perfect case) in which the bridge is still and the crowd is perfectly asynchronous and therefore unable to exert any lateral net force on the bridge itself. In real cases ($\varepsilon > 0$), even if the bridge is still, there is a certain component of lateral force due exclusively to random synchronization phenomena between pedestrians, which is assumed to be $Ng_{average}\tanh(\varepsilon)$ in this model. In general, however, this force is small, so that we assume ε to be small.

Substituting Eq. (41) into $\bar{\beta}_n = \bar{g}_n\sqrt{k_1^2+k_2^2}$, we have:

$$\bar{\beta}_n = \sqrt{k_1^2+k_2^2}\,N g_{average}\tanh\left(\frac{x_n}{\delta}+\varepsilon\right), \tag{42}$$

and we note that, with the previous assumptions, $\bar{\beta}_n$ is no longer an *independent* variable.

Summarizing the previous developments, we have found that our discrete-time model has only two independent variables, x_n and σ_n, and it is described by the two-dimensional map:

$$\begin{cases} x_{n+1} = f_1(x_n,\sigma_n) = \alpha x_n + (1-\alpha)\,\delta\dfrac{N}{N_{cr}}\tanh\left(\dfrac{x_n}{\delta}+\varepsilon\right)\cos(\sigma_n), \\ \sigma_{n+1} = f_2(x_n,\sigma_n) = \dfrac{a}{a+x_n}\sigma_n. \end{cases} \tag{43}$$

The parameter N_{cr} appearing in (43) is of great practical interest and it is defined by:

$$N_{cr} = \frac{(1-\alpha)\delta}{g_{average}\sqrt{k_1^2+k_2^2}} \cong \frac{2\delta\xi}{g_{average}} = \frac{2\xi\delta K}{G_{average}}. \tag{44}$$

Equations (43) and (44) describe the phenomenon of synchronization of pedestrians' motion with the lateral vibrations of footbridges. Note that in this model we consider a global synchronization, without distinguishing between the two different levels of synchronization observed in practice [15]: bridge-pedestrian and pedestrian-to-pedestrian (see Sects. 2.2.7 and 2.2.8).

The model parameters are the damping ξ, the sensitivities a and δ, the imperfection ε, $g_{average}$ and, of course, the number N of pedestrians walking on the bridge.

3.3 Fixed Points

The dynamic aspects of the map (43) of interest for the present chapter are the fixed points

$$\begin{cases} x = f_1(x,\sigma) \\ \sigma = f_2(x,\sigma) \end{cases}, \tag{45}$$

which corresponds to oscillation of the original (physical) system, and their stability. They are investigated in the following by distinguish between the imperfect ($\varepsilon > 0$) and the perfect ($\varepsilon = 0$) cases.

3.3.1 The Imperfect Case

We consider first the real case $\varepsilon > 0$. In this case the system of equations (45) becomes

$$\begin{cases} (1-\alpha)x = (1-\alpha)\,\delta\frac{N}{N_{cr}}\tanh\left(\frac{x}{\delta}+\varepsilon\right)\cos(\sigma) \\ \sigma x = 0 \end{cases}. \tag{46}$$

The second equation of (46) admits two solutions which, inserted in the first equation, give two different solutions:

$$\text{A)}: \begin{cases} x = 0 \\ \sigma = \pm\frac{\pi}{2} \end{cases}, \quad \text{B)}: \begin{cases} \sigma = 0 \\ \frac{x}{\delta} = \frac{N}{N_{cr}}\tanh\left(\frac{x}{\delta}+\varepsilon\right) \end{cases}. \tag{47}$$

The two fixed points $(47)_{\mathrm{A}}$ correspond to the condition of bridge in equilibrium (motionless), with perfect bridge-pedestrians de-synchronization. We guess they are only theoretical solutions, i.e. they are unstable.

The fixed point $(47)_{\mathrm{B}}$ corresponds to perfect bridge-crowd synchronization with non-vanishing oscillations of the bridge. As we will see later, they are stable solutions and they are involved in the phenomenon of synchronous lateral excitation.

It is worth to note that the equilibrium points are independent of the pedestrian sensitivity parameter a; therefore we don't need to determine it experimentally, unlike other models present in the literature [12], [47]. This is a worthy aspect of our model. However, we expect that a will influence the rate of convergence towards the equilibrium solutions.

To discuss the stability of the fixed points Eq. (47), we evaluate the eigenvalues of the Jacobian matrix of the system (43):

$$\mathbf{J} = \begin{bmatrix} \dfrac{\partial f_1}{\partial x_n} & \dfrac{\partial f_1}{\partial \sigma_n} \\ \dfrac{\partial f_2}{\partial x_n} & \dfrac{\partial f_2}{\partial \sigma_n} \end{bmatrix}. \tag{48}$$

For the fixed points A the eigenvalues are both real:

$$\lambda_{1,2} = \frac{1}{2}\left\{1+\alpha \pm \sqrt{(1-\alpha)^2+2(1-\alpha)\delta\frac{\pi}{a}\frac{N}{N_{cr}}\tanh(\varepsilon)}\right\}$$
$$\cong 1-\pi\xi \pm \pi\xi\sqrt{1+\frac{\delta}{a}\frac{N}{N_{cr}}\frac{\tanh(\varepsilon)}{\xi}}. \tag{49}$$

We observe that $\lambda_1 > 1$ and $\lambda_2 < 1$, being $\varepsilon > 0$; therefore the fixed points $(47)_A$ are saddles, and unstable as expected.

In correspondence of the fixed point B, the Jacobian matrix is diagonal and the eigenvalues, both real, are:

$$\lambda_1 = \alpha+(1-\alpha)\frac{N}{N_{cr}}\frac{1}{\cosh^2\left(\frac{x}{\delta}+\varepsilon\right)} = \alpha+(1-\alpha)\left[\frac{N}{N_{cr}}-\left(\frac{x}{\delta}\right)^2\frac{N_{cr}}{N}\right],$$
$$\lambda_2 = \frac{a}{a+x}, \tag{50}$$

where x is the fixed point position, solution of the second equation of system $(47)_B$. We observe that $\lambda_2 < 1$, while the condition $\lambda_1 < 1$ corresponds to the inequality:

$$\left(\frac{x}{\delta}\right)^2 > \frac{N}{N_{cr}}\left(\frac{N}{N_{cr}}-1\right) \tag{51}$$

which is always satisfied. Therefore the solution $(47)_B$ is always stable, as expected. It is shown in red (grey) in Fig. 31 for different values of the parameter ε.

3.3.2 The Perfect Case

Now we consider the unperturbed limit case $\varepsilon = 0$ and we determine likewise the fixed points of the map (43). Again, there are two types of solutions of (45):

$$\text{A)}: \begin{cases} x=0 \\ \forall\sigma \end{cases}, \quad \text{B)}: \begin{cases} \sigma = 0 \\ \dfrac{x}{\delta} = \dfrac{N}{N_{cr}}\tanh\left(\dfrac{x}{\delta}\right) \end{cases}. \tag{52}$$

The solutions $(52)_A$ are a manifold of fixed points non depending on the relative phase crowd-bridge. They are all the points of the straight line $x = 0$. In the presence of perturbations, $\varepsilon > 0$, they are reduced to the only two unstable fixed points $x = 0$, $\sigma = \pm\pi/2$ previously investigated.

The solutions $(52)_B$ are a curve of fixed points consisting of two branches: $x = 0$ and another one, $N = N_{cr}(x/\delta)/\tanh(x/\delta)$, bifurcating from the previous one at $N = N_{cr}$. It is shown in blue (black) in Fig. 31.

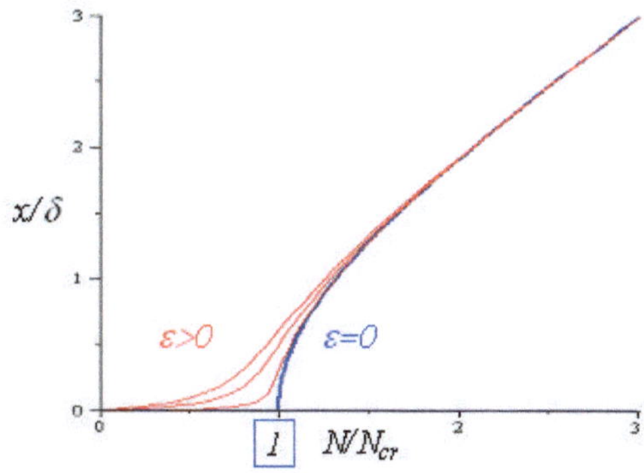

Fig. 31 Synchronized fixed points $x = x(N)$, for different values of parameter ε: $\varepsilon = 0.01,\ 0.05,\ 0.1$ (red (grey) curves); the solution for the unperturbed limit case $\varepsilon = 0$ is also shown (blue (black) curve)

To discuss the stability of the fixed points Eq. (52) we consider the eigenvalues of the Jacobian matrix (48). For the fixed points A the eigenvalues are real and equal to:

$$\lambda_1(\sigma) = \alpha + (1-\alpha)\frac{N}{N_{cr}}\cos(\sigma)\,, \quad \lambda_2 = 1\,. \tag{53}$$

Therefore, all the fixed points of the manifold are non-hyperbolic. As regards their stability, we can only affirm that they are certainly unstable when $|\lambda_1| > 1$, i.e., for

$$N > \frac{N_{cr}}{\cos(\sigma)}, \ \text{for } \sigma \text{ such that } \cos(\sigma) > 0, \tag{54}$$

and

$$N > \frac{N_{cr}}{-\cos(\sigma)}\frac{1+\alpha}{1-\alpha}, \ \text{for } \sigma \text{ such that } \cos(\sigma) < 0. \tag{55}$$

To further investigate the stability, we need to consider the nonlinear terms in the equations. We omit the computations, which are heavy and of little interest for our purposes; we show only the results, which are instead of great interest to improve the understanding of the physical event and to actually use our model. We find that:

1. all the fixed points $x = 0$ and $\sigma \neq 0$ are unstable;
2. the fixed point $x = 0$ and $\sigma = 0$ is stable for $N < N_{cr}$ and unstable for $N > N_{cr}$.

The eigenvalues of the fixed points B in Eq. (52) are given by Eq. (50) and thus they are both real and lesser than 1, so that the corresponding equilibrium position is always stable.

We can conclude that for $N = N_{cr}$, the bifurcating solution $(52)_{\mathrm{B}}$ catches the stability of $x = \sigma = 0$ and triggers the lateral synchronization. This is a pitchfork bifurcation of degenerate type, as the fundamental branch is not made of hyperbolic points (blue (black) lines in Fig. 31).

On the basis of these observations, the fixed points $(47)_{\mathrm{B}}$ can be seen as perturbations of the aforementioned pitchfork bifurcation.

The meaning of N_{cr}, as defined in Eq. (44) is now clear: it is the theoretical point where bifurcation occurs, the practical one being slightly smaller, as we can see from the lift-off point of the red (grey) curves in Fig. 31. It corresponds to the theoretical number of pedestrians triggering the synchronization and thus it is the most important information from an engineering point of view. Its relevance is underlined by the fact that it has a simple analytical formula, which helps in understanding how the various parameters influence the synchronization phenomenon.

In fact, Fig. 31 confirms that for small crowds, N/N_{cr} near zero, the amplitude ratio x/δ is near zero too, as walkers are desynchronized and randomly phased. Then, as more and more people walk on the deck, there is no hint of wide oscillations until the crowd reaches the critical size N_{cr}, after that wobbling and synchrony suddenly emerge simultaneously, as dual aspects of a single instability mechanism (Eq. (41) associates the synchronization with the current state of the oscillator).

3.4 A Case-Study: The London Millennium Footbridge

To test the agreement of our model with the dynamical behaviour of real structures we consider again as a benchmark the London Millennium Bridge, as it is the most well-known and well-documented case of 'lively' footbridge in the literature and in this sense it is the most suitable for a detailed analysis of the model behaviour and of the various aspects of the synchronous lateral excitation phenomenon.

With the data of the Millennium Bridge $M = 113000\,\text{kg}$, $B = 11000\,\text{kg/s}$, $K = 4730000\,\text{kg/s}^2$, $G_{average} = 30\,\text{N}$, $\xi = 0.0075$ and $g_{average} = G_{average}/K = 6.34 \times 10^{-6}\,\text{m}$ (see Sect. 2.1) we have from Eq. (44)

$$N_{cr} = \frac{2\delta\xi}{g_{average}} = 2365\delta. \tag{56}$$

Assuming $N_{cr} = 155$ (see Sect. 2.1) we obtain $\delta = 0.0654\,\text{m} = 6.5\,\text{cm}$; this means that the predicted amplitude of the bridge motion is very close to the observed values of about 5–7 cm on the opening day. There is a good agreement of the model results with the real behaviour of the bridge.

Another consideration can be drawn from Fig. 31 and from the second equation of the system $(47)_B$. They show that for large values of x and N the equation of the bifurcated branch becomes

$$x \cong \delta\frac{N}{N_{cr}}. \tag{57}$$

This linear trend is in good agreement with the SAMEO model results shown in Fig. 32, where the oscillations amplitude versus time curves have a similar linear trend in their final part. This observation allows a further estimate of the ratio δ/N_{cr}: with reference to Fig. 32 we register an amplitude of oscillation $x = 0.41\,\text{m}$ for $N = 1000$ pedestrians, namely, $\delta/N_{cr} = 4.1 \times 10^{-4}$. By assuming $N_{cr} = 155$, we obtain $\delta \cong 6.15\,\text{cm}$, which is comparable with the one previously calculated and confirms its reliability.

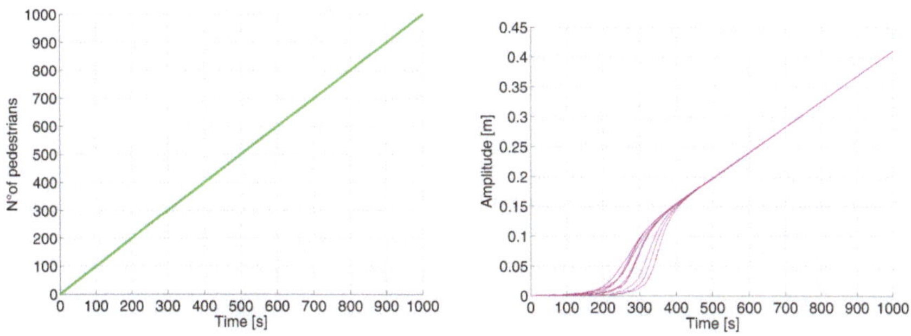

Fig. 32 Numerical results from the SAMEO model: **a** number of walkers on the bridge and **b** amplitude of vibration versus time. The various amplitude curves differ for the initial conditions randomly assigned to pedestrians' motion [30]

Comparing our equilibrium solutions (Fig. 31) with the SAMEO model results (Fig. 32), it is evident that our discrete-time model is able to describe accurately the phenomenon of synchronous lateral excitation, even with a simpler analytical formulation and, more important, without requiring numerical simulations.

4 Conclusions

After a focus on the problem of synchronous lateral excitation in slender footbridges and a critical overview of the existing literature on the topics, in the first section of this chapter a parametric study of the SAMEO model [47] for the pedestrians-induced lateral vibrations of footbridges has been performed and some modifications have been introduced and checked. The application of this model to the case of the London Millennium Bridge provides results in good agreement, both qualitatively and quantitatively, with observations and experiments. The model is able to predict simultaneously both the onset of bridge instability and the onset of crowd synchronization, providing a reliable value of the number of pedestrians which trigger the synchronization, i.e., the critical threshold. The steady state amplitude for bridge motion is also well predicted.

The extended numerical simulations permit to draw the following conclusions.

1. The initial distribution of pedestrians natural frequencies, i.e. their level of typological homogeneity, is decisive both for the trigger point of instability and for the temporal probability of the event to occur (see the relative time scales for onset of large bridge motion in Figs. 7, 9, 10).
2. The shape of the pedestrian loading wave plays an important role. The ideal sinusoidal trend proposed in the original formulation of the model proves to be quite different from the typical time series of lateral forcing as measured by experiments on a treadmill [5]. The square-wave we introduce is more realistic and more conservative, as the critical number of pedestrians decreases.

3. The amplitude of lateral force due to pedestrians can increase dramatically when they change their gate on a moving platform. Numerical results, both for Abrams' relationship and for our bilinear relationship $G(A)$, show the necessity to consider this phenomenon. In fact, it affects the steady-state amplitude of the bridge motion and the speed at which the large amplitude oscillations set up. In particular, our relationship provides a lower critical number of pedestrians. Both the relationships share the same qualitative behaviour of the bridge-crowd system in terms of amplitude versus time curves.
4. The additional pedestrian-to-pedestrian synchronization self-excites the phenomenon, and accordingly the critical threshold considerably reduces. Although the two considered types of synchronization are different in their nature, they usually happen simultaneously and lead to the same result: an increase in the response of the structure. It is clear the necessity to take into account both of them.
5. Different loading paths (i.e. different numbers of pedestrians introduced on the bridge deck per unit time) affect both the time-history of the event and the critical number N_c. However, the phenomenon remains unaltered in its essence, thus emphasizing the robustness of the model.

It is worthy to note that the model depends only on the modal characteristics of the bridge, on the biological dynamics of human walking in a crowd, and on its dynamical interaction with the deck. So, unlike other models described in the literature [12], [11], [33], it is applicable to any bridge where a similar phenomenon is observed or expected to occur. The unique 'weak' point is the determination of the two parameters C (originally introduced to measure pedestrian sensitivity to bridge lateral oscillations) and D (we introduce here to measure the visual sensitivity to the synchronization between a pedestrian and the previous one). However the generality of the model is not invalidated as actual distributions of C and D could be determined through an *ad hoc* experimental campaign on a representative sample of the population. Experiments in this sense would be welcome.

Finally, it is necessary to remark that the SAMEO model is pseudo-stochastic, and it does not consider various (minor in our opinion) aspects. The perfect periodicity of the lateral human-induced load is assumed only with respect to the pedestrians phases (initially randomly assigned), as walking is not a perfectly periodic activity with respect to time, as wide experimental studies confirm [41], [42]; but the adopted approach is deterministic in its evolution and it can be criticized due to the random nature of walking forces, which would suggest the use of a really stochastic approach. Furthermore subharmonic or superharmonic resonances are not considered in the model.

All the previous observations, besides improving the understanding of the underlying physical phenomenon, allow us to state that the SAMEO model, despite some limitations, is sufficiently simple and robust in its previsions; therefore it may prove useful to estimate, e.g., the damping needed to stabilize other exceptionally crowded footbridges against synchronous lateral excitation by pedestrians. In this sense, it could constitute the basis to look for technical solutions aimed at limiting or avoiding the phenomenon of the pedestrians-induced lateral vibrations of footbridges.

As regards to the second section of this chapter, which is dedicated to the analysis of the discrete-time model, the most important conclusions are summarized in the following.

1. Also the proposed discrete-time model is independent of the specific real case-study we have considered (i.e. the London Millennium Bridge). It depends only on the modal characteristics of the bridge, on the biological dynamics of human walking in a crowd, and on its dynamical interaction with the deck.
2. As regards the parameters ε and δ , they do not limit neither the generality of the model nor its easiness and speed of application. In fact ε should be determined experimentally, but its evaluation pertains exclusively the unconstrained human walking dynamics (random synchronization phenomena) and not the footbridge we are studying. δ is a value very close to the bridge maximum lateral displacement and so we usually fix it in stage of project.
3. The stochastic aspects are put into account in the definition of the so-called 'equivalent pedestrian' which resumes the characteristics of the pedestrians in crowd, in terms of maximum exerted force, native frequency and random phase.
4. We assume that all the pedestrians have the same native frequency, even though experiments on a statistical sample of the population reveal a Gaussian distribution [54]. This is not a real problem, because our assumption constitutes only an analytical simplification which doesn't influence the structure of the model and its general validity.
5. Perfect periodicity of the lateral human-induced load is assumed. Actually, the walking force is not perfectly periodic and it could be attenuated due to interaction between the pedestrian and the structure. Moreover, we idealize again the pedestrian forcing as sinusoidal, even if experiments on a treadmill reveal a trend more similar to a square wave [5], as said.

The simple discrete-time model (2D map) is able to explain the main characteristics of the phenomenon of synchronous lateral excitation, without numerical simulations.

A simple analytical formula to compute the critical number of pedestrians triggering the synchronization is proposed. Its application to a real case-study returns reliable values, in good agreement both qualitatively and quantitatively with other consolidated results [30] and with observations [12]. Moreover it shows that N_{cr} depends only on the bridge damping and stiffness, on the average maximum lateral force exerted by walking pedestrians and then on the bridge maximum lateral displacement.

The main result, from a dynamical system point of view, is that the model highlights how the phenomenon is a perturbation of a classical (but degenerate) pitchfork bifurcation, which is the underlying dynamical event. This observation permits an improved understanding of the physical event underlying synchronization.

Finally, we can conclude that by the combined use of continuous- and discrete-time models we achieved a good understanding of the synchronization induced large lateral oscillations of footbridges, and reliable estimations of the triggering number of pedestrians.

References

1. Abrams, D.M.: Two coupled oscillator models: the Millennium Bridge and the chimera state. PhD Dissertation in Theoretical and Applied Mechanics, Cornell University, Ithaca, NY (2006), http://www.mit.edu/ dmabrams/papers/ Abrams_Daniel_Michael_dissertation.pdf (accessed September 28, 2008)
2. Arup videos (2000), http://www.arup.com/MillenniumBridge/indepth/video.html (accessed September 28, 2008)
3. Balthazar, J.M., Mook, D.T., Webwe, H.I., Brasil, R.M.L.R.F., Fenili, A., Belato, D., Felix, J.L.P.: An overview on non-ideal vibrations. Meccanica 38, 613–621 (2003)
4. Bauby, C.E., Kuo, A.D.: Active control of lateral balance in human walking. J. Biomech. 33, 1433–1440 (2000)
5. Belli, A., Bui, P., Berger, A., Geyssant, A., Lacour, J.R.: A treadmill ergometer for three-dimensional ground reaction forces measurement during walking. J. Biomech. 34(1), 105–112 (2001)
6. Blanchard, J., Davies, B.L., Smith, J.W.: Design criteria and analysis for dynamic loading of footbridges. In: Symposium Dyn Behav Bridges, Crowthorne, Berks, TRRL SR275:90-107 (1977)
7. Blekherman, A.N.: Autoparametric Resonance in a Pedestrian Steel Arch Bridge: Solferino Bridge. Paris. J. Bridge Eng. ASCE 12(6), 669–676 (2007)
8. Bodgi, J., Erlicher, S., Argoul, P.: Lateral vibration of footbridges under crowd-loading: Continuous crowd modelling approach. Key Eng. Mater. 347, 685–690 (2007)
9. Borri, C., Höffer, R.: Aeroelastic wind forces on flexible bridge girders. Meccanica 35, 1–15 (2000)
10. Candaten, M., Rinaldi, S.: Peak-to-peak dynamics: a critical survey. Int J. Bif. Chaos 10(8), 1805–1819 (2000)
11. Danbon, F., Grillaud, G.: Dynamic behaviour of a steel footbridge. Characterization and modelling of the dynamic loading induced by a moving crowd on the Solferino Footbridge in Paris. In: Proc Footbridge 2005 (2005)
12. Dallard, P., Fitzpatrick, A.J., Flint, A., Le Bourva, S., Low, A., Ridsdill Smith, R.M., Willford, M.: The London Millennium Footbridge. Struct. Eng. 79(22), 17–33 (2001)
13. Dallard, P., Fitzpatrick, A.J., Flint, A., Low, A., Ridsdill Smith, R.M., Willford, M., Roche, M.: London Millennium Bridge: pedestrian-induced lateral vibration. J. Bridge Eng. 6(6), 412–417 (2001)
14. Eckhardt, B., Ott, E., Strogatz, S.H., Abrams, D.M., McRobie, A.: Modeling walker synchronization on the Millennium Bridge. Phys. Rev. E 75, 021110 (2007)
15. Fib: Guidelines for the design of footbridges. Guide to good practice. Bulletin 32 (2005)
16. Fitzpatrick, A.J., Dallard, P., Le Bourva, S., Low, A., Ridsdill Smith, R.M., Willford, M.: Linking London: The Millennium Bridge. Royal Academy of Engineering, 1–28 (2001)
17. Fujino, Y., Pacheco, B.M., Nakamura, S.I., Warnitchai, P.: Synchronization of human walking observed during lateral vibration of a congested pedestrian bridge. Earthq. Eng. Struct. Dyn. 22, 741–758 (1993)
18. Gallagher, D.P.: Multi-Dimensional Optimization. Numer. Anal. 92, 563 (2006)
19. Gregoriou, G.G., Gotts, S.J., Zhou, H., Desimone, R.: High-Frequency, Long-Range Coupling Between Prefrontal and Visual Cortex During Attention. Science 324, 1207–1210 (2009)

20. Grundmann, H., Kreuzinger, H., Schneider, M.: Dynamic calculations of footbridges. Bauingenieur 68, 215–225 (1993)
21. Hughes, R.L.: A continuum theory for the flow of pedestrians. Transp. Res. Part B: Methodol. 36(6), 507–535 (2002)
22. Hof, A.L., Gazendam, M.G.J., Sinke, W.E.: The condition for dynamic stability. J. Biomech. 38, 1–8 (2005)
23. Hof, A.L., van Bockel, R.M., Schoppen, T., Postema, K.: Control of lateral balance in walking: experimental findings in normal subjects and above-knee amputees. Gait Posture 25, 250–258 (2007)
24. Hoogendoorn, S.P., Bovy, P.H.L.: Pedestrian route-choice and activity scheduling theory and models. Transp. Res. Part B: Methodol. 38(2), 169–190 (2004)
25. Kaneoke, Y.: Magnetoencephalography: In search of neural processes for visual motion information. Progr. Neurobiol. 80(5), 219–240 (2006)
26. Lagarias, J.C., Reeds, J.A., Wright, M.H., Wright, P.E.: Convergence Properties of the Nelder-Mead Simplex Method in Low Dimensions. SIAM J. Optim. 9(1), 112–147 (1998)
27. Lenci, S., Marcheggiani, L.: A discrete-time model for the phenomenon of synchronous lateral excitation due to pedestrians motion on footbridges. In: Proc. Third International Conference Footbridge 2008, Porto, Portugal (2008)
28. Macdonald, J.H.G.: Lateral excitation of bridges by balancing pedestrians. Proc. R. Soc. A 465, 1055–1073 (2009)
29. Matsumoto, Y., Nishioka, T., Shiojiri, H., Matsuzaki, K.: Dynamic design of footbridges. In: Proc. IABSE, Zurich, Switzerland (1978)
30. Marcheggiani, L., Lenci, S.: On a model for the pedestrians-induced lateral vibrations of footbridges. Meccanica 45, 531–551 (2010); doi:10.1007/s11012-009-9269-0
31. McRobie, A., Morgenthal, G., Lasenby, J., Ringer, M.: Section model tests on human-structure lock-in. Bridge Eng. 156(2), 71–79 (2003)
32. Mondada, L.: Emergent focused interactions in public places: A systematic analysis of the multimodal achievement of a common interactional space. J.Pragmat. 41(10), 1977–1997 (2008)
33. Nakamura, S.: Model for lateral excitation of footbridges by synchronous walking. J. Struct. Eng. ASCE 130, 32–37 (2004)
34. Nakamura, S.I., Kawasaki, T.: Lateral vibration of footbridges by synchronous walking. J. Constr. Steel Res. 62, 1148–1160 (2006)
35. Newland, D.E.: Vibration: problem and solution. In: Sudjic, D. (ed.) Blade of light: the story of London's Millennium Bridge. Penguin Books, London (2001)
36. Newland, D.E.: Pedestrian excitation of bridges - recent results. In: Proc. Tenth Int. Congr. Sound Vib., Stockholm, Sweden (2003)
37. Piccardo, G., Tubino, F.: Parametric resonance of flexible footbridges under crowd-induced lateral excitation. J. Sound Vib. 311, 353–371 (2008)
38. Roberts, T.M.: Lateral pedestrian excitation of footbridges. J. Bridge Eng. ASCE 10(1), 107–112 (2005)
39. Roberts, T.M.: Synchronised pedestrian lateral excitation of footbridges. In: Proc. Eurodyn 2005, pp. 1089–1094 (2005)
40. Roberts, T.M.: Probabilistic pedestrian lateral excitation of bridges. Proc. ICE, Bridge. Eng. 158(2), 53–61 (2005)
41. Ricciardelli, F., Pizzimenti, A.D.: Lateral Walking-Induced Forces on Footbridges. J. Bridge Eng. ASCE 12(6), 677–688 (2007)
42. Ricciardelli, F., Pizzimenti, A.D.: Experimental evaluation of the dynamic lateral loading of footbridges by walking pedestrians. In: Proc. Eurodyn. 2005 (2005)

43. Salvatori, L., Spinelli, P.: A discrete 3D model for bridge aerodynamics and aeroelasticity: nonlinearities and linearizations. Meccanica 42, 31–46 (2007)
44. Strogatz, S.H.: Nonlinear Dynamics and Chaos, with Applications to Physics, Biology, Chemistry, and Engineering. Perseus Books, Massachusetts (2000)
45. Strogatz, S.H.: Sync: the emerging science of spontaneous order. Hyperion, New York (2003)
46. Strogatz, S.H.: From Kuramoto to Crawford: exploring the onset of synchronization in populations of coupled oscillators. Physica D 143, 1–20 (2000)
47. Strogatz, S.H., Abrams, D.M., McRobie, F.A., Eckhardt, B., Ott, E.: Crowd synchrony on the Millennium Bridge. Nature 438, 43–44 (2005)
48. Sétra, Footbridges: assessment of vibrational behaviour of footbridges under pedestrian loading. Afgc, Paris (2006), `http://www.setra.equipement.gouv.fr/IMG/pdf/US_0644A_Footbridges.pdf` (accessed September 28, 2008)
49. Trovato, A., Erlicher, S., Argoul, P.: A modified Van der Pol oscillator for modelling the lateral force of a pedestrian during walking. In: Proc. Conference Vibrations, Chocs & Bruit, Lyon, France (2008)
50. Vaughan, C.L.: Theories of bipedal walking: an odyssey. J. Biomech. 36, 513–523 (2003)
51. Venuti, F., Bruno, L.: Crowd-structure interaction in lively footbridges under synchronous lateral excitation: A literature review. Physics of Life Reviews 6, 176–206 (2009)
52. Venuti, F., Bruno, L., Bellomo, N.: Crowd dynamics on a moving platform: Mathematical modelling and application to lively footbridges. Math. Comput. Model. 45, 252–269 (2007)
53. Wheeler, J.E.: Pedestrian induced vibration in footbridges. Technical Rep N°15, Main Roads Dep., Western Australia (1980)
54. Zivanovic, S., Pavic, A., Reynolds, P.: Vibration serviceability of footbridges under human-induced excitation: a literature review. J. Sound Vib. 279, 1–74 (2005)

Applications for Shape Memory Alloys in Structural and Machine Dynamics

Matthew P. Cartmell, Arkadiusz J. Żak and Olga A. Ganilova

Abstract. This chapter presents a review of some of the fundamental science behind applications for shape memory alloys (SMAs) in mechanical engineering structures and machines in the context of an inexhaustive review of the literature. Following this three well known literature models are considered in some detail after which a summary investigation of the effect of SMAs on the dynamics of beams and plates is given. This leads into a discussion of applications in rotor dynamics for which SMA elements are shown to have considerable uses in the modification of resonant behaviour within the rotor. The chapter concludes with further work on plates, and the concept of antagonism as a means for the approximate equalisation of heating and cooling time constants.

Keywords: Shape memory alloys, vibration, dynamics, resonance, nonlinearity.

1 Review of the Literature and Introduction

Shape memory alloys (SMAs) exhibit very interesting and useful mechanical properties, particularly two unique effects known as the *shape memory effect* and *superelasticity*. An SMA which is initially deformed recovers its original low temperature shape during heating, a demonstration of the shape memory effect. This effect can

Matthew P. Cartmell · Olga A. Ganilova
School of Engineering, University of Glasgow, James Watt South Building,
Glasgow G12 8QQ, Scotland, United Kingdom
e-mail: {Matthew.Cartmell,Olga.Ganilova}@glasgow.ac.uk

Arkadiusz J. Żak
Szewalski Institute of Fluid Flow Machinery, Polish Academy of Sciences,
Fiszera 14, 80-952 Gdask, Poland
e-mail: arek@imp.gda.pl

J. Warminski et al. (Eds.): Nonlinear Dynamic Phenomena in Mechanics, SMIA 181, pp. 115–158.
springerlink.com

also generate large internal recovery stresses. In addition, at higher temperatures the superelastic phenomenon may be seen; an effect associated with large nonlinear loading and unloading recovery strains. According to Otsuka and Wayman [47] the first reported steps made in categorizing and understanding the shape memory effect were taken in the 1930s, firstly by Ölander who discovered the pseudoelastic behaviour of Au-Cd alloys, and then by Greninger and Mooradian who observed the formation and disappearance of the martensitic phase by decreasing and increasing the temperature of a Cu-Zn alloy in 1938. The phenomenon of memory as governed by martensitic thermoelasticity was widely reported a decade later by Kurdjumov and Khandros in 1949, and then in 1951 when Chang and Read [15] discovered the shape memory effect in an Au-Cd alloy, after which Buehler et al. [12] found it in the now ubiquitous Ni-Ti alloys. Shape memory properties have since been found in In-Tl [13], [2] Cu-Zn [27] and Cu-Al-Ni [16], and many new forms of alloys, polymers, and ceramics, are all actively being researched. Amongst the many different SMAs that are available the Ni-Ti alloys are of significant practical interest because of their unique and pragmatic mechanical properties, particularly in comparison with Cu-Zn-Al or Cu-Al-Ni alloys [30]. Ni-Ti SMAs exhibit very high ultimate tensile strengths of up to 1000 MPa, with elongations to failure of 50%, recovery stresses up to 800 MPa, and exceptional damping properties [30]. Because the Young's modulus [47], [30], damping characteristics [49], [23] and the capability for large internal forces [51] can all be influenced precisely, SMAs can be applied to various problems in civil and mechanical engineering.

Rapidly emerging engineering applications for SMAs have led to the development of effcient and robust mathematical models for shape memory and superelastic behaviour. In-depth studies of SMA behaviour account for the complexity of the multi-axial stress and strain fields, and also include plasticity due to the shape memory transformations. This phenomenology is discussed in detail by Boyd and Lagoudas [7], Lexcellent et al. [37], Brocca et al. [11], Liew et al. [40], Zhu et al. [65] and Govindjee and Hall [25]. One-dimensional models are commonly used such as those proposed by Tanaka [58], Raniecki et al. [50], Liang and Rogers [39], Boyd and Lagoudas [7] and Brinson [10]. Nishimura et al. [42] investigated macroscopic shape recoverability in a Fe-Cr-Ni-Mn polycrystalline SMA during heating by applying the Tanaka model, whereas Lexcellent et al. [38] studied the pseudoelastic behaviour of a Ni-Ti SMA in connection with the growth of the martensite phase from the parent austenite phase during loading. Wu et al. [64] constructed a simple model using that of Liang and Rogers [39] and studied the repeated stress-temperature cycling of SMAs. Ford and White [22] researched the thermomechanical behaviour of Ni-Ti SMA wires based on the work of Brinson [10], and then Bekker and Brinson [5] looked at the evolution of the martensite fraction during the thermoelastic phase transformation when induced by a generalised thermomechanical load. Bo and Lagoudas [6] undertook an interesting comparison of the Tanaka [58] and Liang and Rogers [39] models and their own model.

A critical study of the literature on the superelastic behaviour of SMAs was completed by Epps and Chopra [21] in which the model developed by Brinson [10] was included. The results obtained from each model were compared against experimental findings.

Applications invariably utilise SMAs in wire or strip form, where such elements are integrated in some manner within composite material structures, and a comprehensive study of the dynamics of Ni-Ti wire reinforced plates has been given by Rogers et al. [52] in which two basic concepts for SMA component operation in situ have been defined. These are the *active modal modification method* (also known as *active property tuning* (APT)) and the active strain energy tuning method (ASET). These theoretical predictions were justified by experiments carried out on Ni-Ti wire-reinforced composite beams. Related work was carried out by Baz et al. [4], who used Ni-Ti SMA wires for adjusting the natural frequencies of composite beams.

The application of integrated SMA wires for affecting the natural frequencies of composite plates has also been proposed by Ostachowicz et al. [45], and Birman [8] who took sandwich panels reinforced by SMA wires to test their thermal buckling. Further work on the thermal bucking and post-bucking behaviour of composite beams and plates when reinforced by SMA components was completed by Choi et al. [19], Thompson and Loughlan [60], Ostachowicz et al. [46] and Tawfik et al. [59]. Lee and Lee [35] discussed the characteristics of certain SMA actuators and Song et al. [55] and Oh et al. [43] also carried out interesting work in which they applied SMA wires to the active shape control of composite beams and plates, respectively. Ostachowicz and Kaczmarczyk [44] investigated the vibrations of delaminated composite plates reinforced by SMA wires and in a gas stream, and Roh and Kim [53] looked at the activation of SMA wires in composite plates undergoing low velocity impacts.

It is important to emphasise that the results summarised above have principally been based on experimentally obtained SMA characteristics for which no specific SMA models were used. Many papers generally discuss various SMA reinforced composite structures, frequently applying the well known literature models of Raniecki et al., Tanaka, Liang and Rogers, and Brinson. As a case in point Chen and Levy used the Tanaka model in their work on beams with SMA layers in which they controlled the vibration [17] and damping characteristics [18]. This model was applied by Birman et al. [9] to test the local stresses found in composite material structures reinforced with Ni-Ti SMAs and subjected to isothermal loading and unloading cycles. Armstrong and Lilgholt [1] presented a summary of superelastic Ni-Ti wire-reinforced polymer matrix composite, and also certain time-dependent responses. Baz et al. [3] made analytical and experimental studies of the capability of SMAs for the shape control of composite beams with integrated Ni-Ti alloy wires, and Epps and Chandra [20] reported on the active frequency tuning of composite beams reinforced by SMA wires, by employing the Liang and Rogers [39] model.

Additionally, Tsai and Chen [61] used the same Liang and Rogers [39] model as a basis for understanding the dynamic stability of composite beams with Ni-Ti SMA wires, and Lee and Lee [34] summarised the thermal buckling and post-buckling behaviour of laminated composite shells with integrated SMA wires. Lau [32] and Lau et al. [33] used the Liang and Rogers [39] model to study the vibration characteristics of composite beams fitted with SMA wires, and Su et al. [56] investigated the thermomechanical behaviour of composite laminates reinforced by Ni-Ti wires by adopting the work of Liang and Rogers [39]. Sun et al. [57] used the Brinson [10] model to examine the thermomechanical deformation of an SMA wire-reinforced elastic beam, and Pae et al. [48] utilised SMA wires to control higher deformation modes in beams. Icardi [28] used the Brinson [10] model to study the deformation of a composite cantilever beam actuated by SMA wires, and Shu et al. [54] applied the Boyd and Lagoudas model to investigate a flexible beam with SMA wires. By using the models developed by Tanaka [58] and Boyd and Lagoudas [7], Kwai [31] usefully uncovered the strain-stress behaviour of Ni-Ti reinforced composites.

This brief summary shows the general activity of the field and confirms that many researchers have used various phenomenological models for SMAs in order to reveal the static and dynamic capabilities of different composite material structures with embedded SMA components. Because of this it is necessary to consider in some detail the differences and similarities between the literature models so as to ensure the accuracy of subsequent simulations.

2 Modelling of the Shape Memory Effect

The main objective of work undertaken by Żak et al. [70] was to generate a comparative study of SMA phenomena. That work is summarised here and is based on the use of the three most popular one-dimensional models, according to Tanaka [58], Liang and Rogers [39], and Brinson [10]. Major differences between these models and their respective accuracies are considered against results from experimental measurements of the performance of a Ni-Ti SMA wire.

The thermomechanical properties of a Ni-Ti alloy were obtained from systematic experimental measurements on a *Flexinol* actuator wire. A Ni-Ti actuator wire of 1 m long, operating at a high temperature of 94 °C was used with properties as given as follows: wire diameter 0.4 mm, martensitic wire resistance 7.87 Ω/m, maximum pulling force of 2 kgf, 4% contraction, 2.75 A activation current at room temperature, 1 s contraction time at room temperature, 10 s off-time. A special experimental test rig was constructed to determine the shape memory and superelastic characteristics, as illustrated in Fig.1.

The test system used two linear bearing carriage plates and a low friction slideway system. An SMA wire was fixed to the carriage plates and electrically insulated from the rest of the mechanical parts so that it could be resistively heated during the tests and then naturally cooled down to room temperature. The upper carriage plate stayed attached to a lead screw throughout the tests, and this enabled the initial displacement, or strain, to be precisely applied and measured. The lower carriage plate was unconstrained during all the tests. In order to investigate the superelastic behaviour of the wire a dead weight load was applied to the lower carriage plate with displacement control over the transformations. The shape memory effect of the wires was investigated by arranging for the lower carriage plate to act against a force sensor and then the internal stress state of the wire was measured as it contracted.

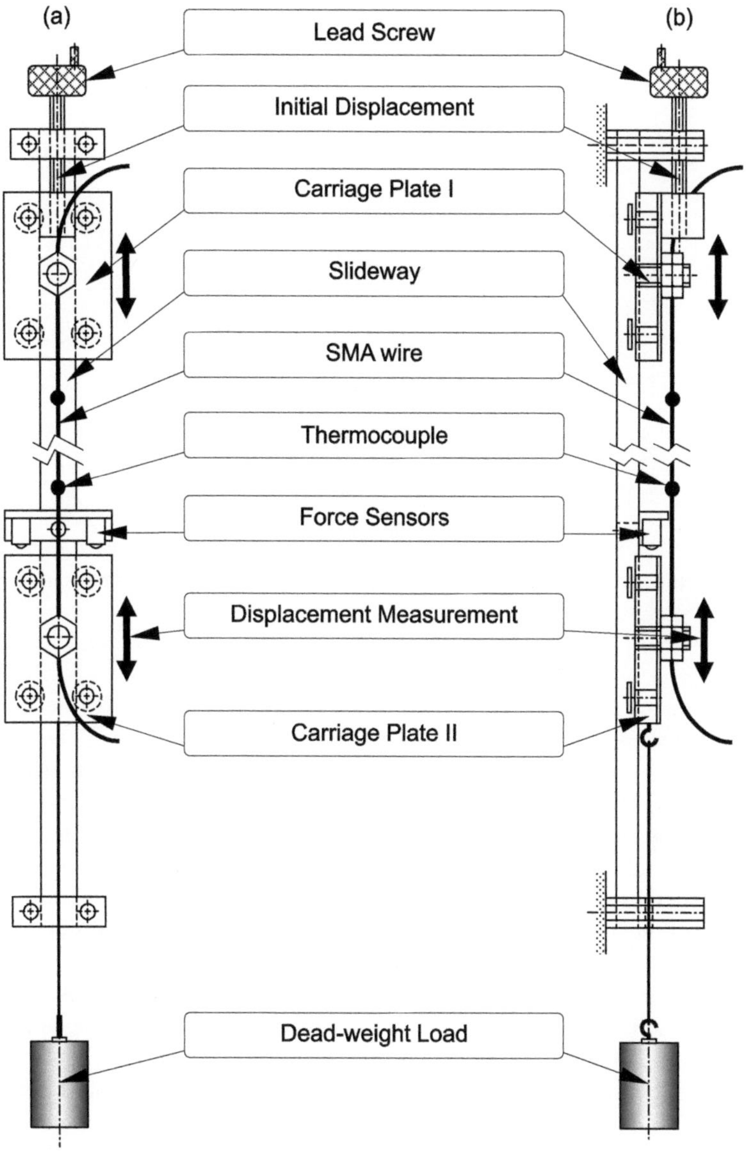

Fig. 1 Experimental test rig schematic, **a** plan, **b** side elevation, after [70]

Thermocouple sensors monitored the temperature of the wire during the tests and the purpose of the initial tests was to determine the start (*S*) and finish (*F*) transformation temperatures using the martensite (*M*) to austenite (*A*) phase of the *Flexinol* wire. It is known that the characteristic transformation temperatures are

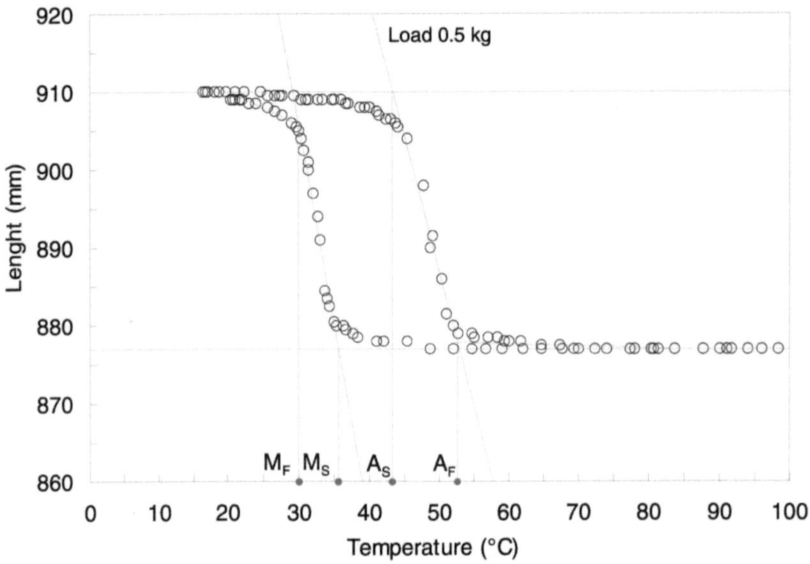

Fig. 2 Experimental results for length as a function of temperature for a *Flexinol* wire, after [70]

stress-dependent, and so they were found at four different initial stress levels associated with dead weight loads of 0.5, 1.0, 1.5 and 2.0 kg respectively. Changes in the length of the wire during each measurement were noted, with these changes being a function of its temperature over the transformations. Therefore, a set of four characteristic transformation temperatures for the wire was found for each load case. The values of the characteristic transformation temperatures M_S, M_F and A_S , A_F could be found by extrapolation for the wire at 'zero' stress conditions, also taking the weight of the lower carriage plate into account. Typical test results are given in Fig.2.

The characteristic transformation temperatures obtained from the results shown in Fig.2 are plotted and so the temperature-stress characteristics of the wire can be determined, and the four lines which indicate changes in the characteristic transformation temperatures of the wire, M_S, M_F and A_S, A_F, as a function of stress, are found, as given in Fig.3. Therefore, the characteristic transformation temperatures of the wire are determined at 'zero' stress conditions by linear extrapolation from the known levels of stresses down to the 'zero' stress conditions.

Further experiments yield other thermomechanical properties of the wire and the martensitic and austenitic Young's modulus of the *Flexinol* wire were extracted from strain-stress characteristics, these being obtained by loading and unloading the wire at constant temperature. Furthermore the critical stresses for the start of the a-m transformation from austenite to martensite σ_{AM}, and the start of the transformation from martensite to austenite σ_{MA}, were deduced and are given in Fig.4.

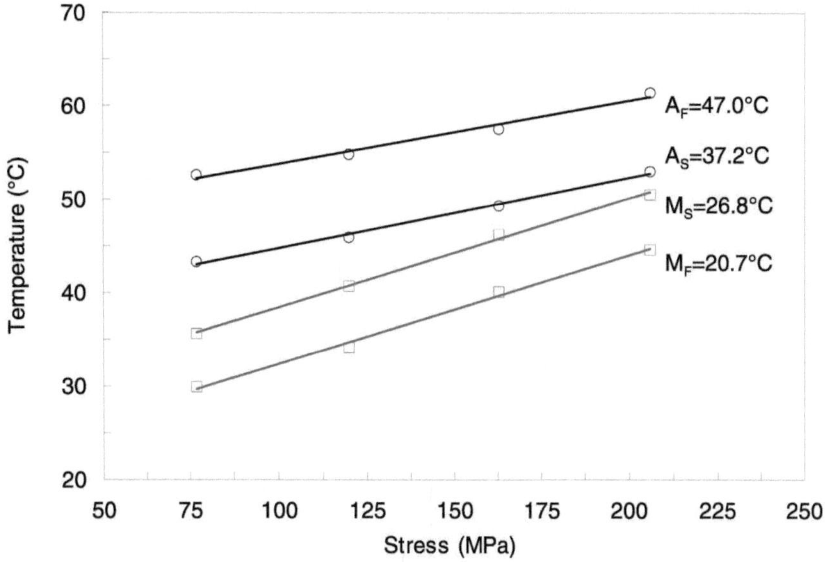

Fig. 3 Experimental temperature - stress characteristics for a *Flexinol* wire, after [70]

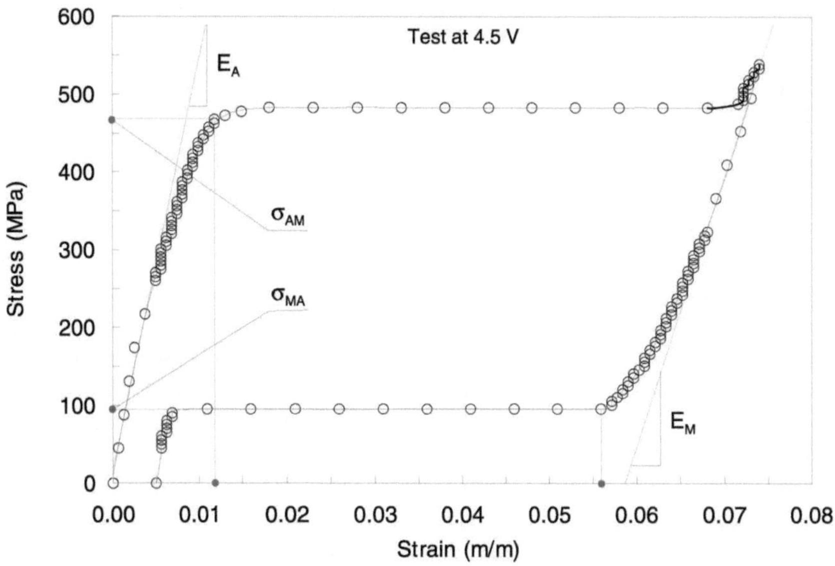

Fig. 4 Experimental stress - strain, 'loading and unloading', results for a *Flexinol* wire, after [70]

Each test was carried out at a constant applied voltage across the SMA wire, so the electrical power was affected primarily by changes in the resistivity of the wire during the transformations, and obviously the temperature of the wire changed correspondingly. The temperature changes were continuously monitored throughout

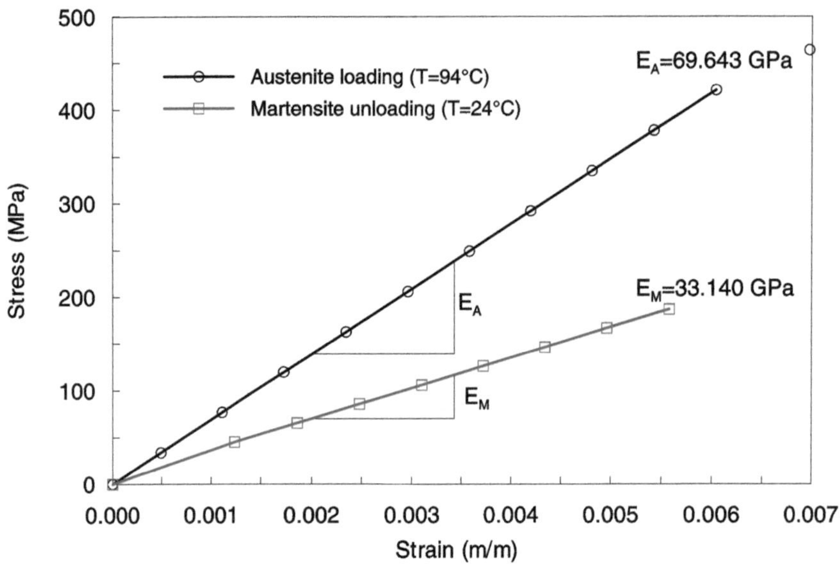

Fig. 5 Experimentally obtained martensitic and austenitic Young's Moduli, after [70]

the tests. Loading and unloading tests were undertaken in order to determine the Young's moduli of the martensite and austenite phases of the wire, and are presented in Fig.5.

A loading test was performed for the wire remaining in the austenite phase at a constant temperature of 94 °C. In the case of the martensite phase the loading and unloading test was carried out at 24 °C and this generated a value for the maximum residual strain ε_L, as well as the critical stresses for the start σ_S and finish σ_F of the transformation from temperature-induced martensite to stress-induced martensite, see [70] for full details. These loading and unloading tests were undertaken for constant voltage levels of 4.0, 4.25, 4.5, 4.75 and 5.0 V. The values of the critical stresses for the start of the transformation from austenite to martensite σ_{AM} were determined from the results obtained, and as a function of wire temperature.

Similarly, the critical stresses for the start of the transformation from martensite to austenite σ_{MA} were recorded throughout the tests, and are reproduced in detail in [70]. The stress influence coefficients c_M and c_A of the wire can be easily found by using the same results in [70]. They are defined by the slopes of the lines for the critical stresses σ_{AM} and σ_{MA} as functions of the temperature of the wire, as shown in Fig.6.

It is important to note that the constitutive equations for the Tanaka [58], Liang and Rogers [39] and Brinson [10] models all assume that the total change in stress within the candidate SMA consists of three components, namely an elastic component proportional to the Young's modulus E and a change in strain ε, also a transformational component proportional to the phase transformation coefficient Ω and a certain change in the martensite volume fraction ξ, and finally a thermal component

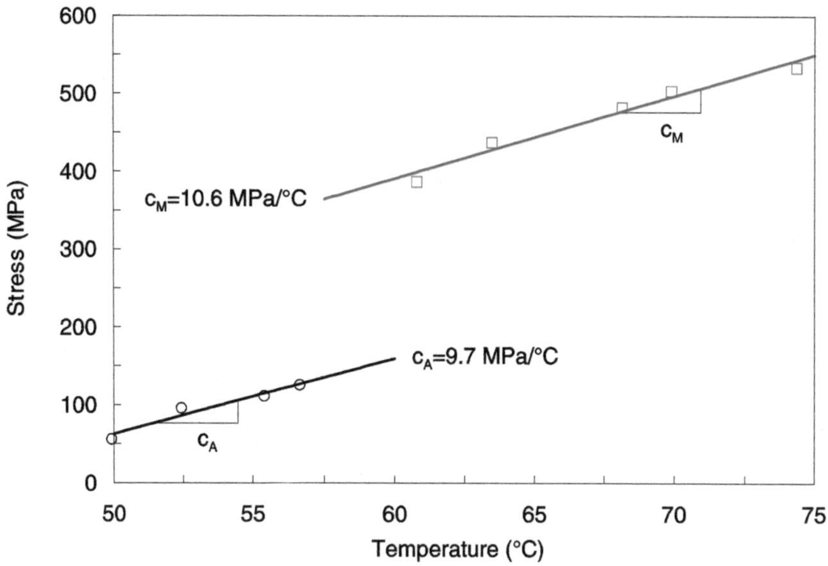

Fig. 6 Experimental results showing the dependence of critical stresses on temperature, after [70]

proportional to the thermoelastic coefficient and the change in temperature T. In the case of the Tanaka [58] and Liang and Rogers [39] models the transformational component is assumed to be proportional to the total martensite volume fraction ξ. The constitutive equation for both these models is given by Eq. (1):

$$(\sigma-\sigma_0)=E(\xi)(\varepsilon-\varepsilon_0)+\Omega(\xi)(\xi-\xi_0)+\Theta(\xi)(T-T_0). \tag{1}$$

In contrast, the Brinson [10] model introduces additional stress-induced ξ_S and the temperature-induced ξ_T martensite volume fractions so that the overall martensite volume fraction ξ during the transformations is defined as being the sum of these two fractions. On this basis the constitutive equation in the Brinson [10] model takes a modified form, as shown in Eq. (2):

$$\begin{aligned}(\sigma-\sigma_0) &= E(\xi)(\varepsilon-\varepsilon_0)+\Omega(\xi)(\xi_S-\xi_{0S})+\Theta(\xi)(T-T_0)\,,\\ \xi &= \xi_S+\xi_T\,.\end{aligned} \tag{2}$$

Generally speaking, the Young's modulus E and the phase transformation coefficient Ω, as well as the thermoelastic coefficient Θ, can all be assumed to be linear functions of the martensite volume fraction ξ in each model:

$$\begin{aligned}E(\xi) &= E_A+\xi(E_M-E_A)\,,\\ \Omega(\xi) &= -\varepsilon_L E(\xi)\,,\\ \Theta(\xi) &= \Theta_A+\xi(\Theta_M-\Theta_A)=\alpha(\xi)E(\xi)\,.\end{aligned} \tag{3}$$

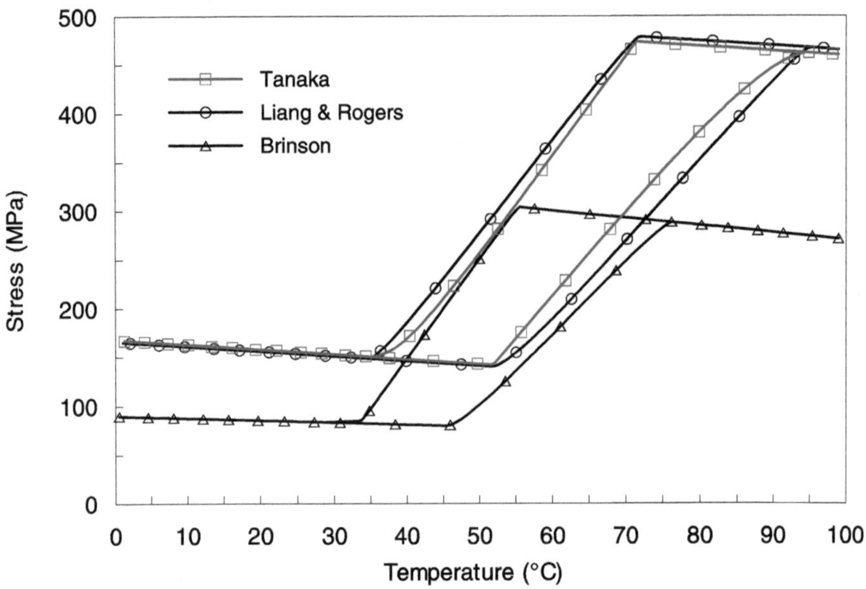

Fig. 7 Stress as a function of temperature for heating and cooling at constant initial strain, after [70]

It should be noted that the functions used to describe the martensite volume fractions ξ, ξ_S, and ξ_T are different from model to model, but are usually chosen to be functions of the stress σ and the temperature T. The effectiveness of the Tanaka [58], Liang and Rogers [39] and Brinson [10] models are demonstrated by means of a comparative study. To this end the shape memory and superelastic behaviour are investigated for a Ni-Ti SMA. All the necessary materials properties are presented in [70] for the wire as previously determined from the experimental testing. The shape memory behaviour of the Ni-Ti alloy is investigated in the first numerical example, assuming that at a constant temperature of T_0 (20 °C) below the martensite finish temperature M_F a sample composed of Ni-Ti alloy is elongated up to its initial strain value ε_0, and then the heating and cooling of the sample occurs. It is assumed in this numerical test that the sample heating continues until completion of the full transformation from martensite to austenite. A similar assumption is made for the cooling in which the cooling of the sample continues until the transformation from austenite to martensite concludes. Numerical results obtained for the Tanaka [58], Liang and Rogers [39], and Brinson [10] models are given in Fig.7 for a constant initial strain of $\varepsilon_0 = 0.005$.

It can be deduced from Fig.7 that considerable differences exist between the results from the Brinson [10] model and the models of Tanaka [58], and Liang and Rogers [39]. This arises from the three constitutive equations. The Tanaka [58], and Liang and Rogers [39] models show in equation (1) that the increase in alloy stress during initial loading is due purely due to the elastic component, and is therefore linearly proportional to the Young's modulus. Because the entire alloy stays in the

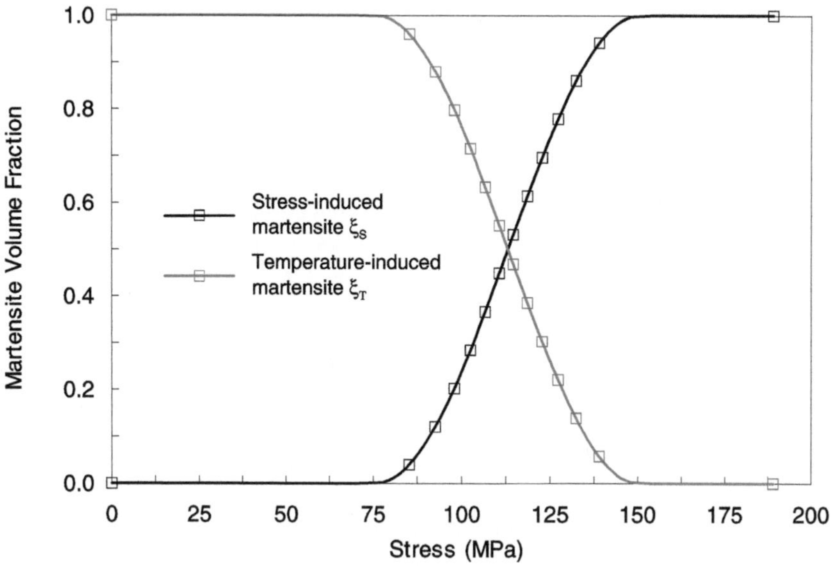

Fig. 8 Martensite volume fraction as a function of loading stress at constant temperature, after [70]

martensite phase during loading no contribution is made by the transformational component of stress. The Brinson [10] model requires that the increase in stress within the alloy during initial loading should consist of two components, elastic and transformational, as defined in equation (2). The stress increases in the same elastic way during initial loading as it does in the Tanaka [58] and Liang and Rogers [39] models. However, this is only until the stress reaches the critical stress value for the start of the transformation from temperature- to stress-induced martensite. At this point the transformational component of stress then increases and reaches its maximum value at the point when the transformation to stress-induced martensite is completed, as illustrated in Fig.8 for a constant temperature of $T_0 = 20\,°\text{C}$ below the martensite finish temperature.

From this we see that the stress within the alloy is lower when compared with the stress calculated from equation (1). Moreover, the total ranges of stress observed for both the Tanaka [58] and Liang and Rogers [39] models are much greater than the corresponding total range of stress seen in the Brinson [10] model. This feature can also be explained by recourse to the constitutive equations for the three models. Equation (1) shows that for the Tanaka [58] and the Liang and Rogers [39] models the increase in alloy stress during heating and cooling is because of the transformational component of stress. Because the entire sample is in the martensite state prior to heating, and then the full transformation from martensite to austenite takes place, we find that the stress proportional to the full residual strain is subsequently recovered during heating. This effect is not observed in the Brinson [10] model despite the fact that the change in stress is also entirely due to the transformational component, as stated in Eq. (2). So, the transformational component in (2) is not proportional to the change in the total martensite volume fraction within the sample, but merely to the stress-induced martensite volume fraction, which is in contrast to the implication of Eq. (1). During the initial loading of the sample prior to heating, it is seen that a fraction of the martensite phase is transformed from temperature-induced martensite to stress-induced martensite, and so the remaining fraction of the stress, proportional to the residual strain, can be recovered during heating.

The changes in stress in the case of the Brinson [10] model are much lower because of that than the changes in stress for the Tanaka [58], and the Liang and Rogers [39] models. It is useful to note an interesting numerical test concerning the superelastic behaviour in the Ni-Ti alloy during loading and unloading of the same sample at constant temperature T_0, below the martensite finish temperature M_F [30], [21]. From this we see that because the entire loading and unloading process of the sample stays in the fully martensite phase no superelastic behaviour can be observed in the case of the Tanaka [58] and Liang and Rogers [39] models during the test. This is because of the fact that changes in stress within the alloy are due purely to the elastic component from equation (1) and this is similar to the results in Fig.7.

Clearly, no transformation either way takes place. Changes in stress within the alloy, according to the Brinson [10] model, are due to the elastic and transformational components as given in Eq. (2). In this case the stress within the alloy increases until its value reaches the value of the critical stress for the start of the transformation from temperature- to stress-induced martensite in a manner similar to the previous explanation. When this point is reached the transformational component of stress increases, and achieves its maximum value when the transformation to stress-induced martensite is complete and the full residual strain has been recovered. Results for changes in the martensite volume fraction within the alloy during loading are given in Fig.9, again for a constant temperature of $T_0 = 20\,°\mathrm{C}$ below the martensite finish temperature.

In the numerical example that follows the superelastic behaviour of a Ni-Ti SMA is also considered. In distinction to the previous numerical test it is assumed now that the Ni-Ti sample is heated up to an initial temperature T_0, above the austenite finish temperature A_F, and then the loading and unloading process for the sample takes place. Results for the Tanaka [58], the Liang and Rogers [39], and the Brinson [10] models are all given in Fig.10, this time for a constant temperature of $T_0 = 50\,°\mathrm{C}$ above the martensite finish temperature.

Figure 10 confirms that the three models agree well in this context and this is a simple consequence of the fact that for each model the changes in stress during loading and unloading are driven by the same processes, under chosen conditions, and constitutive Eqs. (1) and (2) confirm that. The initial changes in alloy stress are driven by the elastic component of stress as long as it remains below the critical stress for the inverse transformation (austenite to martensite). Until this point is reached the entire sample stays in the austenite phase, and then when the critical value of stress for the inverse transformation is reached the transformation from austenite to martensite begins, and the residual strain is gradually recovered because of the transformational components.

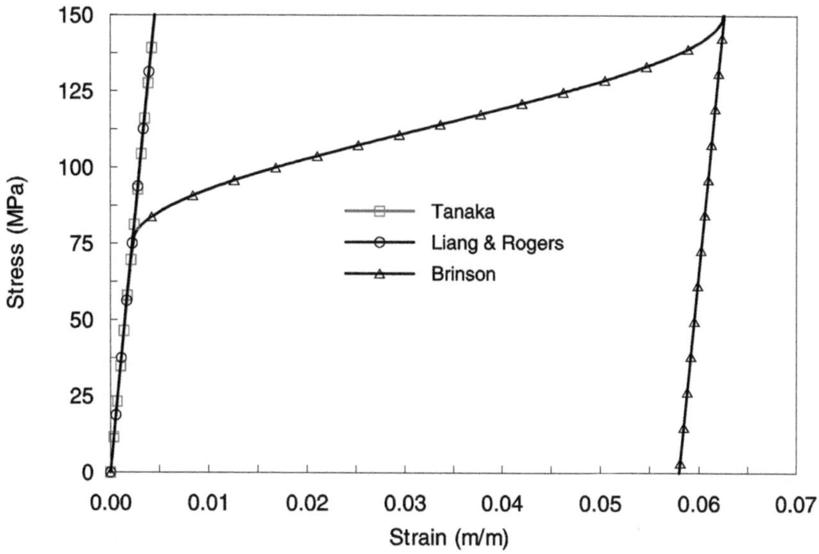

Fig. 9 Stress as a function of loading and unloading strain at constant temperature of 20°C, after [70]

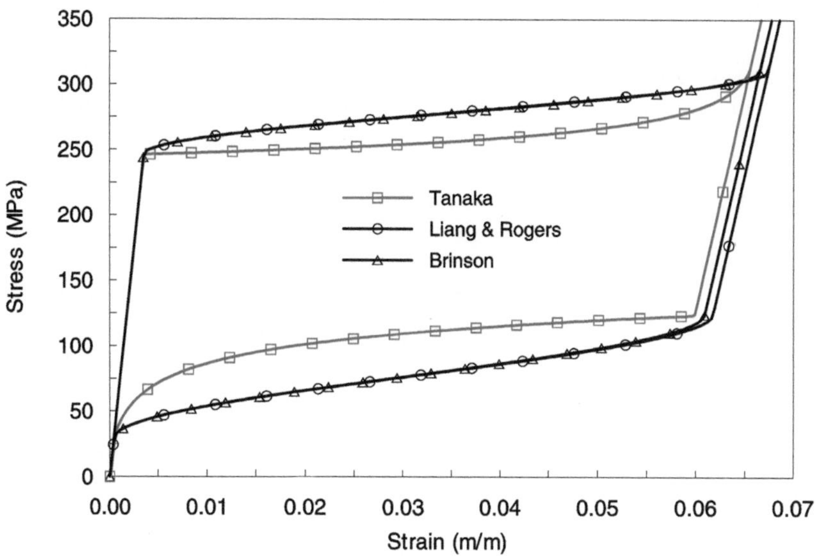

Fig. 10 Stress as a function of loading and unloading strain at constant temperature of 50°C, after [70]

It should be noted, however, that the residual strain recovery for the Brinson [10] model is associated with an increase in stress-induced martensite, whereas the Tanaka [58] and the Liang and Rogers [39] models only recognise one type of martensite. The transformation finishes when the full residual strain is recovered and the sample is fully within the martensite phase. From thereon the martensite phase loading process occurs, and changes in the alloy stress are only due to the elastic components given in the constitutive equations.

Inverse behaviour is observed during the unloading process of the sample. The discrepancies shown in Fig.10 between the results for the models are mostly likely to be due to differences in the actual functions which define the transformation behaviour of the alloy, as well as to some differences in the constitutive equations themselves. An interesting adjunct to this was presented by the authors in which it was found to be advantageous to supplement the Brinson [10] model with temperature dependent critical stresses, and modified equations defining this by means of the temperature-induced martensite volume fraction over the austenite to martensite transformation, are given in [66].

The preceding discussion offers an inexhaustive summary of some of the well-known studies published on the modeling of SMAs and should be used as a basis for further and wider reading.

3 Dynamics of Composite Beams and Plates with Integrated SMA Elements

It has already been proposed that there can be two practical implementations for SMAs within composite hosts, namely Active Property Tuning (APT) and Active Strain Energy Tuning (ASET), and in order to exploit SMA action within structures it is important to distinguish practically between these two approaches. In APT the active SMA element, possibly in the form of a wire, strip, or foil is bonded within the structure in a zero residual strain state. This means that a continuous bond between SMA and host is implied, with no unactivated traction forces exerted by the SMA onto the host. In the case of ASET the SMA is fitted into the host in a nonzero residual strain state, and this requires a fixturing technology which allows the ends of the SMA element to be rigidly attached to the host, and for some means to introduce the required residual strain in the SMA. Various technological solutions for each method can be envisaged, the simplest being uniform adhesion by means of a high-temperature epoxy in the case of APT and sleeved SMAs with end-point attachments in the case of ASET. An integrated SMA foil is shown schematically in Fig.11, noting that this configuration is for APT rather than ASET.

Vinson and Sierakowski [63] derived the governing equation for a composite beam of length L, width B, and thickness H, and with SMA layering due to wires or foils as shown in Fig.11, as follows:

$$BD_{11}\frac{\partial^4 w}{\partial x^4} - P_x\frac{\partial^2 w}{\partial x^2} = p(x)B - \rho S\frac{\partial^2 w}{\partial t^2} \tag{4}$$

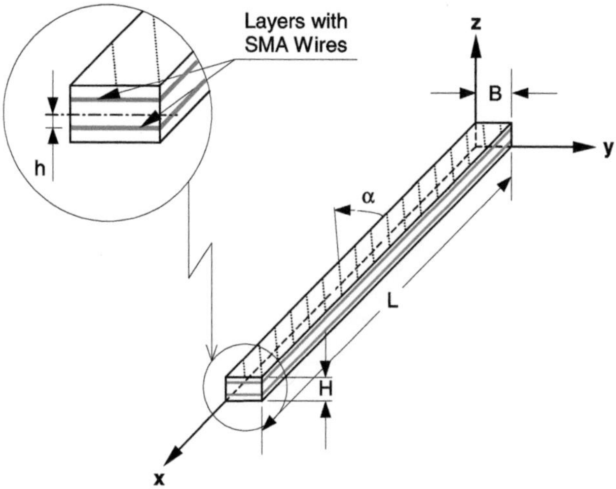

Fig. 11 Schematic of a composite beam with equidistant SMA element layers fitted across the width and along the length and applying the method of Active Property Tuning (APT), after [66]

where B, D_{11}, w, P_x, $p(x)$, ρ, and S are defined as beam width, elastic coefficient, lateral displacement, in-plane beam load (SMA), lateral load, density, and cross-sectional area, respectively. The solution to equation (4) can be expressed in the usual way by means of modal functions based on chosen boundary conditions, and from this the equations for the natural frequency and critical buckling load can be obtained, noting that the critical buckling load is defined at $\omega_n = 0$:

$$w(x,t) = \sum_{n=1}^{\infty} A_n \sin(\frac{n}{L}\pi x)\cos(\omega_n t), \quad n = 1,2,3,\ldots, \tag{5}$$

$$\omega_n = \sqrt{\frac{\pi^2 B D_{11}\left(\frac{n}{L}\right)^4 + P_x\left(\frac{n}{L}\right)^2}{4\rho S}}, \tag{6}$$

$$P_{crit} = -\left(BD_{11}\left(\frac{\pi}{L}\right)^2 + P_x\right). \tag{7}$$

The maximum static deflection can also be calculated by recourse to the Navier method and Fourier series modelling for the deflection of the beam, for some given pressure load. Therefore the maximum static deflection is given by:

$$w(x) = \sum_{n=1}^{\infty} \frac{q_n}{D_n} \sin(\frac{n\pi x}{L}), \quad n = 1,3,5,\ldots \tag{8}$$

where $q_n = 4Bp_0/n\pi$ and $D_n = \pi^4 BD_{11}(n/L)^4 + \pi^2 P_x (n/L)^2$, $n = 1,3,5,\ldots$.

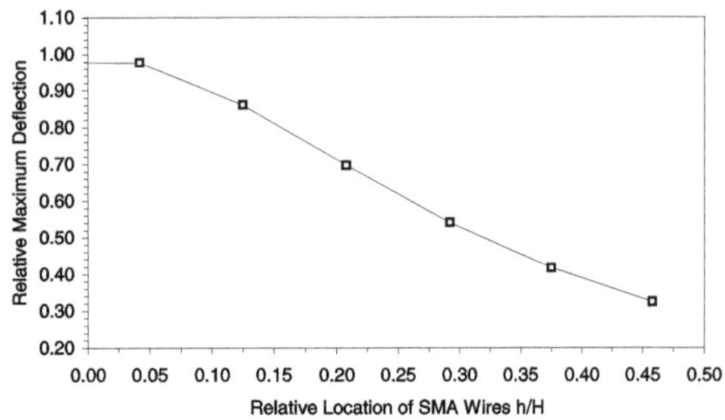

Fig. 12 Relative maximum beam deflection as a function of relative location on the cross-section, after [66], note RVF=30% and $\alpha = 0°$

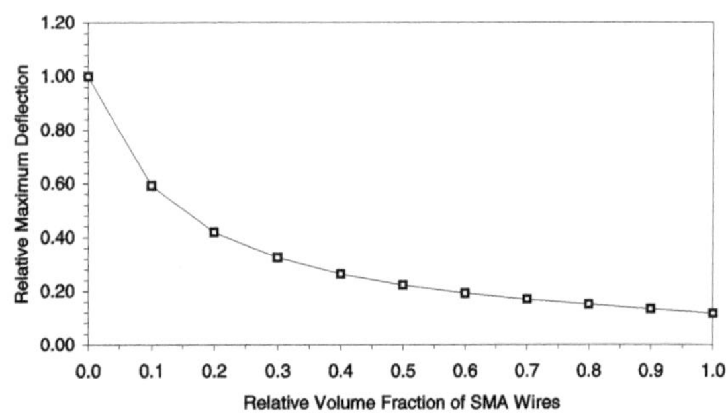

Fig. 13 Relative maximum beam deflection as a function of relative SMA volume fraction, after [66], note $h/H = 0.458$ and $\alpha = 0°$

Żak and Cartmell [66] used Eqs. (5)–(8) to explore the performance of a test beam for a range of test data to show how the SMA can profoundly influence the load bearing capacity of the beam and its overall dynamic performance, for arbitrarily chosen simply supported boundary conditions. The beam is symmetrical and of length 500 mm, width 10 mm and thickness 6 mm, and is composed of 12 layers of an epoxy/nitinol SMA composite for which the epoxy has a Young's modulus of 3.43 GPa, a Poisson's ratio of 0.35, and a density of 1250 kg/m^3. The Nitinol SMA alloy itself has a martensitic Young's modulus of 26.3 GPa and an austenitic Young's modulus of 67.0 GPa, a constant Poisson's ratio of 0.3, and a density of 6448 kg/m^3.

Figure 12 shows that the relative maximum deflection of the composite beam, this being the ratio of the maximum deflection of the beam when the SMA is activated to the maximum deflection of the beam without SMA, (and this definition applies to all subsequent Figures in this section) plotted against the relative cross-sectional location of the two layers each side of the central axis. The effect of the SMA is clearly visible in Figure 12 for which the relative maximum deflection reduces to < 0.4 for $h/H > 0.45$. The effect of relative volume fraction (RVF) of SMA can be seen in Fig.13 and how that this levels off after the RVF exceeds 50%, ending up with a relative maximum deflection of < 0.3 for the an RVF of 50% and above.

Figure 14 shows very clearly indeed the significance of the orientation angle of the active SMA elements from effectively operating along the length of the beam (with an orientation angle of zero) to the case where the wires are oriented across

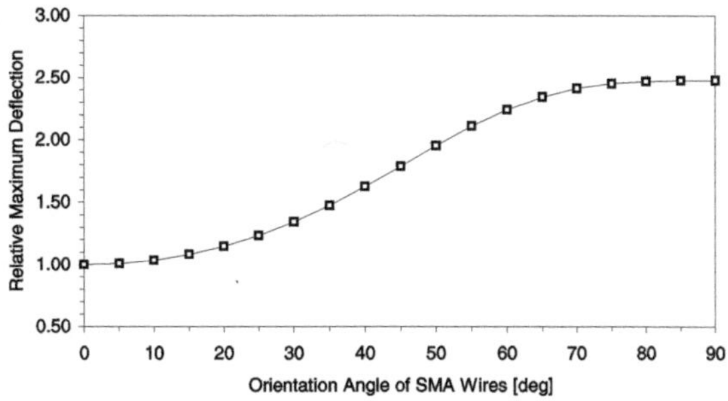

Fig. 14 Relative maximum beam deflection as a function of SMA orientation angle along the length of the beam, after [66], note $h/H = 0.458$ and $\alpha = 0°$

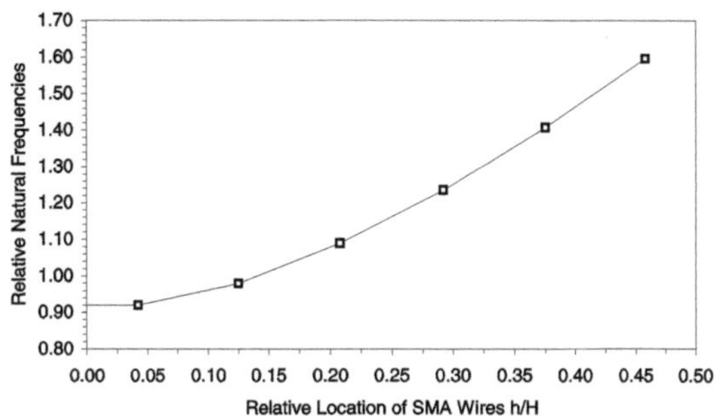

Fig. 15 Relative natural frequency as a function of relative location on the cross-section, after [66], note RVF=30% and $\alpha = 0°$

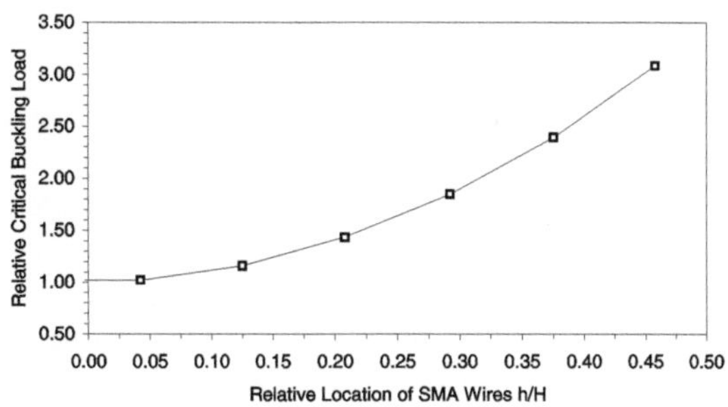

Fig. 16 Relative critical buckling load as a function of relative location on the cross-section, after [66], note RVF=30% and $\alpha = 0°$

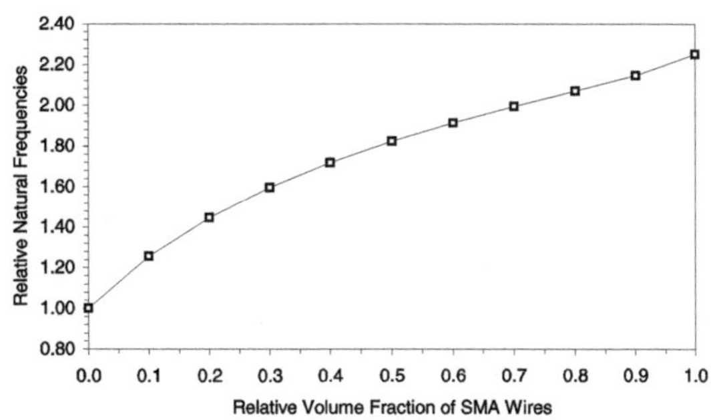

Fig. 17 Relative natural frequency as a function of relative volume fraction, after [66], note $h/H = 0.458$ and $\alpha = 0°$

Fig. 18 Relative critical buckling load as a function of relative volume fraction, after [66], note $h/H = 0.458$ and $\alpha = 0°$

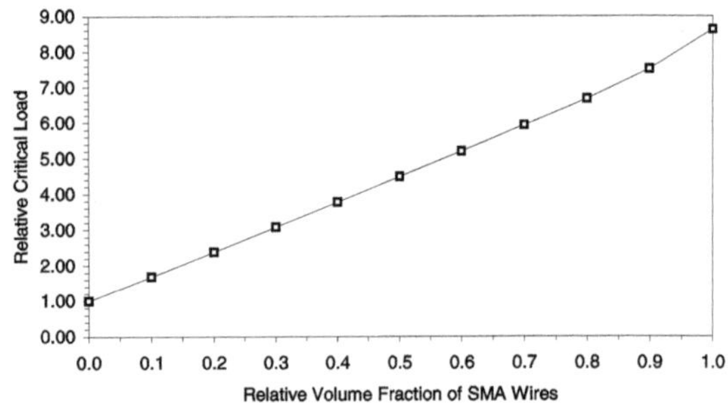

Fig. 19 Relative natural frequency as a function of SMA orientation angle, after [66], note $h/H = 0.458$ and RVF=30%

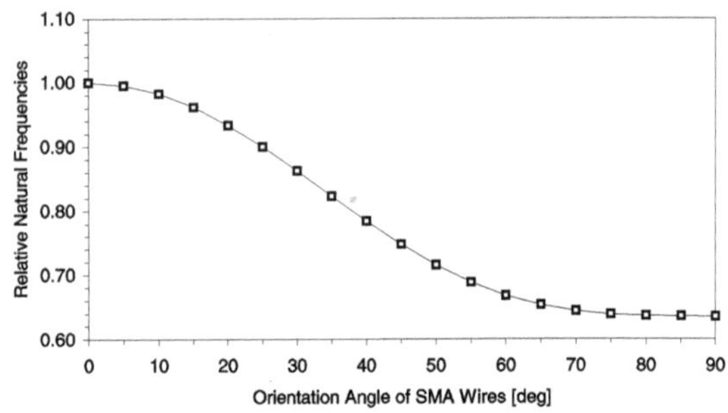

the width of the beam (orientation angle of 90%). In Fig.15 the dynamic case is considered and it can be seen there how the relative natural frequency of the first mode can be affected by the relative location across the cross-section, with an increase of around 1.6 in the natural frequency of this mode when $h/H > 0.45$. Higher modes show similar characteristics [66].

The effect of moving the SMA layers outwards from the central axis on the relative critical bucking load is really emphasised in Fig.16, showing a three-fold increase for $h/H > 0.45$. Figure 17 illustrates the case for the natural frequency of the first mode as a function of RVF, with a doubling of the natural frequency for a RVF exceeding 70%. Lesser, more realistic RVFs still clearly affect the relative natural frequency, as can be seen from the Figure.

Figure 18 shows an almost linear relationship between the relative critical load and the RVF, with an increase of over four times for the critical load for RVFs of 50% and above. In the case of the relative natural frequency of the first mode it can be seen in Fig.19 that this tails off to around 70% of its unactivated value for orientation angles in excess of around 55°. The ASET configuration is more difficult to set up experimentally but can be done with a suitable arrangement of insulated clamps and a calibrated pre-tensioner, with relative changes in the fundamental natural frequency slightly in excess of 2.7 for $h/H = 0.458$, RVF=30%, $\alpha = 0°$, and $P_x = 549.4\,\text{N}$. This consolidates general findings that although somewhat difficult to set up in practice, the ASET configuration for a beam can result in generally superior modification performance of the static and dynamic responses of the beam over the simpler APT methodology. The interested reader is directed to Żak and Cartmell [66] for further details.

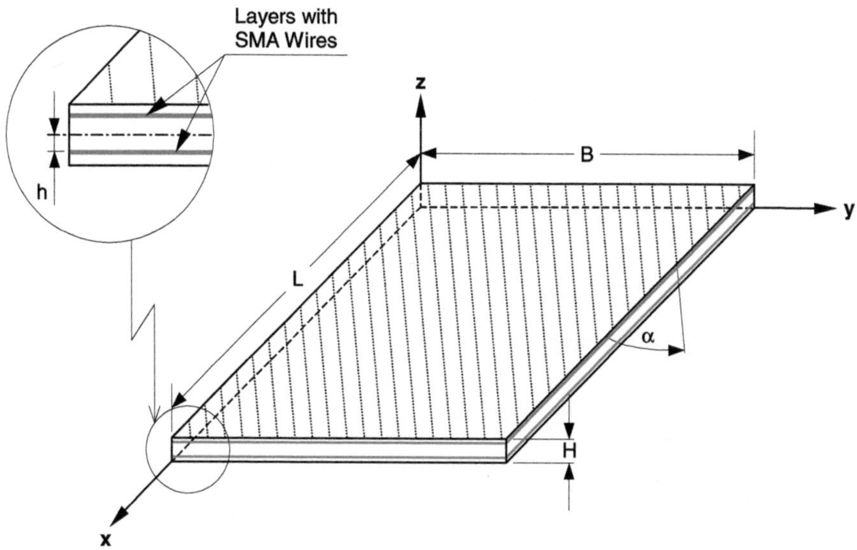

Fig. 20 Schematic of a composite plate with equidistant SMA element layers fitted across the width and along the length and applying the method of Active Property Tuning (APT), after [66]

This approach can also be extended to plates, for which an appropriate governing equation is given as follows [63]:

$$D_{11}\frac{\partial^4 w}{\partial x^4}+2(D_{12}+2D_{66})\frac{\partial^4 w}{\partial x^2 \partial y^2}+D_{22}\frac{\partial^4 w}{\partial y^4}-N_x\frac{\partial^2 w}{\partial x^2}=p(x,y)-\rho H\frac{\partial^2 w}{\partial t^2} \quad (9)$$

noting that L is the length of the plate, B is the width, and H is the thickness, as shown in Fig.20.

Equation (9) can be solved by classical means to lead to a solution for the vibration of the plate in the following form:

$$w(x,y,t)=\sum_{n=1}^{\infty}\sum_{m=1}^{\infty}A_{nm}\sin(\frac{n\pi x}{L})\sin(\frac{m\pi y}{B})\cos\omega_n t\,,\quad n,m=1,2,3,\ldots \quad (10)$$

The natural frequencies and critical buckling load of a plate with arbitrarily chosen simply supported boundary conditions are given by the following relations:

$$\omega_{nm}=\sqrt{\frac{\pi^2 D_{11}\left(\frac{n}{L}\right)^4+2\pi^2(D_{12}+2D_{66})\left(\frac{n}{L}\right)^2\left(\frac{m}{B}\right)^2+\pi^2 D_{22}\left(\frac{m}{B}\right)^4+N_x\left(\frac{n}{L}\right)^2}{4\rho H}}$$
$$n,m=1,2,3,\ldots \quad (11)$$

$$N_{crit} = -\frac{L^2}{\pi^2}(D_{11}\left(\frac{\pi}{L}\right)^4 + 2(D_{12}+2D_{66})\left(\frac{\pi}{L}\right)^2\left(\frac{\pi}{B}\right)^2 + D_{22}\left(\frac{\pi}{B}\right)^4 + N_x\left(\frac{\pi}{L}\right)^2). \quad (12)$$

Once again, the critical buckling load given by Eq. (12) corresponds to the case where, $\omega_n = 0$ and it also relates specifically to the x-direction on the plate. The maximum static deflection can also be obtained analytically for the plate, and this is given by:

$$w(x,y) = \sum_{n=1}^{\infty}\sum_{m=1}^{\infty}\frac{q_{nm}}{D_{nm}}\sin(\frac{n\pi x}{L})\sin(\frac{n\pi y}{B}), \quad n,m = 1,3,5,\ldots \quad (13)$$

where $q_{nm} = 16p_0/nm\pi^2$ and

$$D_{nm} = \pi^4 D_{11}\left(\frac{n}{L}\right)^4 + 2\pi^4(D_{12}+2D_{66})\left(\frac{n}{L}\right)^2\left(\frac{m}{B}\right)^2 + \pi^4 D_{22}\left(\frac{m}{B}\right)^4 + \pi^2 N_x\left(\frac{n}{L}\right)^2,$$

$n,m = 1,3,5,\ldots.$

Equations (9)–(13) are utilized by Żak and Cartmell [66] to investigate the static and dynamic performance of a composite test plate with APT configured SMA element layers in the same basic configuration as the previous beam problem, and for a range of test data, notably where the length and width are equal at 500 mm and the thickness is the same as the beam at 6 mm. The plate comprises exactly the same make-up and mechanical characteristics as the beam, as previously described. The

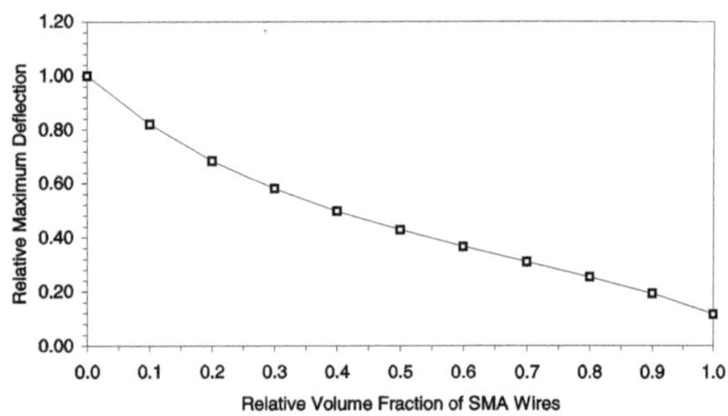

Fig. 21 Relative maximum deflection as a function of relative volume fraction, after [66], note $h/H = 0.458$ and $\alpha = 0°$

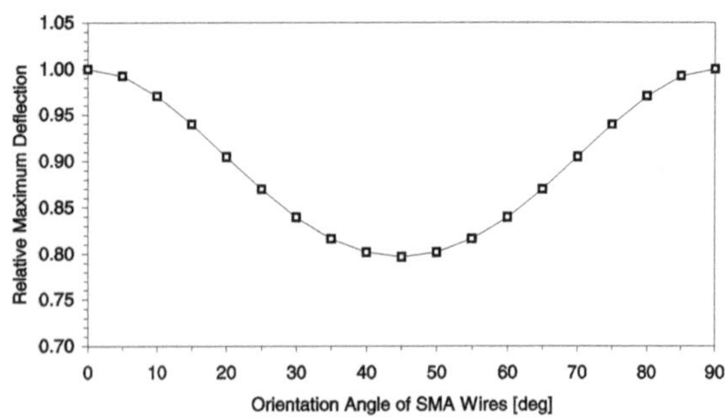

Fig. 22 Relative maximum deflection as a function of SMA orientation angle, after [66], note $h/H = 0.458$ and RVF=30%

Fig. 23 Relative natural frequency of the first plate mode as a function of relative location on the cross-section, after [66], note RVF=30% and $\alpha = 0°$

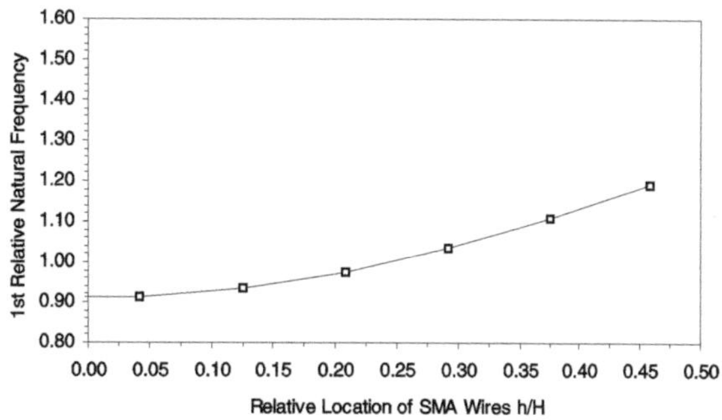

Fig. 24 Relative critical buckling mode as a function of relative location on the cross-section, after [66], note RVF=30% and $\alpha = 0°$

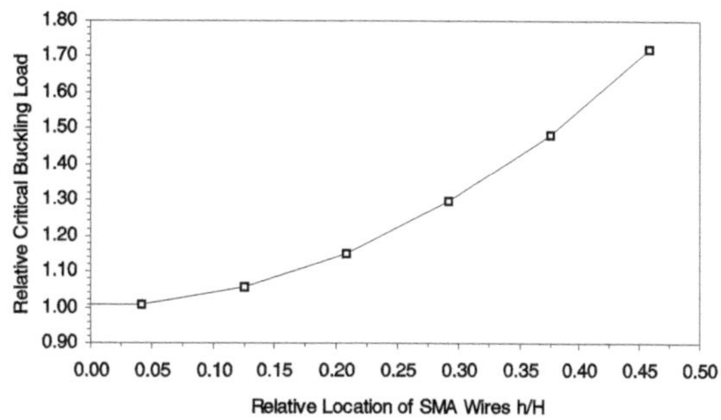

significant influence of the SMA layers on the response of the plate both to static and dynamic conditions could then be confirmed, once again for arbitrarily chosen simply supported boundary conditions.

In the case of the plate the relative maximum deflection for $h/H = 0.458$ and SMA orientation angle $\alpha = 0°$is plotted in Fig.21 as a function of the RVF and it can be seen that this is a shallow nonlinear decline down to around 0.5 for 50% RVF. Figure 22 depicts the relationship between relative maximum deflection and SMA orientation angle for a RVF of 30% and the same cross-sectional location. This interesting result confirms that the relative maximum deflection is symmetrical about a minimum of just over 80% at an orientation angle of 45°.

It is interesting to note in Fig.23 that the relative natural frequency of the first plate mode is reasonably insensitive to the location of the SMA layers on the cross-section of the plate, noticeably less so that than in the case of the beam in Fig.15. This is clearly due to the two dimensional nature of the plate mode as opposed to the one dimensional beam mode which will obviously be very directly influenced by the presence of the SMA layers running in the same direction. Figure 24 illustrates the static case in which relative critical buckling load is plotted also as a function of the relative location of the SMAs on the plate cross-section, and a significant increase in this load is evident, at 1.7, for h/H of around 0.45. This could clearly have important applications in safety critical systems where static buckling in plates and panels needs a maximised margin for control.

Figure 25 summarises the relationship between the relative natural frequency of the first plate mode and the RVF, and this is virtually linear up to approximately

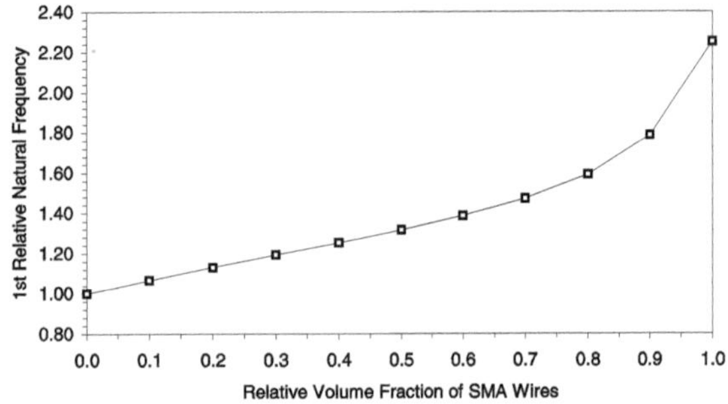

Fig. 25 Relative natural frequency of the first plate mode as a function of relative volume fraction, after [66], note $h/H = 0.458$ and $\alpha = 0°$

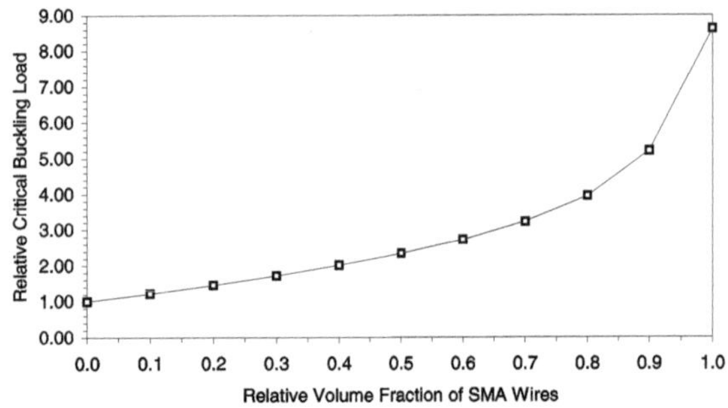

Fig. 26 Relative critical buckling mode as a function of relative volume fraction, after [66], note $h/H = 0.458$ and $\alpha = 0°$

60% RVF, corresponding with a relative natural frequency for the fundamental plate mode of 1.4, after which the effect rises more steeply. The plate is relatively insensitive to more pragmatic lower RVF values up to 30% or so. A similar trend is observed for the relative critical buckling load as a function of RVF in Fig.26 but the range of relative critical buckling load is very much greater than the corresponding range for the same dependence in the relative natural frequency of the fundamental plate mode. Linearity generally prevails up to RVFs of about 45%, but with a large increase in the relative critical buckling load for RVFs above 80%.

Figure 27 confirms that the relative natural frequency of the fundamental mode of the plate is quite insensitive to the SMA orientation angle, with the two-dimensionality of the plate mode clearly at the root of this phenomenon. The relative

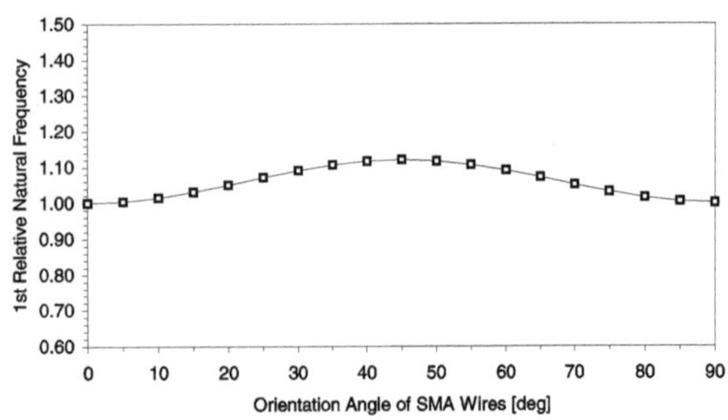

Fig. 27 Relative natural frequency of the first plate mode as a function of SMA orientation angle, after [66], note $h/H = 0.458$ and RVF=30%

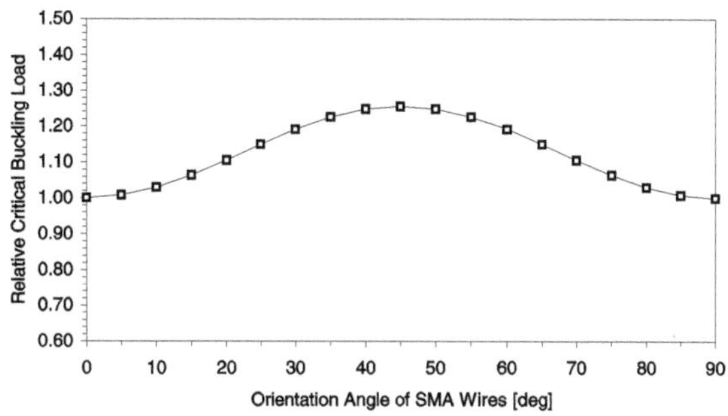

Fig. 28 Relative critical buckling mode as a function of SMA orientation angle, after [66], note $h/H = 0.458$ and RVF=30%

critical buckling load of the plate shows somewhat limited sensitivity to the orientation angle with symmetry about a maximum in excess of 1.2 at an orientation angle of 45°as depicted in Fig. 28.

It should be noted that further details for responses in higher plate modes can be found in Żak and Cartmell [66] and a description of a conjectured ASET configuration for the plate is also given. In this case the results are computed for $h/H = 0.458$, RVF=30%, $\alpha = 0°$, and in-plane load $N_x = 51.8\,\mathrm{kN/m^2}$. This investigation suggested that for the ASET method the relative change in the maximum deflection of the plate would reach 3.7, confirming the superiority of the ASET configuration as a means for modifying the static and dynamic response characteristics of a simply supported plate. It can be confirmed that such performance enhancements are generic across the various possible boundary conditions.

General conclusions are that the greatest modifications to the beam's performance are obtained when the SMA elements run along the longitudinal axis of the beam. The SMA elements should be located as far from the central axis as possible for maximum benefit in both beams and plates, and ideally forming the outer layers. In the case of the plate analysis it has been shown that the most effective SMA orientation angle depends somewhat on the plate's aspect ratio and also on the mode in operation, but in the case of the fundamental mode and the critical buckling mode maximum effect is obtained when the orientation angle is at 45°. Furthermore, the functionality of embedded SMA elements of this sort depends on their relative volume fraction (RVF) and that one of the advantages to an ASET based system is that good performance can be obtained for relatively lower RVFs than for APT, thereby saving on SMA costs.

Finite element modeling can be used to calculate the beam and plate modes for a range of different boundary conditions, and specialised elements were created in order to do this accurately [69]. The beam element, designated 3N4D due to three nodes and four degrees of freedom per node, consists of N layers made of unidirectional composite material, with arbitrarily orientated reinforcing fibres within the layers. The elemental displacement field is constructed on the basis of first order shear deformation theory and the nodal degrees of freedom assumed for the element provide a good approximation for the transverse displacement of the element. The plate element is of the form 8N7D comprising eight nodes and seven degrees

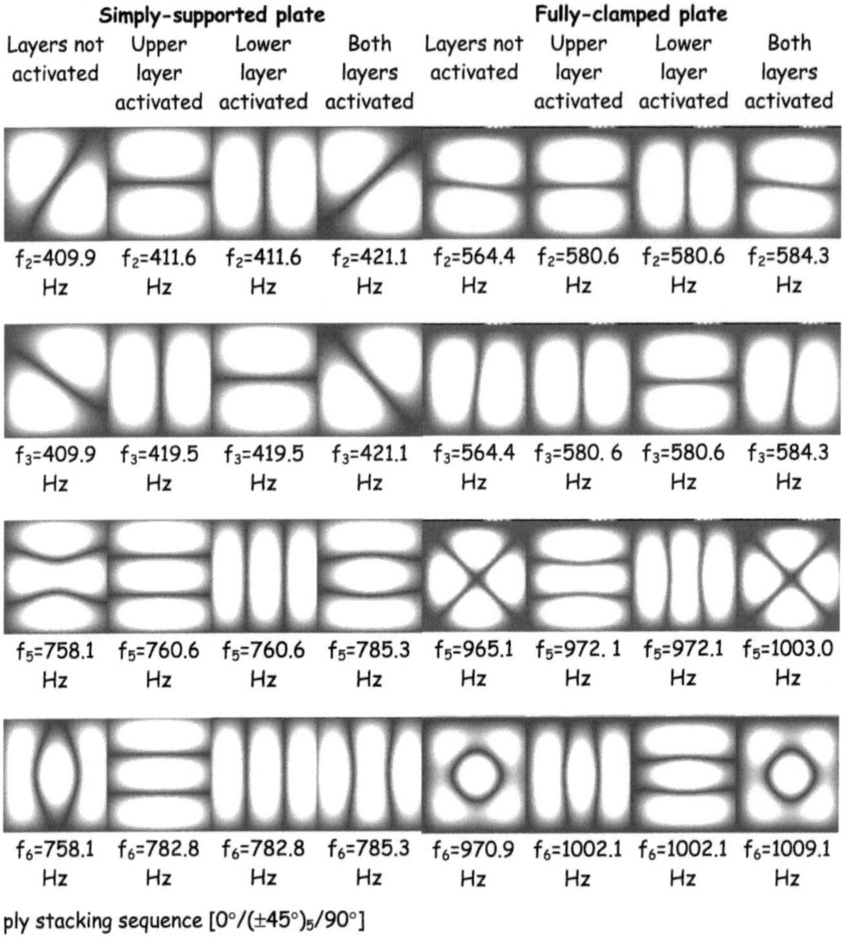

Fig. 29 Calculated modes 2, 3, 5, and 6 for a simply supported and fully clamped square plate with unactivated SMA layers, activated lower and upper layers, and both layers activated, adapted from [69]

of freedom per node, and contains N layers and similarly arranged reinforcing fibres. The nodal degrees of freedom define two longitudinal in-plane displacements, a transverse displacement, two rotations due to the transverse displacement, and two independent rotational corrections due to shearing effects. It is shown in [69] that the displacement and strain fields of the plate element are as given by Kirchhoff plate theory for small thickness to length ratios, whilst the shearing strain influence emerges for higher ratios, and this leads to a good prediction of the static and dynamic behaviour of the plate, with no locking-in effects occurring.

Figure 29 shows some of the calculated modes for a square plate in which two SMA wire layers have been introduced. It can be seen that in both boundary condition configurations the chosen mode shapes corresponding to the unactivated SMA layers and the mode shapes for both layers activated are broadly similar, but the modal natural frequencies for the two cases increase noticeably, by 2.7% in the case of modes 2 and 3 for the simply supported plate to almost 3.6% in the case of modes 5 and 6. The fully clamped plate shows even more natural frequency shift with 3.5% for modes 2 and 3 to 3.9% in the case of modes 5 and 6. Asymmetrical SMA activations show the same tendencies for nodal line reorientations in each boundary condition for modes 2, 3, and 5, but mode 6 shows a reversal of the nodal line structure for the upper and lower layer activation cases, respectively, for the two sets of boundary conditions.

4 Applications to Flexible Rotors

A further interesting application of SMA elements has been explored in some detail by Żak and Cartmell [67], in the form of specialised sub-systems which can be introduced into rotating systems in order to modify their dynamics responses. In this application the sub-system comprises a cylindrical shell with a closed end, onto which SMA elements are integrated so that the stiffness properties of the structure are modified when the SMAs are activated. The shell forms a specialised bearing housing inside which a shaft-end, terminating in a ball bearing, is fitted. The net effect of this is to provide a dynamically variable end condition on the shaft predominantly through the housing stiffness change due to SMA activation. One can apply APT, where the material property changes through activation are exploited, or ASET by which high recovery stresses and material property changes are applied accordingly.

In practice it is much easier to implement APT integrations where the need for calibrated preloads can be avoided, and it can be shown that useful resonant frequency shifts can be achieved this way. This could be particularly important in highly resonant machines that run close to an otherwise unavoidable critical whirl speed for some reason. A controllable bearing housing then offers the possibility of moving the resonance and potentially greatly reducing the associated whirl amplitude. A schematic of the geometry of a bearing housing, shown as a sleeve-ring type of shell component, is given in Fig.30.

Results are given by Żak and Cartmell [67] for aluminium, glass/epoxy and graphite components, al fitted with SMAs as shown in Fig.30, and finite element calculations were carried out for both the APT and ASET configurations under different conditions of static and dynamic load. The greatest changes, defined in terms of relative modifications to static and dynamics characteristics, come from the glass/epoxy component under ASET, and the least effects are predicted for the aluminium version operating when configured with APT. SMA activation reveals considerable modifications to the distributions of the displacement and stress fields, particularly u_z and $\sigma_{\phi z}$, and this is linked to the orientation of the SMA strips within

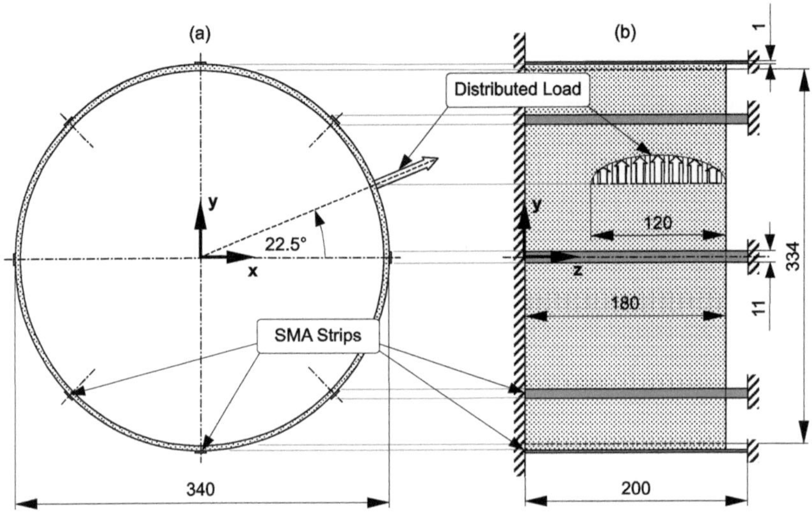

Fig. 30 Geometry of a sleeve-ring shell in the form of a bearing housing for a high speed flexible rotor, **a** front elevation, **b** side elevation, after [67]

the component. It is important to note that in some configurations static loads can generate shear stresses between SMA and host that exceed the elastic limit of the host material particularly when this is composite material. The activation of the SMA strips can significantly alter natural frequencies and response amplitudes under forced vibration. Predictions at the highest levels of performance indicate ASET controlled changes in natural frequency approaching 20% and fundamental vibration mode amplitude reductions exceeding five times, with corresponding APT figures of 10% and up to five times amplitude reduction. On this basis, judicious component design could achieve very high levels of performance from relatively easily configured APT installations. Having conjectured that a suitable shell structure can be designed Żak and Cartmell [68] showed that it could be usefully fitted into an experimental flexible rotor, based along the lines of Fig.31.

The mechanical properties are as shown in Fig.31 with the SMA/Composite sleeve-ring shell component fitted to the right hand end of the rotor shaft. It can be seen that the shaft end fits into a ball-race and that this is press-fitted into the left hand end of the housing. The right hand side of the housing is rigidly attached to the mechanical support. The left hand end of the rotor shaft is belt driven by a speed controlled motor and the vibration of the overall system is measured at the disk in the centre of the shaft.

Predictive modelling assumptions were made on the basis that the mechanical properties (other than the density) for the bearing are the same as for the disk, that the glass/epoxy version's properties were Young's moduli of 65.5/3.43 GPa respectively, Poisson ratios of 0.23/0.35, and densities of 2250/1250 kg/m^3 respectively. The graphite version had a Young's modulus of 275.6 GPa, Poisson ratio of 0.20, and a density of 1900 kg/m^3 for the same grade of epoxy. A finite element model of the whole system was used to determine the possible responses of the system under passive (unactivated) operation and also under SMA activation. Tabulated natural frequencies are given in [68] for unactivated responses of the rotor for both glass/epoxy and graphite/epoxy configurations and different numbers of layers in the sleeve/ring shell. Layering schemes from 2 to 10 layers can affect the natural frequencies by nearly 14% for the glass/epoxy variant and just over 12% in the case of the graphite/epoxy design. This is summarised in Fig.32.

Numerical analysis of the system also shows that the amplitudes of the rotor under forced vibration of the first whirl mode vary considerably dependent on the sleeve/ring material thickness in both material cases, as shown in Figs.33 and 34.

As has been shown in Fig.30 the SMA elements are introduced into the system in the form of strips laid symmetrically onto the outer surface of the sleeve/ring component based on the APT methodology, and that any thermally induced softening of the composite can be neglected, on the basis of short heating times. The Ni-Ti alloy chosen for this study had a martensite finish temperature of 20.7°C, a martensite start temperature of 26.8°C, an austenite start temperature of 37.2°C, an austenite finish temperature of 47.0°C, a martensite stress influence coefficient value of 10.6 MPa/°C, an austenite stress influence coefficient value of 9.7 MPa/°C, a maximum residual strain of 0.058, martensitic Young's modulus of 33.1 GPa, an austenitic Young's modulus of 69.6 GPa, a martensitic coefficient of thermal expansion of $6.6 \times 10^{-6}\,°C^{-1}$, and an austenitic coefficient of thermal expansion of $1.1 \times 10^{-5}\,°C^{-1}$. The finite element model was based on 34000 degrees of freedom and a two-step process was applied in which modes and natural frequencies were initially extracted, followed by a forced vibration frequency sweep.

Modelling experiments were conducted in which the quantities and widths of the SMA strips laid onto the sleeve/ring surface were varied and also where quantities

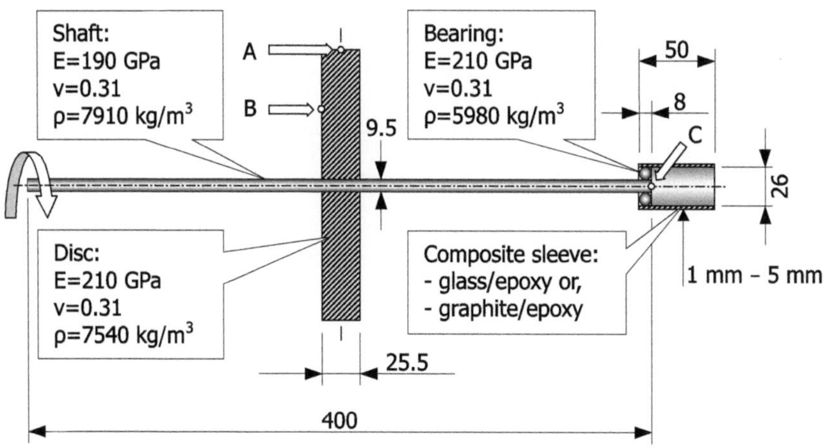

Fig. 31 Schematic of an experimental flexible rotor with SMA/Composite sleeve-ring shell fitted to the right hand end, after [68]

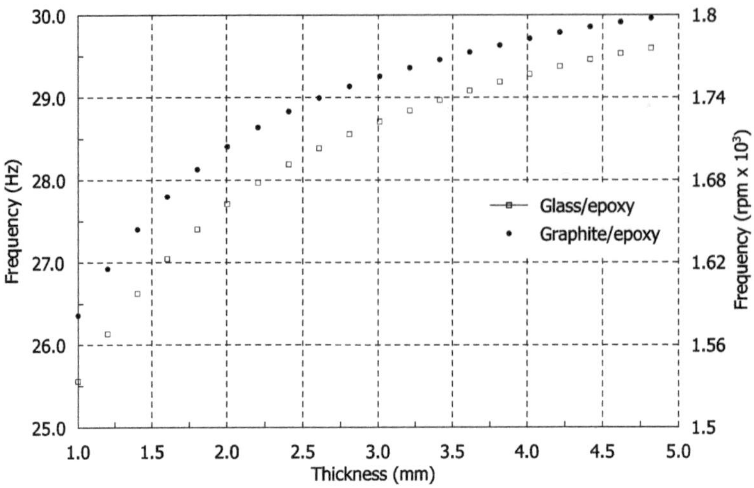

Fig. 32 First mode natural frequencies of the flexible rotor plotted against layers converted to sleeve/ring material thickness, for glass/epoxy and graphite/epoxy variants, with damping 0.05 of critical, after [68]

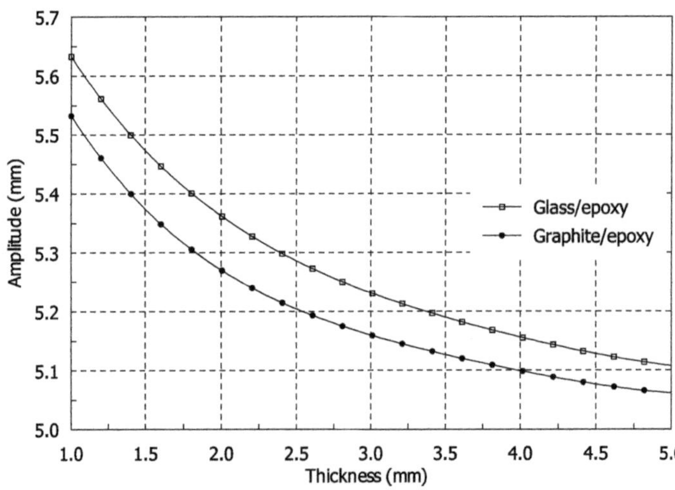

Fig. 33 Displacement amplitude of the rotor measured at the disk against sleeve/ring thickness, for glass/epoxy and graphite/epoxy variants, with damping 0.05 of critical, after [68]

of the host material were removed symmetrically from the lands in between the strips. The differences between the unactivated and activated responses, in terms of natural frequency shift and relative amplitude reduction, were calculated. Broadly speaking the more material that is removed, consistent with a maximum number of elements (noting that four strips of width 5.1 mm are more effective than three strips 6.8 mm wide, despite having the same overall total area for the same length

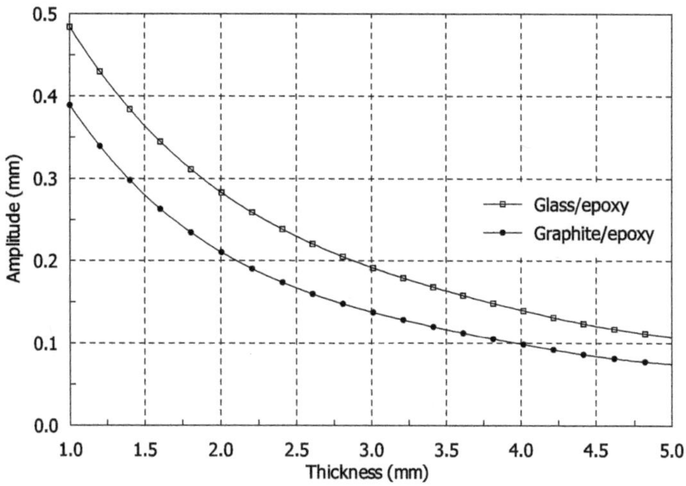

Fig. 34 Displacement amplitude of the rotor measured at the sleeve/ring bearing housing against sleeve/ring thickness, for glass/epoxy and graphite/epoxy variants, with damping 0.05 of critical, after [68]

of 50 mm, indicating that the way in which the SMA elements are distributed is also highly significant) then the higher the general performance in terms of control of frequency and amplitude. Numerical calculations suggest that in the last case a 7% increase in the natural frequency of the first whirl mode is attainable, commensurate with a 45% reduction in the amplitude of the vibration measured at the disk [68]. One has also to bear in mind that the layering/thickness issue continues to be significant in this model, with thinner sleeve/ring host material offering the highest degree of SMA influence, and that shear stresses can readily reach critical levels for the host material, requiring careful design if longevity is the aim and degredation, cracking, and delamination is to be avoided.

Further work on critical speed control in rotors has been reported by Nagaya et al. [41], Viderman and Porat [62], and latterly by Lees et al. [36]. All three works considered the use of SMAs in various guises in different configurations for bearing stiffness control. Lees et al. [36] proposed a pedestal bearing in which an elastomeric o-ring is fitted within a pair of clamped semi-circular clamps put under compression by the action of symmetrically positioned SMA wires on each side. In practice the wires are configured as a single multi-loop wire so that precisely set pre-loads could be set with both sides absolutely symmetrically balanced. The o-ring is fitted within the clamps and surrounds a rolling element bearing, as shown in the prototype test system in Fig.35.

The installation of Fig.35 effectively uses the bracket, o-ring elastomer and the SMA as three stiffnesses in series, such that the overall bearing housing stiffness is given by:

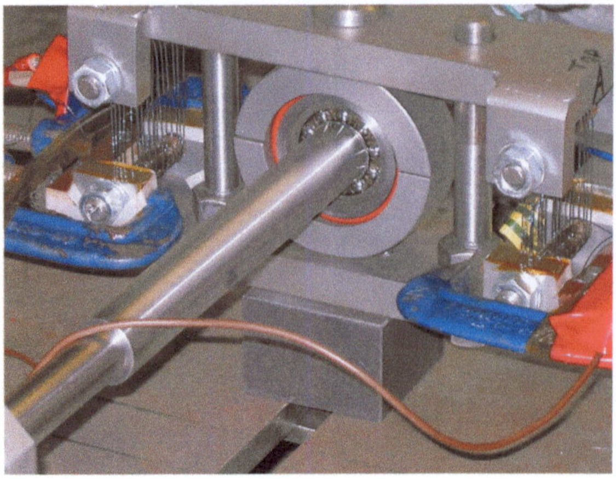

Fig. 35 Prototype SMA loaded bearing housing for a disk-less rotor fitted with a rigid bearing housing 250 mm down the rotor shaft, after [36]

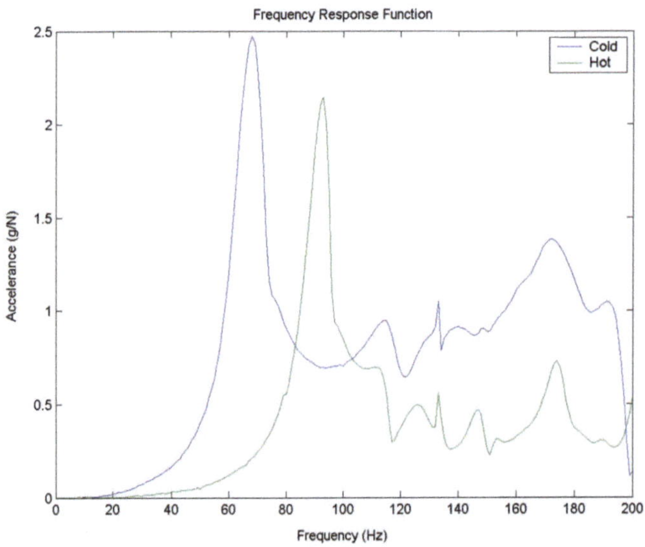

Fig. 36 Experimental impact test results for the o-ring/SMA assembly in the frequency domain, after [36]

$$k = \frac{1}{\frac{1}{k_b} + \frac{1}{k_e} + \frac{1}{k_s}} \tag{14}$$

When this is integrated into a finite element model a very major change in the amplitude response can be achieved during run up when the SMA is activated, for suitably chosen data [36]. Corresponding laboratory tests on the test rig shown in Fig.35 in which impact testing was used to investigate the cross-inertance between the shaft and the bearing system reveal that SMA activation does indeed lead to profound changes in the dynamic response of the system. Impact test results are given in Fig.36, with a change of the order of 45% attainable in the natural frequency of the first whirl mode. This suggests the potential for superior performance to that of the previous sleeve/ring configuration but optimisations of each design still remain to be completed.

5 Antagonistic Actuation Control of Vibration in Plates

Previous sections have already shown that large SMA strains of 8–10% can be achieved, by the thermal shape memory effect or superelasticity. The control of characteristic phase transition temperatures by altering the amount of nickel present, is easily accomplished by exploiting the inverse proportionality between nickel content and the temperature at which the transition takes place. This is highly significant because optimal control of the nickel content means that the temperature threshold can be optimally tailored for the application, also with energy conservation in mind. Degredation over many heating and cooling cycles can readily cause problems but this can be minimized at the outset by raising the level of the volume fraction of SMA used in the application. This can be traded off economically against the cost of energy required to heat (and cool) the SMA across the transformation temperature point.

In this section we consider the problem caused by the very different time constants involved in SMA heating and cooling and pose a possible solution, with an associated analytical model and proposals for experimental verification. The fundamental issue here is that SMAs do no work when cooling, unlike during the heating phase, and so the basic time constants differ dramatically in most installations. In some cases this does not matter but in any installation requiring reasonable levels of adaptivity the cooling phase time constant must really be brought approximately into line with the time constant associated with heating.

One proposal for this has been to harness the idea of *antagonism*, as summarised by Inman et al. [29], whereby a double-acting actuator is proposed so that one side is actively heated while the other passively cools, and vice versa, so that the changes in traction forces during the passive cooling phase are forced to work more at the pace of the adjacent active heating side. This was tested in a further development of the prototype sleeve/ring shell assembly applied to the flexible rotor and discussed previously, and in more detail by Cartmell et al. [14]. The most recent development of this concept has been applied to a composite plate by Ganilova and Cartmell [24] and the rudiments of that work are summarised and discussed henceforth. In the work of Ganilova and Cartmell [24] a special actuator plate is postulated, in which three principal layers are employed in sandwich form. A host layer of bulk composite material, in this case glass-epoxy is located centrally with square SMA plate elements laid out in chequer-board fashion over the surface of the plate, as shown in Fig.37.

The SMA elements are laid out identically on both sides of the composite host and configured for electrically resistive heating in such a way that three different zones are defined, somewhat arbitrarily in the case of this demonstrator system. The

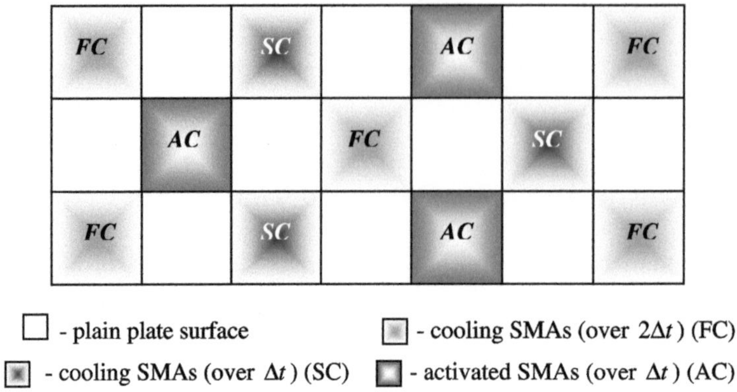

Fig. 37 Glass-epoxy composite and SMA sandwich plate in which the square plate-like SMA elements are laid out and zoned as shown, after [24]

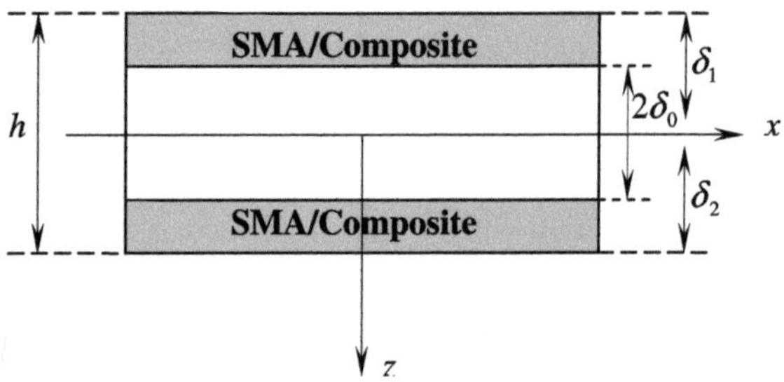

Fig. 38 Side elevation of the glass-epoxy composite and SMA sandwich plate showing the layer thickness dimensions, after [24]

element zones are denoted by *FC*, *SC*, and *AC* and are switched in or out of circuit sequentially in order to demonstrate the antagonistic effect, and also to show that grouped elements may be advantageous for the control of multiple modes of vibration for which highly complex patterns of nodal and antinodal lines can occur. The APT configuration has been investigated to date [24]. In an attempt to demonstrate roughly equalised heating and cooling time constants a typical time constant Δt is taken as the basic unit of time. The acronyms *FC*, *SC*, and *AC* were originally associated with operations denoted by 'first cooling', 'second cooling', and 'activated' labels, and although these meanings have become less significant as the research has developed they are retained here for consistency with [24]. The process works in principle like this: the *FC* elements are heated through the martensite-to-austenite transformation during and then they are cooled whilst the *SC* elements are heated through a further Δt. During the $2\Delta t$ period the *FC* elements have cooled for Δt and the *AC* elements have remained cool for the whole time. At the end of $2\Delta t$ the *AC* elements are heated and the *SC* elements enter the cooling phase for a further period of Δt. This represents one basic system cycle, as shown in Fig.39, during which each zone sequentially heats over Δt whilst the two others are cooling. The antagonism effect is embedded in such a way that two zones are always cooling whilst one is heating, ensuring that continuous useful work is done by the plate in

Fig. 39 Heating and cooling cycle for an antagonistically actuated SMA/Composite plate

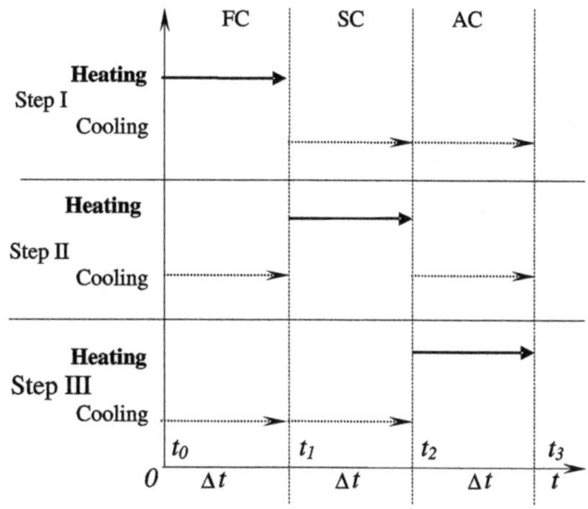

sequentially activated zones and reducing the passive cooling time constant accordingly to somewhere between Δt and $2\Delta t$, dependent on the precise physical configuration. Ganilova and Cartmell [24] also proposed an analytical model for the dynamics of the plate under forced vibration and APT. This is summarised here with some theoretical results shown for chosen physical data.

The governing equation of motion for a plate with integrated SMA layers and any chosen boundary conditions is given by Żak et al. [66]:

$$D_{11}\frac{\partial^4 w}{\partial x^4}+2\left(D_{12}+2D_{66}\right)\frac{\partial^4 w}{\partial x^2 \partial y^2}+D_{22}\frac{\partial^4 w}{\partial y^4}-N_x\frac{\partial^2 w}{\partial x^2}=q(x,y,t)-\rho h\frac{\partial^2 w}{\partial t^2} \quad (15)$$

noting that this is the same as equation (9) except for slightly revised notation on the right hand side. We persevere with the version given as equation (15) in order to maintain consistency with [66]. A generalised excitation is applied to the plate as follows:

$$q(x,y,t)=q_0(t)\sin\frac{\pi x}{a}\sin\frac{\pi y}{b} \quad (16)$$

The density quantity in Eq. (15) is given by $\rho=\sum_{i=1}^{3}\rho_i h_i/h$, in which h_i is the ith layer thickness and h is the whole plate thickness. The in-plane load N_x depends on the SMA elements and therefore on time given that elements switch in and out cyclically, therefore $N_x=N_x(t)$. This has the additional effect of varying the bending stiffnesses, D_{jk}, so that they each assume three different values for the scenario here, effectively leading to $D_{jk}(t)$ instead, but as the transformation time is to be kept arbitrary we drop the time argument henceforth and stay with D_{jk}. A further feature of this analysis is the inclusion of time variant damping which may be a partial side-effect of the phased activation of the zoned SMA elements as well as something additionally included within a generalised version of the problem, as discussed in [66]. This leads to a further re-statement of the governing equation of motion, whereby we obtain:

$$D_{11}\frac{\partial^4 w}{\partial x^4}+2(D_{12}+2D_{66})\frac{\partial^4 w}{\partial x^2\partial y^2}+D_{22}\frac{\partial^4 w}{\partial y^4}-N_x(t)\frac{\partial^2 w}{\partial x^2}$$
$$=q(x,y,t)-2\rho\varepsilon(t)h\frac{\partial w}{\partial t}-\rho h\frac{\partial^2 w}{\partial t^2}. \tag{17}$$

The right hand side of Eq. (17) contains the damping term in which a scaled time-variant damping parameter $\varepsilon(t)$ is included, providing a variable form of classical linear viscous damping. Simply supported boundary conditions are accommodated by means of the following deflection expression in which a and b are the length and width, respectively, of the plate:

$$w(x,y,t)=f(t)\sin\left(\frac{\pi x}{a}\right)\sin\left(\frac{\pi y}{b}\right) \tag{18}$$

Previous work by Żak and Cartmell [66] and Gristchak and Ganilova [26] led to an analytical definition for the bending stiffnesses of a laminated composite plate:

$$D_{jk}=\frac{2}{3}\left[B_{jk}^{N/2}h_{N/2}^3+\sum_{i=1}^{N/2-1}B^i_{jk}\left(h_i^3-h_{i+1}^3\right)\right] \tag{19}$$

where the B^i_{jk} represents the ith layer's elastic coefficients and N is the number of layers. This can be re-stated in a form applicable to the specific case of Fig.38 for which this results:

$$D_{jk}=\frac{2}{3}\left[B^0_{jk}\delta_0^3+B_{jk}\left(\delta^3-\delta_0^3\right)\right]. \tag{20}$$

The outer layers comprise composite/SMA and it is assumed that they are identical dimensionally and physically, leading to $B^1_{jk}=B^2_{jk}=B_{jk}$. The top, middle, and bottom layers are $\delta_1-\delta_0$, $2\delta_0$, and $\delta_2-\delta_0$, respectively, and it should also be noted that i is intentionally defined as zero for the middle layer, 1 for the top, and 2 for the bottom layer, and as top and bottom are identical then $\delta_1=\delta_2=\delta$. In order to apply equation (20) we require the B^i_{jk} value for each layer. For the middle layer the appropriate values are found to be,

$$B^0_{11}=\frac{E^0_{11}}{1-\nu^0_{12}\nu^0_{21}},\ B^0_{22}=\frac{E^0_{22}}{1-\nu^0_{12}\nu^0_{21}},\ B^0_{12}=\frac{\nu^0_{12}E^0_{12}}{1-\nu^0_{12}\nu^0_{21}},\ B^0_{66}=G^0_{12},$$

and in the case of the identical top and bottom layers, for which the superscripts are subsequently dropped, one obtains [66],

$$B_{11}=\frac{E_{11}}{1-\nu_{12}\nu_{21}},\ B_{22}=\frac{E_{22}}{1-\nu_{12}\nu_{21}},\ B_{12}=\frac{\nu_{12}E_{12}}{1-\nu_{12}\nu_{21}},\ B_{66}=G_{12}.$$

It is also necessary to define the Young's moduli appropriately and in [66] these are shown to be in the following forms, in which the relative volume fraction of the ith layer SMA is defined by V, and subscripts f and m describe the SMA fibre and host composite material, respectively,

$$E_{11}=E_fV+E_m(1-V),\ E_{22}=E_m\frac{E_f+E_m+(E_f-E_m)V}{E_f+E_m-(E_f-E_m)V},\ \ G_{12}=G_m\frac{G_f+G_m+(G_f-G_m)V}{G_f+G_m-(G_f-G_m)V},$$
$$\nu_{12}=\nu_fV+\nu_m(1-V),\ \ \hat{\rho}=\rho_fV+\rho_m(1-V).$$

From these forms it is possible to construct an expression which defines the density of the whole plate, and this is given by [66]:

$$\rho=\sum_{i=1}^{3}\frac{\rho_i h_i}{h}=\frac{2}{h}\left[\hat{\rho}(\delta-\delta_0)+\rho_0\delta_0\right] \tag{21}$$

noting that ρ_0 is the density of the middle layer in which there is no SMA.

Equation (17) can now be re-expressed in the following form after the substitution of Eq. (18), and the use of Eqs. (19)–(21):

$$\lambda^2 f''(t)+2\bar{\varepsilon}(t)f'(t)+\tilde{\bar{D}}(t)f(t)=\tilde{q}(t) \tag{22}$$

Further definitions are required such that $\lambda^2=h/a$, $\bar{\varepsilon}(t)=(h/a)\,\varepsilon(t)$, and

$$\tilde{\bar{D}}(t)=\frac{1}{\rho a}\left[\tilde{D}+\bar{N}(t)\right],\ \tilde{q}(t)=\frac{q(t)}{\rho a},$$

and noting that the following are also required,

$$\tilde{D}=\pi^4\left(\frac{D_{11}}{a^4}+2\frac{D_{12}+2D_{66}}{a^2b^2}+\frac{D_{22}}{b^4}\right),$$

and $\bar{N}(t)=(\pi/a)^2 N_x(t)$.

Equation (22) is a nonhomogeneous linear ordinary differential equation with variable coefficients and is solved in [66] by means of the two-stage hybrid Wentzel-Kramers-Brillouin-Galerkin method [26]. The first stage determines the complementary solution of the homogenous problem and this is of the form,

$$f(t,\lambda)=\exp\left[\int_a^t\left(\frac{1}{\lambda}f_0(t)+f_1(t)\right)dt\right] \tag{23}$$

where

$$f_0=\pm i\left[2\int\frac{e^2\int\frac{\tilde{\bar{D}}(t)}{\bar{\varepsilon}(t)}dt}{\bar{\varepsilon}(t)}dt\right]^{-1/2},\quad\text{and}\quad f_1=-\frac{1}{2}\frac{d}{dt}(\ln f_0).$$

Determination of the appropriate particular integral is summarised in detail by Ganilova and Cartmell [24] for these chosen forms for the physical and scaled damping and SMA in-plane load terms, respectively, $\varepsilon(t) = Ct$ leading to $\bar{\varepsilon}(t) = (h/a)Ct$, and $N_x(t) = st$ leading to $\bar{N}(t) = (\pi/a)^2 st$. The quantities s and C are constants chosen to accord with a specific experimental system design and these relations clearly define simple linear proportionalities with time for both damping and in-plane loading, both of which are predominantly attributable to SMA activation over Δt. The general solution to Eq. (22) is given in [24] as:

$$f(t) = e^{-\delta_{02}\int_{\bar{a}}^{t}\bar{f}_0(t)dt}\left[(s_1+c_1)\sin\left(\delta_{01}\int_{\bar{a}}^{t}\bar{f}_0(t)dt\right)+(s_2+c_2)\cos\left(\delta_{01}\int_{\bar{a}}^{t}\bar{f}_0(t)dt\right)\right] \tag{24}$$

for which we require forms for s_1 and s_2, as follows,

$$s_1 = \int \frac{\tilde{q}(t)e^{\delta_{02}\int_{\bar{a}}^{t}\bar{f}_0(t)dt}\cos\left(\delta_{01}\int_{\bar{a}}^{t}\bar{f}_0(t)dt\right)}{\lambda^2\delta_{01}\frac{d}{dt}\left(\int_{\bar{a}}^{t}\bar{f}_0(t)dt\right)}dt \text{ and } s_2 = -\int \frac{\tilde{q}(t)e^{\delta_{02}\int_{\bar{a}}^{t}\bar{f}_0(t)dt}\sin\left(\delta_{01}\int_{\bar{a}}^{t}\bar{f}_0(t)dt\right)}{\lambda^2\delta_{01}\frac{d}{dt}\left(\int_{\bar{a}}^{t}\bar{f}_0(t)dt\right)}dt.$$

Definitions of the quantities δ_{01}, δ_{02}, and $\bar{a}$ are given in full by Gristchak and Ganilova [26] and quantities c_1 and c_2 are obtained from initial conditions [26]. A numerical example is considered next from which an experimental test system has evolved in parallel. Although the experimental system is undergoing preliminary testing at the time of writing it can be seen in Fig.43. The plate is of dimensions $a = 0.5$ m and $b = 0.214$ m, with an epoxy middle layer of thickness 9 mm and identical top and bottom layers of 1.5 mm in thickness. The top and bottom layers are composed of the chequer-boarded SMA squares filled in with epoxy composite. From these layer thicknesses we can calculate that $\delta_0 = 4.5$,mm and $\delta_1 = \delta_2 = \delta = 6$ mm. The Young's moduli, bending stiffnesses, Poisson's ratio, and density are defined and calculated for the epoxy middle layer, as appropriate, to yield the following, $E_{11}^0 = E_{22}^0 = E_{12}^0 = 3.43$ GPa, $B_{66}^0 = G_{12}^0 = 1.27$ GPa, $\nu = 0.35$, $\rho_0 = 1250$ kg/m^3.

The activated, austenitic phase, SMA elements have the following properties $E_f = E_A = 69.6$ GPa, $G_f = G_A = 26.77$ GPa, $\nu = 0.3$, $\rho_f = \rho_{SMA} = 6448.1$ kg/m^3. The density of the top and bottom layers is given by $\hat{\rho}$ for a volume fraction of V=0.5238, hence $\hat{\rho} = 3972.8$ kg/m^3, and from this the density of the whole plate can be evaluated, in this case as $\rho = 1930.7$ kg/m^3. In the process already described the FC, SC, AC phases comprise 5, 3, and 3 elements, as shown in Fig.37, and so the relative volume fractions change accordingly to be 0.24, 0.14, and 0.14, requiring updated calculations each time for E_{11}, E_{22}, G_{12}, ν_{12}, and $\hat{\rho}$, as needed for the top and bottom layers.

In order to provide some validation of the analytical solution the results from this are compared with those obtained by numerically integrating Eq. (22). Initially $N_x(t)$ and $\varepsilon(t)$ are held constant at $0.1N$ and unity, respectively, with initial conditions

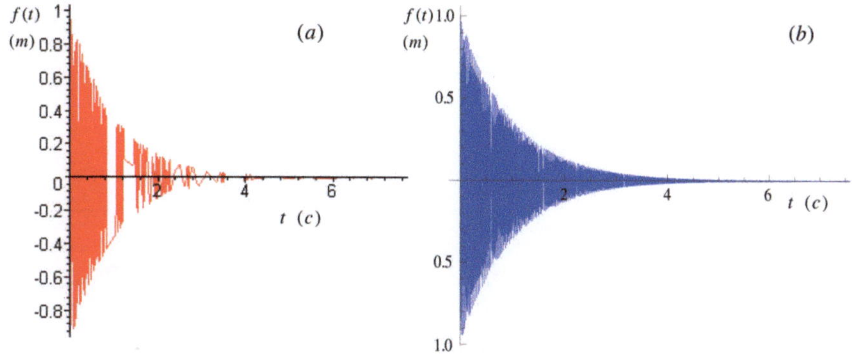

Fig. 40 Analytical-hybrid responses **a** and numerical responses **b** for $N_x(t) = 0.1N$, $\varepsilon(t) = 1$, $f(0.00001) = 0.1$, $\dot{f}(0.00001) = 1$, $q(t) = 0$, after [24]

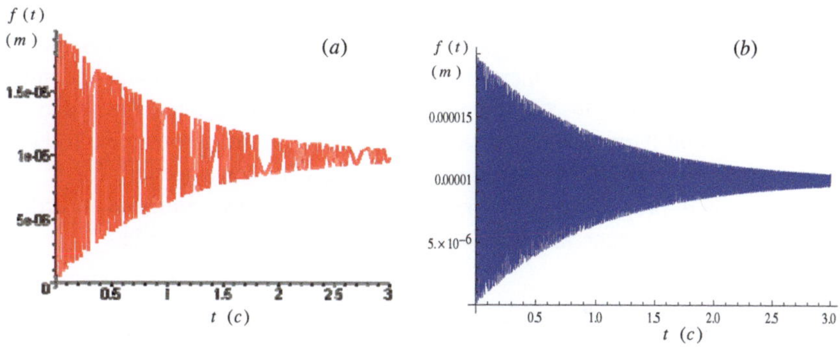

Fig. 41 Analytical-hybrid responses **a** and numerical responses **b** for $N_x(t) = 0.1$ N, $\varepsilon(t) = 1$, $f(0) = 0$, $\dot{f}(0) = 0$, $q(t) = 500N$, after [24]

$f(0.00001) = 0.1$ and $\dot{f}(0.00001) = 1$ and $q(t) = 0$, with analytical hybrid and numerical responses as given in Fig.40 in the time domain.

It is immediately evident that despite the sparseness at some points in the plot in Fig. 40(a), the dynamics of the system are captured to a remarkably similar extent for in both cases. Note that the initial conditions are given for time close to zero in order to ease the burden of numerical calculation. A second case is shown in Fig. 41 in which the in-plane SMA force and damping are maintained at the same constant values but the excitation force is increased from zero to 500 N, and the initial conditions are set to zero.

Once again the dynamics of the system are faithfully reproduced by the hybrid analytical solution and show very similar transient trends about a closely similar dc offset, representative of the particular integral part of the hybrid-analytical solution which, in turn, is commensurate with the constant excitation load. A final case is given in Fig.42 for which time variant damping is introduced, along with an oscillating excitation, such that $\varepsilon(t) = t$, and $q(t) = \cos t$ N, again with zero initial conditions.

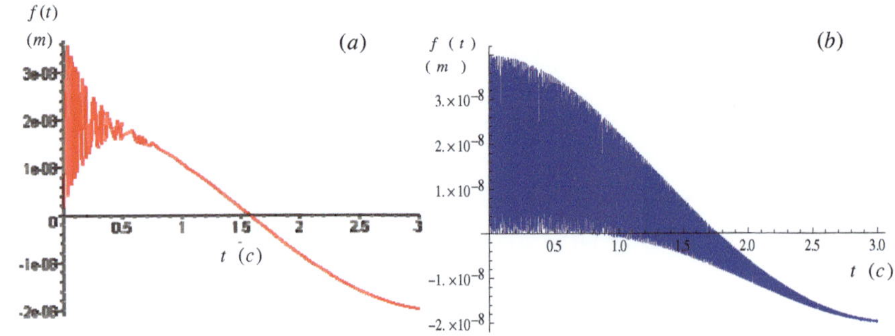

Fig. 42 Analytical-hybrid responses **a** and numerical responses **b** for $N_x(t) = 0.1$ N, $\varepsilon(t) = t$, $f(0) = 0$, $\dot{f}(0) = 0$, $q(t) = \cos t$ N, after [24]

The two solutions show similar tendencies for the mean of the response, with identical drift properties in time, but a far more persistent envelope in the case of the numerical solution which suggests that the hybrid-analytical model may not consolidate the dissipative properties of the system in the same way as the numerical solution. This is because of the numerical calculation required for s_1 and s_2 in equation (24) and the fact that for some data cases simplifications have to be implemented in order to reach convergent solutions, with consequences for solution accuracy. A new programme of experimental work has been initiated in order to validate the model further, and a precursor test system is shown in Fig.43.

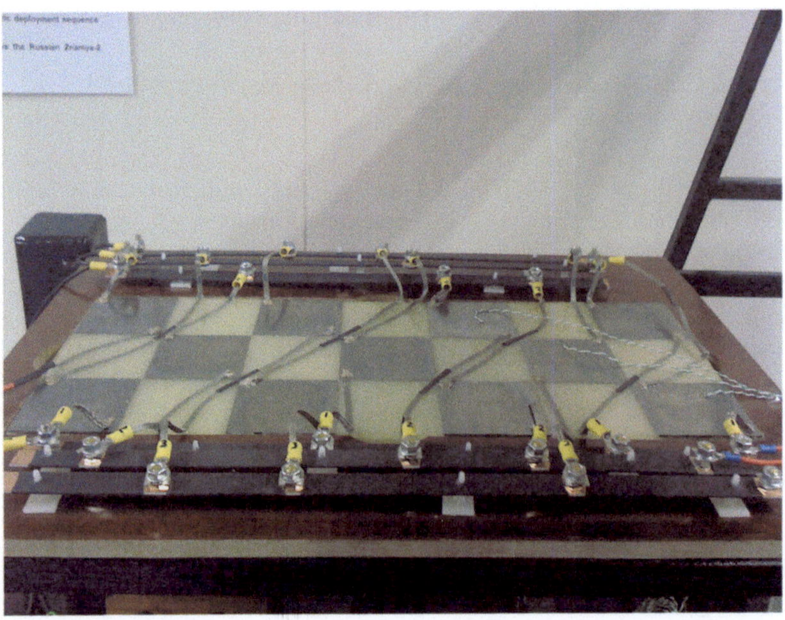

Fig. 43 Laboratory test rig system, showing the composite/SMA plate and resistive heating wiring

The plate is simply supported, and associated wiring and high current bus bars for resistive heating of zones FC, SC, and AC can be seen. Three thermocouples can also be seen fitted to right hand side elements associated with all three zones. Sequential connections are provided so that the heating current can be applied, sequentially, and in a controlled manner, to all the elements in each zone. The experiment has been designed and built in order to test further the theoretical results and results will be presented separately in a forthcoming publication.

6 Conclusions

The main aim of this chapter has been to give a reasonable grounding in SMA modelling and immediately to place that within the context of the literature so that the main models are clarified for the reader. The chapter then summarises appropriate notions of structural integration, where SMA elements can be embedded within composite hosts, initially in the form of beams and plates, then in cylindrical shells. This is tackled by assuming two different embedding technologies as a starting point, Active Property Tuning (APT) and Active Strain Energy Tuning (ASET), and then including appropriate terms, including an ASET-based in-plane SMA load, so that natural frequencies, critical loads, and modal response amplitudes can be calculated, firstly for beams, and then for plates, each with equi-positioned layers of SMA each side of the neutral axis.

Extensive numerical results are provided for both generic structures for variations in the principal physical parameters principally for APT installations because these are simpler to set up in practice than ASET configurations which require arrangements for ensuring precise in-plane preloading, but still give reasonable to good performance. In the case of the plate a finite element model is used to calculate four different modal responses for two different boundary conditions (SSSS and CCCC) and for four different activation scenarios (both layers activated, upper layer activated, lower layer activated, neither layer activated), and this shows clearly the profound qualitative changes that are introduced by active elements embedded in this way, and it also complements the previously discussed quantitative results for responses under different conditions.

The chapter continues with a section on flexible rotors in which a similar technology extended to cylindrical shells is exploited to create active bearing housings in order to modify the vibrational response of a flexible rotor when whirling due to mass unbalance. Finite element models are used to predict the changes in both the first whirl mode natural frequencies and the disk amplitude in the first whirl mode for different bearing housing thickness for two composite materials, and the effects of SMA activation are also discussed within the text.

Further designs for active rotor bearings are presented, and once again the effectiveness of SMA activation is demonstrated. The chapter concludes with a novel proposal for larger scale integration of active SMA elements in a composite plate so that a repeating chequerboard structure is obtained. This is utilised both for so-called antagonistic action, where the cooling phase of one subset of elements is speeded up by direct mechanical action from a neighbouring subset undergoing activation heating, and also as an effective means for reducing the amplitude of two-dimensional plate modes. A candidate three layer plate comprising an inner glass-epoxy host plate with chequerboard distributions of SMA on the upper and lower surfaces is modelled and a generalised differential equation of motion is obtained for such a structure, with an approximate analytical solution obtained by recourse to a two stage hybrid Wentzel-Kramers-Brillouin-Galerkin methodology. Time domain responses are given for some pragmatic cases. This section concludes the chapter with proposals for an extensive experimental investigation of this system with a view to obtaining operational data from which further nonlinear models may be tested in the future.

Acknowledgements. Financial support from EPSRC, Rolls-Royce plc, and the University of Glasgow is acknowledged. Contributions from colleagues and students are recognised, in particular L.Atepor, V.Z.Gristchak, D.J.Inman, S.Jana, S.Kaczmarczyk, M.Krawczuk, A.W.Lees, W.Ostachowicz, and M.Wiercigroch.

References

1. Armstrong, W.D., Lilgholt, H.: The time dependant, super-viscoelastic behavior of NiTi shape memory alloy fiber reinforced polymer matrix composites. Mater. Sci. Eng. B 68, 149–155 (2000)
2. Basinski, Z.S., Christian, J.W.: Experiments on the martensitic transformation in single crystals of Indium-Thallium alloys. Acta Metallurgica 2, 148–166 (1954)
3. Baz, A., Chen, T., Ro, J.: Shape control of NITINOL-reinforced composite beams. Composites B 31, 631–642 (2000)
4. Baz, A., Poh, S., Ro, J., Gilheany: Control of natural frequencies of NITINOL-reinforced composite beams. Journal of Sound and Vibration 185, 171–185 (1995)
5. Bekker, A., Brinson, L.C.: Phase diagram based description of the hysteresis behavior of shape memory alloys. Acta Materialia 46, 3649–3665 (1998)
6. Bo, Z., Lagoudas, D.C.: Comparison of different thermomechanical models for shape memory alloys. Adaptive Structures and Composite Materials: Analysis and Application AD-45/MD-54 9-15 (1994)
7. Boyd, J.G., Lagoudas, D.C.: A theromdynamical constitutive model for shape memory materials. Part I. The monolithic shape memory alloy. International Journal of Plasticity 12, 805–842 (1996)
8. Birman, V.: Stability of functionally graded shape memory alloy sandwich panel. Smart Materials and Structures 6, 278–286 (1997)
9. Birman, V., Saravanos, D.A., Hopkins, D.A.: Micromechanics of composites with shape memory alloy fibers in uniform thermal fields. AIAA Journal 30, 1905–1912 (1996)
10. Brinson, L.C.: One-dimensional constitutive behavior of shape memory alloys: Thermomechanical derivation with non-constant material functions. Journal of Intelligent Material Systems and Structures 4, 229–242 (1993)

11. Brocca, M., Brinson, L.C., Bažant, Z.P.: Three-dimensional constitutive model for shape memory alloys based on microplane model. Journal of the Mechanics and Physics of Solids 50, 1051–1077 (2002)
12. Buehler, W.J., Gilfrich, J.V., Wiley, R.C.: Effect of low-temperature phase change on the mechanical properties of alloys near composition. TiNi Journal of Applied Physics 34, 1475–1477 (1963)
13. Burkart, M.W., Read, T.A.: Diffusionless phase change in the Indium-Thallium system Transactions of AIME. Journal of Metals 197, 1516–1524 (1953)
14. Cartmell, M.P., Inman, D.J., Lees, A.W., Ganilova, O.A., Atepor, L.: Smart materials applications to structural dynamics and rotating machines. In: Proc. Euromech., Kazimierz Dolny, Poland, May 21-24, vol. 498 (2008)
15. Chang, L.C., Read, T.A.: Plastic deformation and diffusionless phase change in metals - the Gold-Cadmium beta phase. Transactions of AIME Journal of Metals 191, 47–52 (1951)
16. Chen, C.W.: Some characteristics of the martensite transformation of Cu-Al-Ni alloys. Journal of Metals 9, 1202–1203 (1957)
17. Chen, Q., Levy, C.: Active vibration control of elastic beam by means of shape memory alloy layers. Smart Materials and Structures 5, 400–406 (1996)
18. Chen, Q., Levy, C.: Vibration analysis and control of flexible beam by using smart damping structures. Composites: Part B 30, 395–406 (1999)
19. Choi, S., Lee, J.J., Seo, D.C., Choi, S.W.: The active buckling control of laminated composite beams with embedded shape memory alloy wires. Composite Structures 47, 679–686 (1999)
20. Epps, J., Chandra, R.: Shape memory alloy actuation for active tuning of composite beams. Smart Mater. Struct. 6, 251–264 (1997)
21. Epps, J.J., Chopra, I.: Comparative evaluation of shape memory alloy constitutive models with test data. In: Proceedings of the 38th AIAA Structures Structural Dynamics and Materials Conference and Adaptive Forum, Kissimmee Florida, April 7-10 (1997)
22. Ford, D.S., White, S.R.: Thermomechanical behavior of 55Ni45Ti. Nitinol Acta Metallurgica 44, 2295–2307 (1996)
23. Gandhi, F., Wolons, D.: Characterization of the pseudoelastic damping behavior of shape memory alloy wires using complex modulus. Journal of Smart Material Structures 8, 49–56 (1999)
24. Ganilova, O.A., Cartmell, M.P.: An analytical model for the vibration of a composite plate containing an embedded periodic shape memory alloy structure. Composite Structures 92, 39–47 (2010)
25. Govindjee, S., Hall, G.J.: A computational model for shape memory alloys. International Journal of Solids and Structures 37, 375–760 (2000)
26. Gristchak, V.Z., Ganilova, O.A.: A hybrid WKB-Galerkin method applied to a piezoelectric sandwich plate vibration problem considering shear force effects. Journal of Sound and Vibration 317, 366–377 (2008)
27. Hornbogen, E., Wassermann, G.: Über den Einfluβ von Spannungen und das Auftretten von Umwandlungsplastizitt bei Beta$_1$-Beta-Umwandlung des Messings. Zeitschrift fr Metallkunde 47, 427–433 (1956)
28. Icardi, U.: Large bending actuator made with SMA contractile wires: theory numerical simulation and experiments. Composites B 32, 259–267 (2001)
29. Inman, D.J., Cartmell, M.P., Lees, A.W., Leize, T., Atepor, L.: Proposals for controlling flexible rotor vibration by means of an antagonistic SMSA/composite smart bearing. In: Proc. MPSVA 2006, Bath, UK (September 2006)

30. Kustov, S., Van Humbeeck, J.: Damping properties of SMA Advances in Shape Memory Materials. Materials Science Forum 583, 85–109 (2008)
31. Kwai, M.: Effects of matrix inelasticity on the overall hysteretic behavior of TiNi-SMA fiber composite. Int. J. Plast. 16, 263–282 (2000)
32. Lau, K.-T.: Vibration characteristics of SMA composite beams with different boundary conditions. Mater. Des. 23, 741–749 (2002)
33. Lau, K.-T., Zhou, L.-M., Tao, X.-M.: Control of natural frequencies of a clamped-clamped composite beam with embedded shape memory alloy wires. Compos. Struct. 58, 39–47 (2002)
34. Lee, H.J., Lee, J.J.: A numerical analysis of the buckling and postbuckling behavior of laminated composite shells with embedded shape memory alloy wire actuators. Smart Mater. Struct. 9, 780–787 (2000)
35. Lee, H.J., Lee, J.J.: Evaluation of the characteristics of a shape memory alloy spring actuator Smart. Materials and Strucutres 9, 817–823 (2000)
36. Lees, A.W., Jana, S., Inman, D.J., Cartmell, M.P.: The control of bearing stiffness using shape memory. In: Proc. ISCORMA-4, Calgary, Canada, August 27-30 (2007)
37. Lexcellent, C., Leclercq, S., Garbry, B., Bourbon, G.: The two way shape memory effect of shape memory alloys: an experimental study and a phenomenological mode. International Journal of Plasticity 16, 1155–1168 (2000)
38. Lexcellent, C., Moyne, S., Ishida, A., Miyazaki, S.: Deformation behaviour associated with the stress-induced martensitic transformation in Ti-Ni thin films and their thermodynamical modelling. Thin Solid Films 324, 184–189 (1998)
39. Liang, C., Rogers, C.A.: One-dimensional thermomechanical constitutive relations for shape memory material. Journal of Intelligent Materials and Structures 1, 207–234 (1990)
40. Liew, K.M., Kitipornachai, S., Ng, T.Y., Zou, G.P.: Multi-dimensional superelastic behavior of shape memory alloys via nonlinear finite element method. Engineering Structures 24, 51–57 (2002)
41. Nagaya, K., Takeda, S., Tsukui, Y., Kumaido, T.: Active control method for passing through critical speeds of rotating shafts by changing stiffnesses of the supports with the use of memory metals. Journal of Sound and Vibration 113(2), 307–315 (1987)
42. Nishimura, F., Watanabe, N., Tanaka, K.: Back stress and shape recoverability during reverse transformation in an Fe-based shape memory alloy. Materials Science and Engineering A 247, 275–284 (1998)
43. Oh, J.T., Park, H.C., Hwang, W.: Active shape control of a double-plate structures using piezoceramics and SMA wires. Smart Materials and Structures 10, 1100–1106 (2001)
44. Ostachowicz, W.M., Kaczmarczyk, S.: Vibrations of composite plates with SMA fibres in a gas stream with defects of the type of delamination. Composite Structures 54, 305–311 (2001)
45. Ostachowicz, W., Krawczuk, M., Żak, A.: Natural frequencies of a multilayer composite plate with shape memory alloy wires. Finite Element in Analysis and Design 32, 71–83 (1999)
46. Ostachowicz, W., Krawczuk, M., Żak, A.: Dynamics and buckling of a multilayer composite plate with embedded. SMA wires Composite Structures 48, 163–167 (2000)
47. Otsuka, K., Wayman, C.M.: Shape Memory Materials. University Press Cambridge, Cambridge (1998)

48. Pae, S., Lee, H., Park, H., Hwang, W.: Realization of higher-mode deformation of beam using shape memory alloy wires and piezoceramics. Smart Mater. Struct. 9, 848–854 (2000)
49. Piedboeuf, M.C., Gauvin, R., Thomas, M.: Damping behaviour of shape memory alloys: strain amplitude frequency and temperature effects. Journal of Sound and Vibration 214, 885–901 (1998)
50. Raniecki, B., Lexcellent, C., Tanaka, K.: Thermomechanical models of pseudoelastic behaviour of shape memory alloys. Archives of Mechanics 44, 261–284 (1992)
51. Rogers, C.A., Baker, D.K., Jaeger, C.A.: Introduction to smart materials and structures Smart Materials Structures and Mathematical Issues, pp. 17–28. Technomic Publishing Company Inc. (1989)
52. Rogers, C.A., Liang, C., Baker, D.K.: Dynamic control concepts using shape memory alloy reinforced plates Smart Materials Structures and Mathematical Issues, pp. 39–62. Technomic Publishing Company Inc. (1989)
53. Roh, J.-H., Kim, J.-H.: Hybrid smart composite plate under low velocity impact. Composite Strucutres 56, 175–182 (2002)
54. Shu, S.G., Lagoudas, D.C., Hughes, D., Wen, J.T.: Modeling of a flexible beam actuated by shape memory alloy wires. Smart Mater. Struct. 6, 265–277 (1997)
55. Song, G., Kelly, B., Agrawal, B.N.: Active position control of a shape memory alloy wire actuated composite beam. Smart Materials and Structures 9, 711–716 (2000)
56. Su, Z., Mai, H., Lu, M., Ye, L.: Thermo-mechanical behaviour of shape memory alloy reinforced composite laminate (Ni-Ti/glass-fibre/epoxy). Compos. Struct. 47, 705–710 (1999)
57. Sun, G., Sun, S., Wu, X., Wu, J.: A study on thermomechanical deformation of elastic beam with embedded shape memory alloy wires. Mater. Des. 21, 525–528 (2000)
58. Tanaka, K.: A thermomechanical sketch for shape memory effect: One-dimensional tensile behavior. Res. Mechanica 18, 251–263 (1986)
59. Tawfik, M., Ro, J.J., Mei, C.: Thermal post-buckling and aeroelastic behaviour of shape memory alloy reinforced plates. Smart Materials and Structures 11, 297–303 (2002)
60. Thompson, S.P., Loughlan, J.: The control of the post-buckling response in thin composite plates using smart technology. Thin-walled Structures 36, 231–263 (2000)
61. Tsai, X.-Y., Chen, L.-W.: Dynamic stability of a shape memory alloy wire reinforced composite beam. Compos. Struct. 56, 235–241 (2002); a double-plate structures using piezoceramics and SMA wires Smart Mater. Struct. 10,1100-1106
62. Viderman, Z., Porat, I.: An optimal control method for passage of a flexible rotor through resonances. Journal of Dynamic Systems Measurement and Control 109(3), 216–223 (1987)
63. Vinson, J.R., Sierakowski, R.L.: The behavior of structures composed of composite materials. Martinus Nijhoff Publishers, Dordrecht (1986)
64. Wu, X., Grummon, D.S., Pence, T.J.: Modeling phase fraction shakedown during thermomechanical cycling of shape memory materials. Materials Science and Engineering A, 273-275, 273–275 (1999)
65. Zhu, J., Liang, N., Huang, W., Liew, K.M., Liu, Z.: A thermodynamic constitutive model for stress induced phase transformation in shape memory alloys I. International Journal of Solids and Structures 39, 741–763 (2002)
66. Żak, A.J., Cartmell, M.P.: Analytical modelling of shape memory alloys and flat multilayered composite beams and plates with shape memory alloy wires. 1st Research Report, Department of Mechanical Engineering, University of Glasgow (2000)
67. Żak, A.J., Cartmell, M.P.: Statics and dynamics of a sleeve-ring component with SMA strips. 3rd Research Report, Department of Mechanical Engineering, University of Glasgow (2002)

68. Żak, A.J., Cartmell, M.P.: Dynamics of a rotor system with a smart SMA-embedded sleeve-ring component. 4th Research Report, Department of Mechanical Engineering, University of Glasgow (2002)
69. Żak, A.J., Cartmell, M.P., Ostachowicz, W.M.: Dynamics of multi-layered composite beams and plates with SMA wires. 2nd Research Report, Department of Mechanical Engineering, University of Glasgow (2001)
70. Żak, A.J., Cartmell, M.P., Ostachowicz, W.M., Wiercigroch, M.: One-dimensional shape memory alloy models for use with reinforced composite structures. Smart Mater. Struct. 12, 338–346 (2003)

Theoretical and Experimental Nonlinear Vibrations of Sagged Elastic Cables

Giuseppe Rega

Abstract. The chapter presents a comprehensive overview of recent advancements in the theoretical and experimental research on modelling, analysis, response, and nonlinear/nonregular phenomena in the finite amplitude, resonant, forced dynamics of sagged, horizontal or inclined, elastic cables. Asymptotic solutions and a rich variety of features of nonlinear multimodal interaction occurring in various resonance conditions are comparatively discussed. Dynamical and mechanical characteristics of some main experimentally observed responses are summarised, along with the relevant robustness, spatio-temporal features, and dimensionality. Challenging issues arising in the characterisation of involved bifurcation scenarios to complex dynamics are addressed, and hints for proper reduced-order modelling in cable nonlinear dynamics are obtained from both asymptotic solutions and experimental investigations, in the perspective of a profitable cross-validation of the observed nonlinear phenomena.

Keywords: Suspended cable, resonant nonlinear dynamics, asymptotic solution, experimental analysis, bifurcation and chaos, reduced-order modelling.

1 Introduction

Suspended cables are lightweight, flexible structural elements used in several applications in mechanical, civil, electrical, ocean and space engineering, due to their capability of transmitting forces, carrying payloads and conducting signals across large distances. At the same time, the sagged cable is a basic element of theoretical interest in applied mechanics, as well as an archetypal model of various phenomena

Giuseppe Rega
Department of Structural and Geotechnical Engineering, Sapienza University of Rome, Via Antonio Gramsci 53, 00197 Roma, Italy
e-mail: giuseppe.rega@uniroma1.it

J. Warminski et al. (Eds.): Nonlinear Dynamic Phenomena in Mechanics, SMIA 181, pp. 159–210.
springerlink.com

in the nonlinear dynamics of elastic systems with initial curvature, for being prone to large amplitude vibrations.

Classical analyses of cable linear vibrations ([25], [74]) have been recently complemented by further achievements, which include, e.g., experimental validation of theoretical phenomena [60], derivation of approximate formulae useful for design purposes [77], consideration of the effects of damage [38] or bending stiffness [59], unified treatment of shallow and non-shallow cables [31]. However, major research effort has been devoted in the last few decades to cable nonlinear dynamics, with the attention being paid to finite amplitude vibrations of sagged elastic cables under a variety of conditions of planar and nonplanar, internal, external and/or parametric, resonances. Variably refined theoretical models, and purely analytical, numerical or mixed treatments have been considered in handling the problem. In fact, owing to the inherent combination of system quadratic and cubic nonlinearities, occurring for sagged cables, their finite amplitude response exhibits an extremely rich variety of nonlinear dynamic phenomena, as highlighted by many research achievements. The relevant state of art and the available results on modelling, solutions, and ensuing dynamic phenomena are comprehensively documented in a few recent review articles concerned with both deterministic ([50], [51]) and stochastic [24] regimes of cable nonlinear dynamics.

Yet, in the last few years, further meaningful research has been accomplished, both theoretically and experimentally. It has been aimed at highlighting, and possibly overcoming, a number of existing modelling and analysis limitations which affect the description and understanding of finite amplitude dynamics, as well as at throwing light on some specific, yet interesting, features of system nonlinear response. Limitations can be schematically summarised as follows:

- consideration (i) of only approximate continuous cable models, (ii) of quite low-dimensional finite representations and analysis of such models, and (iii) of solely shallow horizontal or nearly taut inclined cables, with the ensuing incomplete and/or unsatisfactory description of actual cable dynamics;
- solely partial description of the many involved (in-plane or in-plane/out-of-plane) interaction phenomena possibly occurring in various internal/external/parametric resonance conditions, along with limited knowledge on possible transition scenarios to non-regular dynamics;
- incomplete cross-validation of analytical and numerical solutions, and lack of experimental description and understanding of the actual nonlinear response of sagged cables.

This chapter does not aim at comprehensively reporting on all advances recently made on the above mentioned issues. Yet, it is intended at providing an overview on a number of meaningful theoretical and experimental achievements obtained on some of them by the writer and co-authors, in the last few years. They are concerned with both modelling and analysis, as well as with the description of the diverse phenomenology of system response occurring in different dynamical situations.

The chapter is organised as follows. Modelling and theoretical analysis are addressed in Sect. 2, by discussing different approximate continuous models, by summarising the distinguishing dynamic features of horizontal versus inclined sagged cables, and by introducing the multimode discretization aimed at obtaining low-dimensional cable models to be tackled via asymptotic techniques in different internal resonance conditions. A summary of the ensuing information is provided as regards the actual resonance activation, the expectable response features – with special reference being paid to the extent of nonlinear modal interaction –, the contribution of non-resonant modes to cable response, and the reduced-order models to be possibly formulated for reliably highlighting the main aspects of resonant nonlinear dynamic response of the underlying infinite-dimensional system. Typical nonlinear phenomena occurring in cable forced dynamics are exemplified in Sect. 3, by distinguishing and comparing the responses of horizontal and inclined cables in different internal resonance conditions, by showing cases of non-regular multi-harmonic responses, and by discussing features of nonlinear dynamic displacements and tension as obtained with different cable models. In turn, Sect. 4 is entirely devoted to the experimental characterisation of cable nonlinear dynamics. Upon generally dwelling on the issue of dimensionality of system nonlinear response – to be addressed via rather sophisticated techniques of nonlinear experimental analysis and dynamical system theory –, attention is focused on different bifurcation scenarios of transition to complex dynamics observed for a reference experimental cable/mass system. They include quasi-periodic transition to chaos and a global codimension-two bifurcation which organises involved regimes of regular nonlinear response and the ensuing homoclinic chaos. The possibility of a profitable feedback between experiments and theory is also highlighted by reporting features of on-going research aimed at formulating low-dimensional phenomenological models of sagged cables suitable for theoretical/numerical investigations and aimed at overcoming some persisting limitations of cable modelling, which may prevent us from correctly reproducing the experimentally observed features of response scenario. The chapter ends (Sect. 5) with a short summary on other topics of cable nonlinear dynamics, which have been meaningfully addressed in the last few years within the relevant scientific community, and on further expected research developments.

2 Cable Modelling and Theoretical Analysis

2.1 Continuous Modelling

The continuous model most widely used in the literature for analysing large amplitude, forced, damped, three-dimensional vibrations of a suspended cable refers to a perfectly flexible, homogeneous, linearly elastic system, with negligible torsional, bending and shear rigidities. It is based on the assumption of quasi-static axial stretching (also called static *condensation*), and considers a *horizontally* hanging *shallow* cable with two supports at the same level ([50], [51]). Any of the previous assumptions is somehow relaxed in more recent continuous formulations.

Still keeping the elastic material assumption and accounting for the sole axial rigidity, a more general model of suspended cable considers an *arbitrarily sagged* [70] and possibly *inclined* [69] system, and is based on a *refined* kinematical description of the cable element deformation. Figure 1 displays a suspended cable with arbitrary inclination angle θ in a fixed Cartesian coordinate (X, Y, Z) system. Three different configurations of the cable element in the natural ($\mathrm{d}s_n$), static ($\mathrm{d}s$) and final dynamic ($\mathrm{d}s_f$) states are considered, with the function $y(x)$ describing the planar static equilibrium of the cable under its own gravity g. Keeping the horizontal span X_H fixed, the vertical span Y_H is varied to attain a desired inclination, and the horizontal component H of cable static tension is adjusted to attain a desired cable sag value d. The cable in-plane (out-of-plane) dynamics is described by the horizontal u and vertical v (w) displacement components measured from the static configuration. Here, x is the spatially independent variable, and t denotes time.

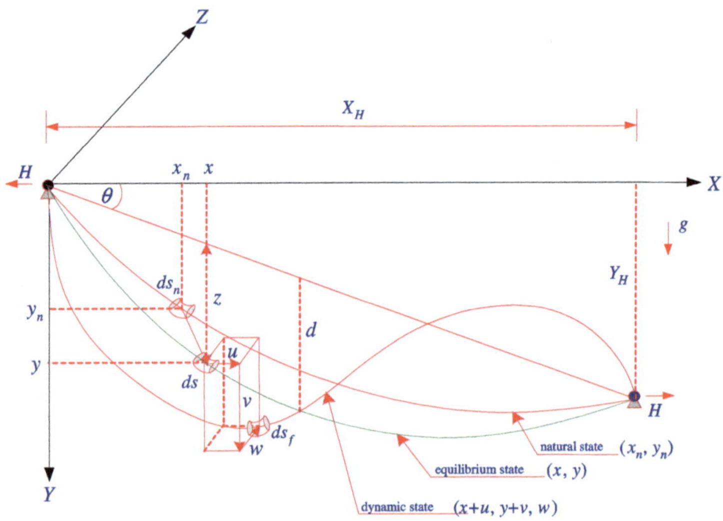

Fig. 1 Different configurations of an arbitrarily sagged and inclined cable

Based on the so-called engineering strain measure, the total strain of an infinitesimal cable element is given by:

$$e_f = \frac{ds_f}{ds_n} - 1 = \frac{1+e}{\sqrt{1+y'^2}}\sqrt{(1+u')^2+(y'+v')^2+w'^2} - 1, \tag{1}$$

where $e = (\mathrm{d}s - \mathrm{d}s_n)/\mathrm{d}s_n$ is the initial static strain. The governing *exact* nonlinear partial differential equations (PDEs) of 3D coupled, forced, damped motion of the cable about static equilibrium read:

$$\left(\frac{EA+EA(1+e)u'}{\sqrt{1+y'^2}}-\frac{EA(1+u')}{\sqrt{(1+u')^2+(y'+v')^2+w'^2}}\right)' = \left(\frac{w_C}{g}\ddot{u}+c_u\dot{u}\right)\frac{\sqrt{1+y'^2}}{1+e}-F_u\sqrt{1+y'^2},$$

$$\left(\frac{EAy'+EA(1+e)v'}{\sqrt{1+y'^2}}-\frac{EA(y'+v')}{\sqrt{(1+u')^2+(y'+v')^2+w'^2}}\right)' = \left(\frac{w_C}{g}\ddot{v}+c_v\dot{v}\right)\frac{\sqrt{1+y'^2}}{1+e}-F_v\sqrt{1+y'^2}$$

$$\left(\frac{EA(1+e)w'}{\sqrt{1+y'^2}}-\frac{EAw'}{\sqrt{(1+u')^2+(y'+v')^2+w'^2}}\right)' = \left(\frac{w_C}{g}\ddot{w}+c_w\dot{w}\right)\frac{\sqrt{1+y'^2}}{1+e}-F_w\sqrt{1+y'^2}. \quad (2)$$

E, A and w_C are the cable Young modulus, uniform cross-sectional area and self-weight per unit unstretched length, respectively, c_u, c_v, c_w are viscous damping coefficients, and F_u, F_v, F_w are components of uniformly distributed harmonic external forcing. A prime (dot) represents partial differentiation with respect to the horizontal space coordinate x (time t).

In conditions of moderately large vibration amplitudes and assuming small static strain ($ds \approx ds_n$), the total strain (Eq. (1)) becomes

$$e_f = e + e_d = e + \frac{1}{1+y'^2}\left(u' + y'v' + \frac{1}{2}\left(u'^2+v'^2+w'^2\right)\right), \quad (3)$$

where e_d is the extensional dynamic strain expressed through its Lagrangian measure. The ensuing, *approximate*, third-order nonlinear PDEs of motion read, in non-dimensional form [71]:

$$\rho\ddot{u}+c_u\rho\dot{u} = \left\{u' + \frac{\alpha}{\rho^3}\left(u'+y'v'\right) + \frac{\alpha}{\rho^3}\left(u'^2+y'u'v'+\frac{1}{2}\left(u'^2+v'^2+w'^2\right)\right) + \frac{\alpha}{2\rho^3}\left(u'^3+u'v'^2+u'w'^2\right)\right\}' + \rho F_u,$$

$$\rho\ddot{v}+c_v\rho\dot{v} = \left\{v' + \frac{\alpha}{\rho^3}\left(y'u'+y'^2v'\right) + \frac{\alpha}{\rho^3}\left(u'v'+y'v'^2+\frac{y'}{2}\left(u'^2+v'^2+w'^2\right)\right) + \frac{\alpha}{2\rho^3}\left(u'^2v'+v'^3+v'w'^2\right)\right\}' + \rho F_v,$$

$$\rho\ddot{w}+c_w\rho\dot{w} = \left\{w' + \frac{\alpha}{\rho^3}\left(u'w'+y'v'w'\right) + \frac{\alpha}{2\rho^3}\left(w'u'^2+w'v'^2+w'^3\right)\right\}' + \rho F_w, \quad (4)$$

where $\alpha = EA/H$, $\rho = \sqrt{1+y'^2}$. The corresponding homogeneous boundary conditions are $u(0,t) = u(1,t) = v(0,t) = v(1,t) = w(0,t) = w(1,t) = 0$.

System (4) exhibits quadratic and cubic geometrical nonlinearities associated with cable initial curvature (only the former) and axial stretching; however, due to the interaction between longitudinal and transverse dynamics – which ensues from accounting for the overall inertia effects – quadratic nonlinearities occur even in the absence of initial sag, namely, in the taut string case. The extensional dynamic strain (see Eq. (3)) turns out to be spatially non-uniform and, accordingly, the nonlinear dynamic tension ($T_d = EAe_d$) exhibits a spatio-temporal variation along the cable. As the exact model, the approximate one nonlinearly couples the longitudinal and transverse (in-plane or out-of-plane) cable dynamics, thus being referred to as a kinematically *non-condensed* model to distinguish it from the above mentioned *condensed* model, typically considered in the cable literature ([50], [66]). Besides further kinematical assumptions, the latter basically results from neglecting the inertia and viscous damping effects in the longitudinal PDE of motion (4_1), which corresponds to assuming that the cable nonlinearly stretches in a quasi-static manner in the absence of longitudinal external loading. Accounting for the cable boundary conditions, the dynamic strain in Eq. (3) becomes

$$e_d(t) = \int_0^1 \left(y'v' + \frac{1}{2}v'^2 + \frac{1}{2}w'^2 \right) \mathrm{d}x \tag{5}$$

and entails a space-independent dynamic tension, averaged over the spatial integral. Accordingly, the governing integro-PDEs of motion accounting for the solely transverse (vertical and out-of-plane) cable dynamics read

$$\begin{aligned} \ddot{v} + c_v\dot{v} &= v'' + \alpha\left(y'' + v''\right)\int_0^1 \left(y'v' + \frac{1}{2}\left(v'^2 + w'^2\right)\right)\mathrm{d}x + F_v, \\ \ddot{w} + c_w\dot{w} &= w'' + \alpha w''\int_0^1 \left\{ y'v' + \frac{1}{2}\left(v'^2 + w'^2\right)\right\}\mathrm{d}x + F_w \end{aligned} \tag{6}$$

with the u displacement component depending on v and w.

The *exact* cable model described by Eq. (2) is solely referred to in *purely numerical* treatments of a given nonlinear dynamic problem based on, e.g., the space-time finite difference method coupled with a predictor-corrector iteration algorithm ([69], [70]) for the inclined and horizontal cable, respectively) or on a finite element procedure. The major advantage of a numerical treatment stands in allowing to capture the spatial richness of cable nonlinear response and its time-varying content, and in obtaining information about the possibly significant involvement of higher-order modes which is allowed by the considered multidegree-of-freedom model. In contrast, if the interest is in highlighting the characterising features of system nonlinear dynamics in different external/internal resonance conditions, anyone of the solely handable *approximate* models, whose analysis can be pursued via analytical (symbolic) or *mixed analytical-numerical* approaches, are referred to. The ensuing static and nonplanar/planar linear and nonlinear dynamic results can be thoroughly validated against those of the exact model via numerical techniques (Runge-Kutta/shooting, finite elements, finite differences, see [71]), thus allowing for a proper approximate continuous model selection in different technical situations.

2.2 *Static Equilibrium and Planar Linear Free Dynamics*

As far as cable sag is concerned, shallowness is usually assumed in the nonlinear dynamics literature, by considering the non-dimensional parabolic, *symmetric*, static equilibrium profile $y(x) = 4dx(1-x)$ – which adequately describes cables with values of the sag-to-span ratio up to about 1/8 – both in the classical case of *small sag* horizontal cables [50] and for the nearly taut inclined cables addressed in recent studies [79]. In the linearised planar dynamics [25], denoting with S the cable equilibrium length, this entails the occurrence of the classical *crossover* phenomenon of the in-plane natural frequencies ω depending on the Irvine parameter $\lambda/\pi = (1/\pi)\sqrt{(w_C S)^2 EA/H^3}$ which accounts for the cable elastic and geometric properties: see the points $\lambda/\pi \approx 2n$ ($n = 1,2,\ldots$) in the frequency spectrum of Fig. 2a, which displays, based on the non-condensed model, the first six planar frequencies ω/π (dimensionalised with respect to the fundamental frequency of the corresponding taut string) versus λ/π. With increasing λ, the i-th crossover ($i = 1,2,\ldots$) marks the transition of the corresponding couple of i-th symmetric and antisymmetric in-plane modes from a sequence where the former is lower to a sequence where the former is higher, along with the increase by two of the number of nodal points of the symmetric modes along the cable span [26].

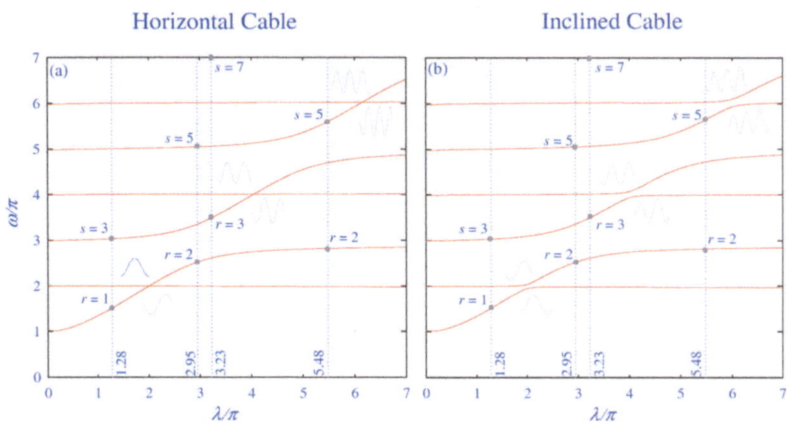

Fig. 2 Planar frequency spectrum of sagged, horizontal **a** or inclined **b**, cables. Several combinations of 2:1 internally resonant high-frequency (s) and low-frequency (r) modes are identified

In contrast, for arbitrarily inclined cables with *moderate sag* (i.e., with values of the sag-to-span ratio somehow larger than the limiting value of the parabolic profile), a closed-form solution of the cable static equilibrium configuration, approximate up to cubic order, allows us to account for the actual system *asymmetry* [25]. In the linearised planar dynamics, to be addressed via a numerical procedure (e.g., Galerkin or finite elements [71]), this entails a qualitative modification from the crossover phenomenon occurring in the frequency spectrum of symmetric cables to the frequency *avoidance* (or veering) phenomenon [37] occurring in the spectrum of inclined cables (Fig. 2b) for increasing values of the generalised cable parameter $\lambda/\pi = (1/\pi)\sqrt{(w_C S \cos\theta)^2 EA/T_a^3}$ ([73], [74]), where T_a is the static tension at the cable point where the local inclination angle is approximately equal to θ.

A veering phenomenon of the same nature actually characterises the linear free dynamics of other asymmetric mechanical systems [49]. Avoidance is associated with the occurrence of cable hybrid, i.e. asymmetric, modes [74], which result from a mixture of symmetric and antisymmetric modal shapes, as exemplified in Fig. 3, whose underlying linear eigenvalue problem of free undamped planar motion, derivable from Eqs. ($4_{1,2}$), has been solved via the Galerkin method based on an assumed sine-based series [71].

The distinguishing and contrasting linear dynamics features of crossover vs avoidance, or symmetric/antisymmetric vs hybrid modes, also somehow affect the system nonlinear dynamic behaviour, as it will be discussed in Sect. 3.

Of course, when dealing with actually non-shallow horizontal or inclined cables, the exact catenary solution of the static equilibrium configuration based on the hyperbolic function [25] has to be used, and the ensuing linear frequencies can be again obtained via a numerically-based (Galerkin or finite element) procedure (see, e.g., [62]).

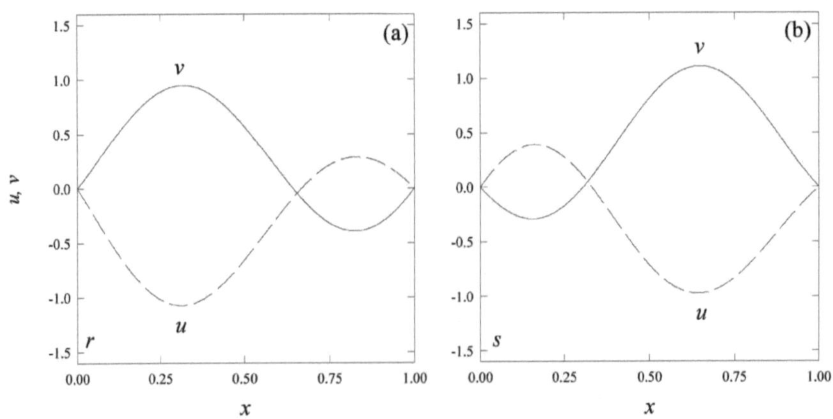

Fig. 3 Low-frequency (r) **a** and high-frequency (s) **b** modes for $\theta = 45^o$ and $\lambda/\pi \approx 2$: u (dashed) and v (continuous) displacement components

2.3 Multimode Discretization for Nonlinear Dynamics

Aiming at obtaining a low-dimensional reduced cable model, suitable for analytical solutions of the sole nonlinear temporal problem, a multimode discretization of the PDEs of the non-condensed approximate infinite-dimensional model is pursued, and then an asymptotic analysis of the ensuing system of nonlinearly coupled ODEs is developed through any perturbation technique, typically, the multiple time scales method [43].

Focusing on the solely planar nonlinear cable vibrations, for the sake of simplicity, and casting the PDEs of motion in state-space (displacement and velocity) first-order format, *multimode discretization* involves two main steps ([33], [71]).

(i) Projecting the first-order PDEs onto the infinite-dimensional orthonormal eigenbasis of cable linear modes, as follows,

$$U^J(x,t) = \sum_{m=1}^{\infty} f_m(t)\zeta_m^J(x), \qquad V^J(x,t) = \sum_{m=1}^{\infty} p_m(t)\zeta_m^J(x), \tag{7}$$

where $J = 1,2$, $U^1 = u$, $U^2 = v$, $V^1 = \dot{u}$, $V^2 = \dot{v}$, $\zeta_m^1 = \phi_m$, $\zeta_m^2 = \varphi_m$, f_m and p_m are the displacement and velocity modal coordinates associated with both the horizontal φ_m and vertical ϕ_m shape functions of the m-th in-plane mode;

(ii) obtaining the infinite set of nonlinearly coupled ODEs in the unknown time dependent displacement and velocity coordinates via the Galerkin method, as follows,

$$\begin{aligned} &\dot{f}_m - p_m = 0, \\ &\dot{p}_m + 2\mu_m p_m + \omega_m^2 f_m = \sum_{i=1}^{\infty}\sum_{j=1}^{\infty} \Lambda_{mij} f_i f_j + \sum_{i=1}^{\infty}\sum_{j=1}^{\infty}\sum_{k=1}^{\infty} \Gamma_{mijk} f_i f_j f_k + Z_m \cos\Omega t, \end{aligned} \tag{8}$$

In Eq. (8), ω_m is the m-th natural frequency, μ_m and Z_m are modal damping and external forcing terms, respectively, Λ_{mij} and Γ_{mijk} are quadratic and cubic nonlinear coefficients depending on the modal shape functions and governing the (m,i,j,k) overall displacement coupling: see [71], where the qualitative difference between the sets of nonlinear coefficients as obtained with the non-condensed and condensed models is also highlighted.

Of course, for all practical computations, a finite number of terms has to be considered in the infinite-dimensional discretization, and a reduced-order model properly suited to deal with the specific dynamic problem has to be identified via convergence analyses. In this respect, it is worth noting that, provided enough modes are retained in the discretization, the relevant outcomes turn out to be substantially equivalent to those furnished by the direct application of the asymptotic method to the original approximate PDEs, with no a priori assumptions of the displacement solution form ([27], [33]).

2.4 Internal Resonances and Asymptotic Solutions

Depending on the generalised Irvine elasto-geometric parameter governing the dynamic behaviour of horizontal/inclined cables and due to the coexistence of quadratic and cubic nonlinearities in the cable equations of motion, a very rich pattern of *internal resonance* conditions (1:1, 2:1, and/or 3:1) involving different modes occurs at both crossover (avoidance) frequencies of horizontal (inclined) cables and away from them.

Internal resonances can be (i) single or multiple, respectively involving a couple or a combination of modes, as well as (ii) planar or non-planar, with the latter involving both in-plane and out-of-plane modes; moreover, (iii) they can involve either mixed (symmetric and antisymmetric) or solely symmetric or antisymmetric modes, as well as hybrid modes (for inclined cables). 1:1 planar or multiple internal resonances typically, but not exclusively, occur at crossover/avoidance points, the former involving symmetric and antisymmetric modes (horizontal cable) or hybrid modes (inclined cable), the latter possibly involving a variety of modes with different combinations of resonant frequencies. Yet, all internal resonances can take place over the whole in-plane/out-of-plane frequency spectrum of the cable, with a variable involvement of modes and frequency ratios: by way of example, in Fig. 2, a number of planar 2:1 resonances involving different couples of high-frequency (s) and low-frequency (r) modes are identified for both the horizontal and the inclined cable.

It is important to note that not all of the nominally activable internal resonance conditions are actually activated, because the involved modes may be nonlinearly orthogonal with each other, and the vanishing nonlinear orthogonality of modes may represent a necessary and sufficient condition for actual *activation* of an internal resonance [29]. This automatically precludes a number of internal resonance situations from actually taking place, as discussed in [29]. Anyway, if internal resonances are activated, they entail a strong modal interaction and energy exchange between the involved modes, to an extent which depends on the specific resonance condition and on the nature of modes (as it will be discussed in Sect. 2.5.2). These phenomena eventually result in harmful undesirable motions of the system, mostly if the internal resonance condition already affecting the system nonlinear free dynamics is complemented by the *external*, primary or secondary, resonance of anyone of the involved modes possibly occurring in a forced dynamics problem, or by the fundamental or principal resonance associated with a parametric excitation.

In the framework of a *multiple scale analysis* ([43], [44]) of the weakly nonlinear periodic solutions of Eq. (8), attention has to be paid to a suitable ordering of the internal resonance relationships specifically concerned, as well as to the ordering of external resonance, damping and excitation. Moreover, the analysis has to be accomplished up to the second order (ε^3-order, ε being a small bookeeping parameter), with the aim of capturing the combined effects due to higher-order quadratic and cubic nonlinearities [33]. Primary resonance of a high-frequency (s) mode and its 1:1 or 3:1 internal resonances with a low-frequency (r) mode are suitably dealt with via the frequency relationship orderings

$$\Omega = \omega_s + \varepsilon^2 \sigma_f, \quad \omega_s = \omega_r + \varepsilon^2 \sigma \quad \text{or} \quad \omega_s = 3\omega_r + \varepsilon^2 \sigma,$$

whereas primary resonance and 2:1 internal resonance are expressed as

$$\Omega = \omega_s + \varepsilon \sigma_f, \quad \omega_s = 2\omega_r + \varepsilon \sigma,$$

where σ_f and σ are external and internal detuning parameters, respectively. At the same time, damping and excitation amplitude have to be ordered in such a way to balance with the nonlinearities and the internal resonance at the same ε^3-order, in both resonance cases, namely $\mu_m \to \varepsilon^2 \mu_m$ and $Z_m \to \varepsilon^2 Z_m$.

Referring to various internal resonance cases and to the set of perturbation equations typically occurring in the multiple scale analysis of the sole *free* vibration problem at various ε-orders, in the 1:1 and 3:1 resonances – which are associated with cubic nonlinearities – the secular effects directly appear at the ε^3-order, whereas in the 2:1 resonance – associated with quadratic nonlinearities – they are split between the ε^2-order (due to quadratic nonlinearities only) and the ε^3-order (due to higher-order quadratic as well as cubic nonlinearities). Upon imposing the solvability conditions at various ε-orders of the perturbation analysis, using the method of reconstitution of the complex amplitudes ([44], [57]) and moving to the amplitudes polar form, the multiple scale analysis provides the amplitude and phase modulation equations (APMEs) of the involved resonant modes, whose fixed points correspond to periodic motions of the ODEs (8) governing the dynamics of the reduced-order model.

Analysis of the APMEs with the associated nonlinear interaction coefficients, as provided by the multiple scale analyses of various (1:1, 2:1, 3:1) internal resonance conditions ([29], [78], [66], [55]) allows to draw a general description of various possible resonant solutions occurring for horizontal (symmetric) and inclined (asymmetric) cables, in the case of *forced* planar vibrations under uniformly-distributed vertical harmonic excitation at primary resonance with some internally resonant mode. Depending on (i) the generalised elasto-geometric parameter λ/π, (ii) the kind of activated internal resonance, and (iii) the primary resonance of either a high-frequency (s) or low-frequency (r) mode, a summary on the existence of *uncoupled* (UC) and/or *coupled* (C) mode planar solutions for horizontal (inclined) cables at crossover (avoidance), or away from it, is reported in Table 1.

In whatever internal resonance condition, the uncoupled (UC) solutions only involve the directly excited resonant mode whereas coupled (C) solutions also involve the non excited mode which is indirectly driven into the response via an internal resonance enhanced mechanism of energy transfer. Both UC and C solutions coexist for 1:1 resonant crossover cables, whereas only C solutions exist for avoidance cables, regardless of the mode being directly excited. This also occur for other asymmetric systems, e.g. [30]. In turn, as regards 2:1 and 3:1 resonances, UC and C (only C) solutions exist when directly exciting a high-frequency (low-frequency) mode, irrespective of the considered horizontal (inclined) cable being at crossover (avoidance) or away from it. This implies that 2:1 and 3:1 (1:1) resonant solutions do (do not) depend on the mode being directly excited, while they do not (do) depend on the cable geometry being symmetric or asymmetric. It is thus evident how moving from horizontal to inclined cables – with the associated modification from purely symmetric/antisymmetric modes at crossover to hybrid modes at avoidance – is expected to entail a different scenario of nonlinear response at 1:1 internal resonance [55].

Table 1 A summary of regular planar solutions in horizontal/inclined cables at various internal resonances. UC (C), uncoupled-mode (coupled-mode) solution, CR (AV), crossover (avoidance)

			horizontal (sym.) cables		inclined (asym.) cables	
$s:r$	λ/π	n	$\Omega=\omega_s+\varepsilon^n\sigma_f$	$\Omega=\omega_r+\varepsilon^n\sigma_f$	$\Omega=\omega_s+\varepsilon^n\sigma_f$	$\Omega=\omega_r+\varepsilon^n\sigma_f$
1:1	CR vs. AV	2	UC/C	UC/C	C	C
2:1	CR vs. AV	1	UC/C	C*	UC/C	C
	non-CR vs. non-AV					
3:1	non-CR vs. non-AV	2	UC/C	C	UC/C	C

* At CR, only with non-vanishing excitation on the low-frequency mode.

The multiple scale analysis in various internal resonance conditions also furnishes the cable coupled dynamic configurations associated with their *nonlinear normal modes*, which meaningfully account for the spatial corrections, with respect to the reference linearly resonant modes, due to the quadratic nonlinearity effects of all infinite modes of the discretization or, in practical terms, of the nonresonant modes considered in a finite-dimensional discretization ([33], [65]). By way of example, the second-order coupled dynamic configurations associated with the u ($J=1$) and v ($J=2$) displacement components of a 1:1 resonant horizontal/inclined cable are expressed as

$$\begin{aligned} U^J(x,t) \approx\; & a_r\cos(\Omega t-\gamma_r)\,\zeta_r^J(x)+a_s\cos(\Omega t-\gamma_s)\,\zeta_s^J(x)\\ &+\frac{1}{2}a_s^2\left[\cos(2\Omega t-2\gamma_s)\,\psi_{ss}^J(x)+\kappa_{ss}^J(x)\right]\\ &+\frac{1}{2}a_r^2\left[\cos(2\Omega t-2\gamma_r)\,\psi_{rr}^J(x)+\kappa_{rr}^J(x)\right]\\ &+a_sa_r\left[\cos(2\Omega t-\gamma_r-\gamma_s)\,\psi_{rs}^J(x)+\cos(\gamma_r-\gamma_s)\,\kappa_{rs}^J(x)\right] \end{aligned} \tag{9}$$

whereas those of a 2:1 resonant horizontal/inclined cable read

$$\begin{aligned} U^J(x,t) \approx\; & a_r\cos\left(\frac{\Omega t}{2}-\frac{\gamma_s}{2}-\frac{\gamma_r}{2}\right)\zeta_r^J(x)+a_s\cos(\Omega t-\gamma_s)\,\zeta_s^J(x)\\ &+\frac{1}{2}a_s^2\left[\cos(2\Omega t-2\gamma_s)\,\psi_{ss}^J(x)+\kappa_{ss}^J(x)\right]\\ &+\frac{1}{2}a_r^2\left[\cos(\Omega t-\gamma_s-\gamma_r)\,\psi_{rr}^J(x)+\kappa_{rr}^J(x)\right]\\ &+\frac{1}{2}a_sa_r\left[\cos\left(\frac{3}{2}\Omega t-\frac{3}{2}\gamma_s-\frac{1}{2}\gamma_r\right)\psi_{rs}^J(x)+\cos\left(\frac{1}{2}\Omega t-\frac{1}{2}\gamma_s+\frac{1}{2}\gamma_r\right)\kappa_{rs}^J(x)\right] \end{aligned} \tag{10}$$

In Eqs. (9) and (10), a_r and a_s are real-value amplitudes of the low-frequency and high-frequency mode, respectively, γ_r and γ_s are corresponding phases, κ_{ss}^J and κ_{rr}^J are static drift effects due to quadratic nonlinearities, and the second-order shape functions ψ_{ss}^J, ψ_{rr}^J, ψ_{rs}^J, κ_{rs}^J govern the spatial displacement corrections to the horizontal ($J = 1$) and vertical ($J = 2$) shape functions ζ_r^J, ζ_s^J of the two resonant (r, s) modes, due to quadratic effects of all resonant/non-resonant modes. These spatial corrections are also of major importance as regards the evaluation of cable nonlinear dynamic tension and of its actual space-time modifications (see Sect. 3.3 forward).

Of course, the accomplished multiple scale analytical predictions can be validated against the numerical outcomes of finite element [17] or finite difference ([65], [68]) solutions of the original, exact or approximate PDEs of motion, under specified initial conditions.

2.5 Modal Interaction Coefficients as Predictive Tools for Reliable Nonlinear Dynamic Response

Still focusing on planar nonlinear dynamics, the analysis of the nonlinear interaction coefficients of multiple scale solution at various ε-orders highlights a number of important phenomenological aspects which play a different role for diverse, horizontal or inclined, cables.

2.5.1 Internal Resonance Activation

As already mentioned, there are restrictions as to the nature of the involved cable modes (e.g., symmetric and/or antisymmetric) in order for their internal resonance to be actually activated [29]. In fact, activation of a specific resonance depends on the condition that the governing nonlinear interaction coefficients be different from zero, which is in turn associated with the involved normal modes being nonlinearly non-orthogonal. A systematic discussion on the conditions for actual activation of 2:1, 3:1 and 1:1 internal resonances for a wide class of shallow symmetric structural systems also including horizontal cables is made in [29].

When considering also asymmetric systems, differences occur as regards vanishing or non-vanishing of some nonlinear interaction coefficients for horizontal and inclined cables, respectively, just based on the existence or non-existence of nonlinear orthogonality properties.

The relevant effects become apparent in the 2:1 internal resonance case. Indeed, for horizontal cables, 2:1 internal resonance is activated only when the involved high-frequency (s) mode is symmetric. This occurs regardless of the low-frequency (r) mode being symmetric or antisymmetric, so that, e.g., the 2:1 resonance of all non-crossover cables ($\lambda/\pi \approx 1.28$, 2.95, 3.23, 5.48) in Fig. 2a is indeed activated, along with that of the second crossover cable ($\lambda/\pi \approx 4$) which involves the high-frequency symmetric mode – out of the two therein coexisting – and the low-frequency (1st antisymmetric) mode. In contrast, for inclined cables, owing to the asymmetry effects of inclined sagged configurations which entail modal hybridity and non-vanishing of the governing first-order quadratic coefficient [55], the 2:1 resonance with the low-frequency mode is activated for the second avoidance cable ($\lambda/\pi \approx 4$, see Fig. 2b) irrespective of the involved high-frequency (3rd or 4th) mode, since both of them are now hybrid. Depending on frequency-tuning and hybridity capacity, such activation feature persists over a rather wide range of system parameters. Overall, this also gives broad hints about the most likely involvement of a larger number of modes within a multiple internal resonance for avoidance cables than for crossover cables, owing to the non-satisfied nonlinear orthogonality of the relevant modes.

In contrast, for the 1:1 internal resonance cases at crossover/avoidance – whose activation is governed by the non-vanishing of a few second-order nonlinear interaction coefficients – no practical differences exist between horizontal and inclined cables in this respect, even though the former exhibit a larger number of vanishing coefficients due to the nonlinear orthogonality of eigenfunctions of mixed modal type at crossovers. Accordingly, nearly tuned 1:1 resonances are activable near both crossovers and avoidances.

2.5.2 Resonant/Non-resonant Modal Contributions and Reduced-Order Modelling

A major issue is concerned with the evaluation of the variable contribution of *resonant* and non-negligible *non-resonant* modes to cable overall response, with the involvement of the latter (of either low-order or higher-order) strongly depending on the role played by the second-order effects of quadratic nonlinearities, which come into play just owed to the accomplished second-order perturbation analysis ([65], [66], [55], [32]).

The importance of quadratic modal contributions depends on the combination of the effects (i) of the elasto-geometric parameter and the static equilibrium configuration, (ii) of the modal characteristics of horizontal/vertical displacements (see, e.g., Fig. 3), and (iii) of the system frequencies commensurability. Some or all of these dependencies also affect the outcomes of cubic interaction coefficients, as well as those of the second-order displacement shape functions given by Eqs. (9) and (10). In the sequel, some distinguishing items of non-resonant modal contributions for horizontal and inclined cables are discussed by addressing 2:1 and 1:1 resonances separately. In view of a finite-dimensional analysis of cable nonlinear dynamics, which is the solely pursuable numerical treatment of the infinite-dimensional discretization of the original PDEs, comparison of contributions from various resonant and non-resonant modes to the interaction coefficients, as determined by the governing nonlinearities, allows us to get important *predictive* hints about the proper identification of variable *reduced-order models* (ROMs) needed to reliably describe the nonlinear dynamics of various resonant cables.

2:1 internal resonance: horizontal vs inclined cables

Analysis of second-order quadratic modal contributions in 2:1 resonance shows that, besides the two resonant modes, only *symmetric* non-resonant modes affect the solution of horizontal cables away from crossover ([8], [70]), whereas *all* non-resonant modes – irrespective of their order or spatial character – do contribute for inclined cables. For cables with a quite low sag below crossover/avoidance, i.e. $\lambda/\pi \approx 1.28$ (see Fig. 2), the quadratic effects due to non-resonant modes are very small compared with those produced by the two resonant modes. Therefore, in the nonlinear dynamic analysis of these cables, it makes sense to consider the minimal (two-degree-of-freedom) reduced-order model accounting for the sole two resonant modes.

However, this is no more the case for larger-sagged cables with $\lambda/\pi \approx 5.48$ (Fig. 2), for which the higher-order effects of quadratic nonlinearities become pronounced, and a number of non-resonant modes, e.g., the intermediate-order 4th (3rd and 4th) and higher-order 7th, 9th, 11th (6th, 7th, 9th and 11th) ones play a meaningful role, too, in all quadratic coefficients of the horizontal (inclined) cable.

In other cases, some non-resonant modes play a role even greater than the resonant ones. This may happen, for instance, for inclined cables at avoidances (Fig. 2b), where multiple internal resonances are actually realised when considering, e.g., the coupled 1st and 4th modes or the coupled 1st and 3rd modes nearly below ($\lambda/\pi \approx 3.84$) or above ($\lambda/\pi \approx 4.14$) second avoidance, respectively. In fact, in these cases, the major influence of non-resonant (i.e., non-modelled) modes substantially results from the contribution of the 3rd (4th) hybrid mode nearly coexisting with the modelled 4th (3rd) one near avoidance for $\lambda/\pi \approx 3.84$ ($\lambda/\pi \approx 4.14$). Yet, other non-resonant modes, of either intermediate or higher order, can also play a significant role, as it occurs, e.g., for the horizontal cable with $\lambda/\pi \approx 5.48$).

1:1 internal resonance: horizontal vs inclined cables

For horizontal cables at crossover, the 1:1 resonance plays the major role and the effects of the second-order quadratic modal contributions are substantially in agreement with those due to 2:1 resonances away from crossovers. They confirm the importance of accounting for both resonant and non-resonant (*higher-order*, mostly *symmetric*) modes in the asymptotic solution of even *relatively shallow* cables. In turn, symmetric non-resonant modal contributions become as more important as higher-order crossovers are considered, whereas antisymmetric non-resonant contributions are nearly always negligible in view of a proper reduced-order model selection.

For inclined cables at avoidances, due to the system high modal density and strong coupling, the non-modelled hybrid mode – out of the two modes therein coexisting with nearby frequencies – may contribute to the response even greater than the directly-modelled hybrid mode, as already mentioned. This highlights the need to account for both of them and the possible involvement of a larger number of coupled modes in avoidance cables than in crossover cables. For cables at first ($\lambda/\pi \cong 2$) or second ($\lambda/\pi \cong 4$) avoidance with different inclinations (θ), the bar charts in Fig. 4 schematically show the percent contributions of each resonant (encircled) and non-resonant modes to the second-order quadratic terms embedded in various nonlinear interaction coefficients K of the asymptotic solution [55]. All charts in Fig. 4 highlight significant contributions of both resonant and non-resonant (lower- and/or higher-order) modes for 1:1 resonant avoidance cables, confirming the already summarised outcomes of the companion analyses of 1:1 (2:1) resonant crossover (non-crossover/non-avoidance) cables ([66], [65]). For instance, the increasing importance of non-resonant modal contributions with increasing sag is confirmed by the higher order modal truncation (up to $m \approx 9$) required for $\theta = 30°$ by the second-avoidance cable, which involves higher-order ($r = 3$, $s = 4$) hybrid modes (Fig. 4c), with respect to the first-avoidance cable in Fig. 4a (up to $m \approx 5$).

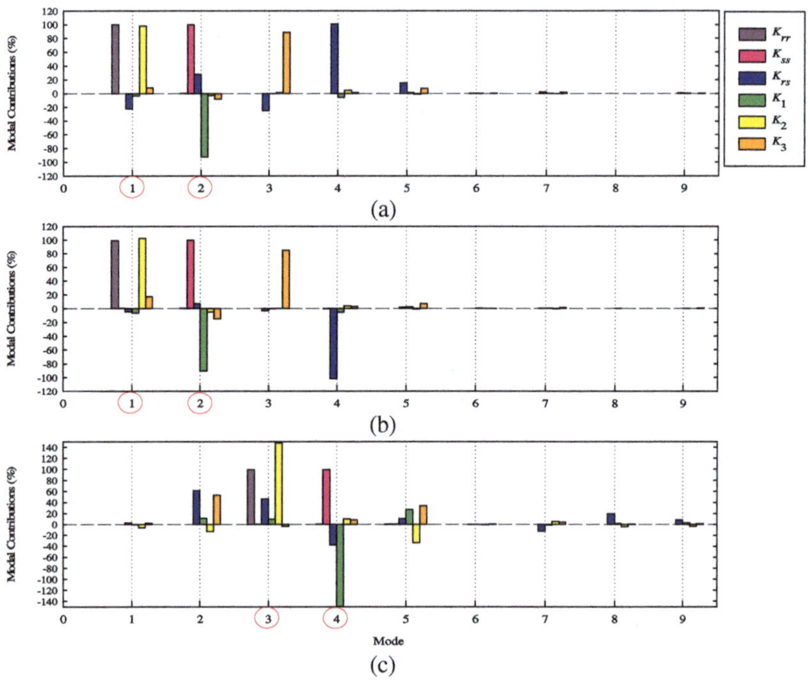

Fig. 4 Mode contributions to various nonlinear interaction coefficients K: $\theta = 30°$ **a**, **c**, 45° **b**; $\lambda/\pi \cong 2$ **a**, **b**, $\lambda/\pi \cong 4$ **c**

However, in contrast with a 1:1 resonant first-crossover cable, whose non-trivial contributions of non-resonant antisymmetric modes ($m = 4, 6, \ldots$) to some coefficients (K_{rs} or K_3) are negligible [66], the 1:1 resonant first-avoidance cables in Figs. 4a ($\theta = 30°$) and 4b ($\theta = 45°$) reveal meaningful contributions of the $m = 4$ mode to K_{rs}, which are mainly due to the high modal density of asymmetric resonant/non-resonant shape functions entering that coefficient. The remarkable role played by asymmetric modal contributions of non-resonant modes is also apparent in the K_3 contributions of 3rd and 5th modes in Figs. 4a and 4b, as well as in the contributions of 5th and higher-order modes in Fig. 4c for the second-avoidance cable.

Overall, comparison of the effective interaction coefficients reveals quantitative as well as qualitative differences in the nonlinear contributions (values and/or sign) to different avoidance cables, which affect in a variable way their nonlinear dynamics and stability under given control parameters.

In general, it can be stated that in the 1:1 resonance at avoidance (crossover), the asymmetry (symmetry) features of inclined (horizontal) cables make the contribution from non-resonant modes greater (lower).

Overall, based on the results of both considered internal resonances, the quadratic modal contributions put into evidence the significance of accounting for both resonant and non-resonant (higher-order) modes in the resonant dynamic solutions of cables with significant sags and/or remarkable asymmetry features due to inclination. Meaningful effects of non-resonant modes have been overlooked in several papers that investigated modal interactions near crossovers by accounting for only resonant modes in the second-order perturbation analysis or by constraining the analysis to just the first order [51]. Based on the method of multiple scales, the lowest-dimensional discretisation may yield quantitatively-inaccurate and/or qualitatively-crude results with respect to the coinciding infinite-dimensional discretisation [65] or direct ([47], [57]) perturbation analysis.

The selection of the proper *reduced-order model* to be referred to for capturing the main response features of the actual underlying infinite-dimensional system depends on system parameter values and coupled amplitudes with a variety of inherent options. For instance, as regards 2:1 resonance, it has been said that the minimal (two-degree-of-freedom) model involving only the resonant modes shows capable of providing reliable results for very low-sagged cables. When increasing the sag, for relatively low-sagged cables it may be sufficient to account for also non-resonant modes in the sole evaluation of the nonlinear amplitudes and frequencies, thereby developing an improved first-order multiple scale solution [65]. In contrast, non-resonant modes have to be accounted for in also the evaluation of nonlinear dynamic displacements (thus developing the full second-order multiple scale solution) – where they provide possibly meaningful spatial corrections – as the cable sag and/or inclination become significant.

Of course, the higher-order modal contributions become less important when the order of modal truncation is increased up to finally yielding converging results. In this respect, sample results in terms of resonantly coupled amplitudes and frequencies can be found in [65]. Anyway, it is worth noting how a two-mode resonant solution is also capable of properly signaling the breakdown of the minimal reduced-order modelling, whether embedded within an infinite-dimensional Galerkin expansion. In such a case, it is mandatory to account in the reduced-order model for also the modes identified as responsible for the breakdown of the minimal one.

3 Nonlinear Phenomena in Forced Dynamic Response

Features of cable nonlinear modal interaction and ensuing finite amplitude vibrations primarily depend on whether considering solely planar or actually non-planar cable dynamics. In general, the latter can be hardly addressed via analytical or mixed analytical-numerical treatments because of the need to consider reduced-order models accounting for a proper number of in-plane and out-of-plane modes possibly involved in internal/external resonance conditions such as those occurring at crossover/avoidance points or away from them. Accordingly, just a relatively small number of studies have been devoted to analysing cable non-planar dynamics via asymptotic solutions of cable reduced-order models. These include classical discretised models, like the two-d.o.f. and three-d.o.f. ones by ([34], [36]) and the four-d.o.f. model by [11] used to analyse the nonlinear vibrations of horizontal cables at first crossover under multiple internal resonance, or the companion four-mode model obtained by directly applying the asymptotic treatment to the original PDEs ([57], [46]). In contrast, actual multidegree-of-freedom models of large-order, to be dealt with via purely numerical (finite element, finite difference) space-time treatments, have been used in, e.g., ([80], [69], [70], [17]). Yet, the great amount of information on nonlinear dynamic response made available by these models somehow prevents the interested reader from a clear understanding of their most basic features associated with the occurrence of specific conditions of mechanical/dynamical interest (small vs larger sag, horizontal vs inclined, crossover vs avoidance, comparative role of internal/external resonances).

Thus, most of the research effort in the recent literature on cable nonlinear dynamics has been devoted to analysing solely planar nonlinear vibrations in various possible situations, allowing us to attain a relatively general and systematic understanding of a few basic features of cable response, which are common to different situations or distinguish them from each other. A sample is represented by the comparative summary of regular nonlinear solutions possibly exhibited by the sagged cable in different technical conditions, already reported in Table 1.

Accordingly, we restrict ourselves to considering *planar* nonlinear dynamics, where the accomplished second-order analysis allows us to refer to minimal reduced-order models accounting for the sole two internally resonant modes, if these are well tailored to describe specific situations of mechanical/dynamical interest. These minimal ROMs can be deemed reliable to describe the actual system dynamics, at least in principle, based on the information provided by the asymptotic analysis (see Sect. 2.5.2) as regards the comparative importance of resonant/non-resonant modal contributions and the frequency ranges where they are expected to fail. Yet, notwithstanding the restricted framework of solely planar cable dynamics, the features of the highlighted regular and non-regular nonlinear vibrations strongly depend on a large number of items, which include considering (i) horizontal or inclined cables, as well as crossover (avoidance) or non-crossover (non-avoidance) frequencies, and comparing (ii) different conditions of internal/external resonance, with the associated cable modes, as well as (iii) different approximate continuous models underlying the reduced-order ones.

In the sequel, based on non-condensed/condensed models, multimode discretization and second-order multiple scale analyses, the nonlinear interactions and nonlinear/complex phenomena characterising the resonantly forced planar vibrations of various horizontal/inclined cables under different internal resonance conditions are summarised. Reference is made to a number of paradigmatic cases of resonant response, obtained by considering the non-condensed continuum, which is the more general approximate model. However, comparisons with the response furnished by the condensed model are also reported. In all of the results, fixed-point (i.e., *equilibrium* or steady-state) solutions of the APMEs (written in Cartesian form, see, e.g., [55]), corresponding to periodic responses of the underlying ODEs of motion, are evaluated by the Newton-Raphson procedure, whereas their limit cycle (i.e., *dynamic*) solutions, corresponding to periodically modulated responses of the ODEs, are evaluated by the shooting application. Overall response paths are traced out via the continuation approach ([61], [45] upon quasi-statically varying a control parameter (external frequency detuning σ_f or excitation amplitude F) within a small incremental and adaptive step-size. *Stability* of fixed points, and the ensuing *bifurcations*, are examined based on the eigenvalues of the relevant (4x4) Jacobian matrix, whereas those of limit cycles are based on Floquet multipliers.

Depending on initial conditions, direct numerical simulation of the APMEs via the fourth-order Runge-Kutta scheme is utilised to validate continuation results and to characterise the post-bifurcation dynamics in terms of possibly *non-regular* responses, by also using phase-plane projections, Fourier power spectral density (PSD) and Poincaré maps (sampled at each forcing period) to characterise responses after decayed transient states. Note that if one is interested in validating the ROM ensuing from the perturbation analysis, simulations should actually affect the original PDEs (4) governing the actual nonlinear system.

In all continuation diagrams, solid lines indicate stable fixed points, whereas dashed and dotted lines indicate unstable fixed points whose stability is exchanged through saddle-node (SN) or pitchfork (PF), and Hopf (HF) bifurcations, respectively. The latter play a crucial role in the onset of periodic, quasi-periodic and chaotic responses, which are established via different mechanisms including period doubling (PD) and intermittency. Filled (open) symbols indicate stable (unstable) limit cycles of amplitude and phases.

3.1 Multimodal Interaction and Resonant Vibrations

Representative cases of multimodal interaction and resonant vibrations of horizontal and inclined cables are selected to properly highlight the effects of a number of items on the system response. Sample frequency-response diagrams are presented. More complete scenarios of multimodal interaction and resonant vibrations can be found in ([66], [55]) in terms of both frequency-response and force-response diagrams.

3.1.1 Horizontal Cables

First, we consider a horizontal cable at nearly crossover in either 1:1 or 2:1 internal resonance with $X_H = 850\,\text{m}$, $A \approx 0.1159\,\text{m}^2$, $w_C \approx 9.48\,\text{kN/m}$, $E = 1.794 \times 10^8\,\text{kN/m}^2$ [66] and a relatively large sag. Second crossover ($\lambda/\pi \approx 4.03$) is considered, where $EA/H = 1018.26$ and the sag-to-span ratio is about 1/20. Therein, both internal resonances involve a higher-order symmetric and a lower-order anti-symmetric mode, being the 4th ($\omega_s = 12.544$) and 3rd ($\omega_r = 12.537$) modes in the 1:1 resonance, and the 4th and 1st ($\omega_r = 6.223$) modes in the 2:1 resonance, respectively (see Fig. 2a). The external excitation is assumed to put energy into the system via the symmetric higher-order (s) mode, with the companion mode being driven into the response via the relevant internal resonance. As already reported in Table 1, both coupled (a_r, a_s) and uncoupled (a_s) solutions exist in both 1:1 and 2:1 resonance cases. Small damping c such that $\mu_r = 0.005$ and $\mu_s = 0.006$ is considered in both cases, along with perfect internal tuning $\sigma = 0$.

1:1 internal resonance

The frequency-response curve obtained with the non-condensed continuous model in the 1:1 internal resonance case is shown in Fig. 5a for a small value of the harmonic forcing amplitude $F = 0.002$ [66]. It exhibits stable/unstable coupled solution (a_r, a_s) curves on both sides of the zero-σ_f axis, which ensue from the uncoupled a_s branches via PF_1 and PF_2 bifurcations. In particular, the diagram exhibits a softening-type nonlinearity, with the stable coupled solutions persisting over a wide σ_f range. The directly excited a_s amplitude prevails left of perfect external tuning ($\sigma_f = 0$), whereas the internally-resonant excited a_r amplitude slightly prevails right of it.

The dynamic characteristics of the overall response are similar to those occurring at first crossover, under a comparable 1:1 resonance involving the first two

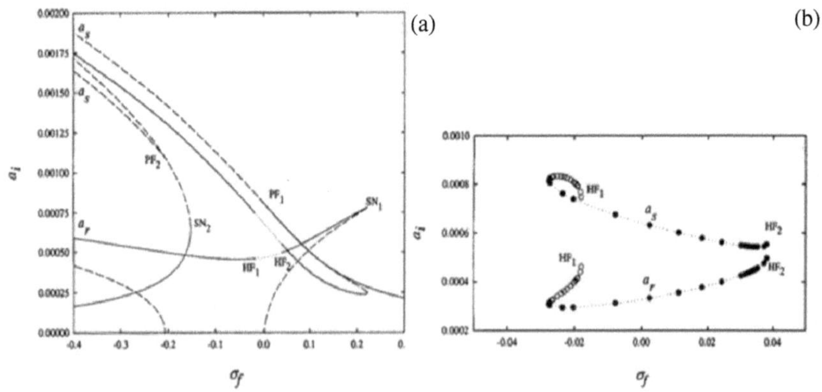

Fig. 5 Frequency-response curves **a** and dynamic solution branches after Hopf bifurcations **b** of 1:1 resonant second crossover cable ($\lambda/\pi \approx 4.03$). $\sigma = 0, F = 0.002$

antisymmetric (lower-order) and symmetric (higher-order) modes, as also predictable by the analysis of the nonlinear interaction coefficients [66]. Minor quantitative differences are herein represented by the larger range of coupled solution with no coexisting stable uncoupled solution (in between $SN_2 - PF_1$ in Fig. 5a), and by some major extent of energy transfer from the directly excited a_s mode to the driven a_r mode. Yet, there are also some qualitative differences. At first crossover, there are branches of coupled solution always stable, whereas at second crossover (Fig. 5a) two Hopf bifurcations occur in the coupled branches before they experience the jump via SN_1 to the uncoupled branches, when σ_f is swept up. The HF_1 and HF_2 points occurring around $\sigma_f = 0$ are responsible for coupled-mode periodic solution. A comparison of dynamic solution branches emerged from HF_1 and HF_2 is depicted in Fig. 5b, where HF_1 is seen to be sub-critical because the ensuing limit cycle is unstable, and a jump from the stable branch onto the steady response via cyclic-fold bifurcation occurs when sweeping σ_f down, whereas HF_2 is super-critical because the ensuing limit cycle is stable.

To verify the existence of amplitude-steady or -modulated solutions and to better characterise the modal interactions, as distinguished by the stability, continuation results are complemented by numerical integrations of the APMEs with initial conditions of the fixed points. Associated with Fig. 5b in which $F = 0.002$, the time histories showing the coupled a_r (solid lines) and a_s (dashed lines) amplitudes are illustrated in Figs. 6a and 6b for $\sigma_f = -0.1$ and $\sigma_f = 0$, respectively. Considering still $\sigma_f = 0$ but a larger F (0.005), the associated time histories are shown in Figs. 6c.

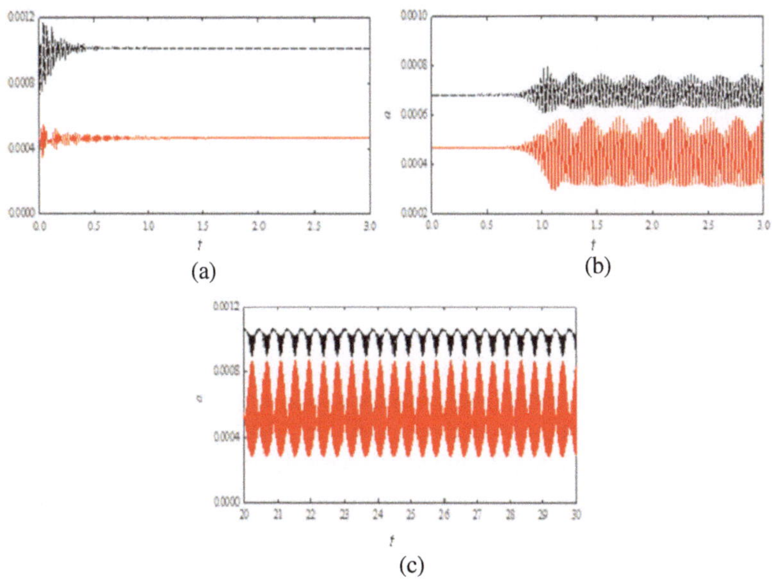

Fig. 6 Amplitude time histories from numerical integration of 1:1 resonant second crossover cable ($\lambda/\pi \approx 4.03$). $\sigma = 0$ and **a** $\sigma_f = -0.1$, $F = 0.002$, **b** $\sigma_f = 0$, $F = 0.002$, **c** $\sigma_f = 0$, $F = 0.005$. Solid (dashed) lines denote a_r (a_s)

For $\sigma_f = -0.1$, it is seen how, after experiencing initial short-interval transient dynamics, the time response settles down onto a steady-state coupled response, as predicted by continuation. Analogously, for $\sigma_f = 0$, the predicted coupled-mode dynamic solution after the Hopf bifurcation HF_1 (Fig. 5a) is confirmed by the periodically amplitude-modulated time histories established in Fig. 6b upon imposing the unstable fixed-point solution as initial conditions. However, when increasing F at $\sigma_f = 0$, with unchanged initial conditions, more marked amplitude-dependent limit cycles and stronger interaction features are seen to occur in Fig. 6c with respect to Fig. 6b. Overall, numerical results validate the continuation outcomes, yet providing a further comparison in terms of nonlinear temporal behaviours.

It is worth commenting on whether and how the response changes when using the condensed model [66] instead of the non-condensed one. The overall pattern of the frequency-response diagram is substantially confirmed apart from a more softening nonlinear response (also predictable by looking at the relevant interaction coefficients), which entails differences of the a_r, a_s amplitude values, slight shifts of the SN, PF and HF bifurcation points, quantitative differences in the stationary and non-stationary amplitudes, and a few qualitative differences of non-regular responses (e.g., chaoticity instead of quasi-periodicity for the case corresponding to Fig. 6c).

2:1 internal resonance

Let's now consider the two-d.o.f model suitable to describe the 2:1 internal resonance dynamics of the same cable at second crossover. The relevant frequency-response curve obtained with the non-condensed model is shown in Fig. 7 [64] for the same values of forcing amplitude ($F = 0.002$), modal damping ($\mu_r = 0.005$ and $\mu_s = 0.006$) and perfect internal detuning ($\sigma = 0$). Of course, the mechanical meaning of the two internally resonant modes is different with respect to the 1:1 resonance case, i.e., the higher-order (s) mode is still the 4th one whereas the lower-order (r) mode is now the 1st one, however still being the former symmetric and the latter antisymmetric, as requested by the conditions to be satisfied for actual internal resonance activation.

There is an overall resemblance of the frequency-response diagram with that at 1:1 resonance (Fig. 5a) as regards the softening behaviour, the onset of the coupled (a_r, a_s) solution in the neighbourhood of $\sigma_f = 0$, and the occurrence of a small range of dynamic solutions along the relevant stable branches. However, the coupled solution is seen to persist (stable) in a range of forcing frequencies wider than that in the 1:1 case and, mostly, to exhibit indirectly excited a_r amplitudes definitely prevailing with respect to the directly excited a_s amplitudes, due to modal interaction effects and energy transfer from higher-order to lower-order mode. These dynamic characteristics are in qualitative agreement with the cable response results shown in ([48], [34]) for the 2:1 internal resonance involving the first symmetric in-plane (higher-order) and out-of-plane (lower-order) modes at first crossover and, more generally, they appear to be typical of many other two-d.o.f systems at 2:1 internal resonance. Again, using the condensed model leads to results with quantitative and some minor qualitative differences [64].

If one considers a non-crossover cable ($\lambda/\pi \approx 2.95$) in between first and second crossover points, with the symmetric 2:1 internally resonant modes being the 2nd and 5th, respectively (see Fig. 2a), considerable differences occur with respect to the response of the 2:1 resonant cable at second crossover, as shown by the frequency-response diagram in Fig. 8. The uncoupled a_s solution now exhibits a hardening nonlinearity, while the indirectly excited a_r amplitudes play a dominant – or comparable – role in the coupled responses, with respect to the directly excited a_s amplitudes, only in a small range around perfect detuning.

For all considered cases, the analyses and comparisons can of course be extended to force-response diagrams, where quantitative – and also some qualitative (as regards stability and bifurcations) – differences may occur when varying the internal resonance, the involved modes, the crossover/non-crossover condition, and the continuous model in the background.

3.1.2 Inclined Cables

For inclined cables, attention is solely focused on the *1:1 resonance* at avoidance [55] in order to highlight the differences due to the occurrence of hybrid (instead of symmetric and antisymmetric) modes, with respect to the 1:1 resonance at crossover of companion horizontal cables. Indeed, the nonlinear response at 2:1 internal resonance does not exhibit meaningful differences with respect to what already observed for horizontal cables. In particular, uncoupled and coupled responses do coexist both at avoidances (as in Fig. 7 for the second crossover cable) and away from them (see Table 1).

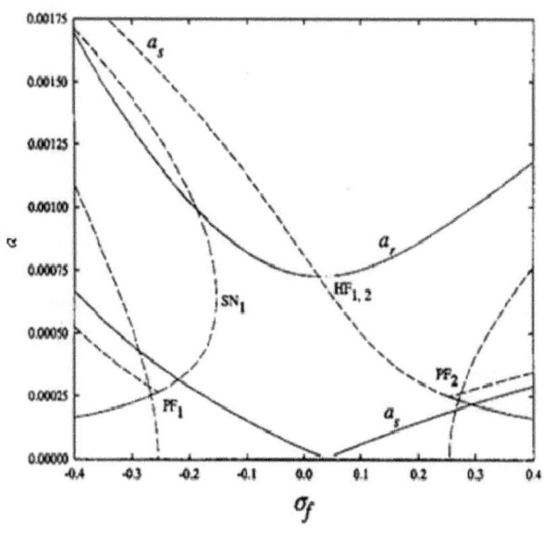

Fig. 7 Frequency-response curves of 2:1 resonant second crossover cable ($\lambda/\pi \approx 4.03$), $\sigma = 0, F = 0.002$

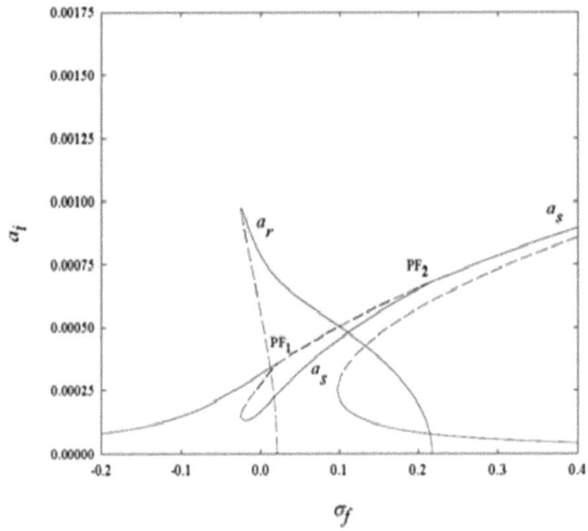

Fig. 8 Frequency-response curves and bifurcations of 2:1 resonant non-crossover cable ($\lambda/\pi \approx 2.95$), $\sigma = 0$, $F = 0.002$

This is no more the case for 1:1 resonant avoidance cables, as also summarised in Table 1, where the solely coupled (a_r, a_s) mode solution exists. The same values of excitation amplitude ($F = 0.002$) and damping coefficients ($\mu_r = 0.005$ and $\mu_s = 0.006$) as those considered for the horizontal cables are referred to, whereas herein the internal detuning σ is only allowed to be positive, and the value $\sigma = 0.04$ is considered. An additional parameter is here represented by the inclination angle.

The first avoidance cable with $\theta = 45°$ is considered, whose frequency-response curves of a_r and a_s amplitudes are reported in Fig. 9. Overall, the responses are seen to exhibit two coupled-mode solutions isolated from each other: the main one (B1) entirely spanning the frequency σ_f range in the neighbourhood of nearly perfect primary resonance, with predominant a_r amplitudes, whereas the other one (B2) exhibiting the so-called *frequency island* phenomenon, with predominant a_s amplitudes. Note that such kind of coupled-mode isolated solutions also occur at lower inclinations or higher avoidances [55], even though the range of existence of B1 becomes larger and that of B2 variably moves towards left of the zero-σ_f axis. Generally speaking, these dynamic characteristics notably differ from those of 1:1 resonant horizontal cables at crossovers (see Fig. 5 as well as [66]), wherein the B1 and B2 branches merge together giving rise to both uncoupled- and coupled-mode responses, with the latter originating from the former via pitchfork bifurcations and with Hopf bifurcations being found only in the coupled-mode branch. It is worth mentioning that coupled-mode responses and frequency islands have been documented also for other systems exhibiting avoidances, e.g. in [30], who considered

1:1 resonant imperfect beams at first avoidance with primary resonance of a low-frequency mode.

In Figs. 9a (a_r) and 9b (a_s), overall jump phenomena are seen to occur through turning points, and there are three Hopf bifurcations, HF_1 and HF_2 located on the main B1 branch, and a single HF_3 occurring in the B2 island along with a connection of unstable foci and saddles (having two positive real eigenvalues). The two isolated solutions overlap with each other, with meaningful ranges of two unstable foci coexisting in a particular σ_f range nearly centred at the zero-σ_f axis.

The solutions emerged from each HF point are traced out in Figs. 9c (a_r) and 9d (a_s), which reveal individual routes of continuation results. All HF_{1-3} bifurcations are supercritical because the born limit cycles are stable. All dynamic solution branches lose stability via period doubling (PD_{1-3}) bifurcations, each of which having one critical Floquet multiplier that crosses the unit circle at -1 [45]. Such PD bifurcations pave the way to quasi-periodic or chaotic oscillations, depending on given σ_f values and initial conditions (see Sect. 3.2).

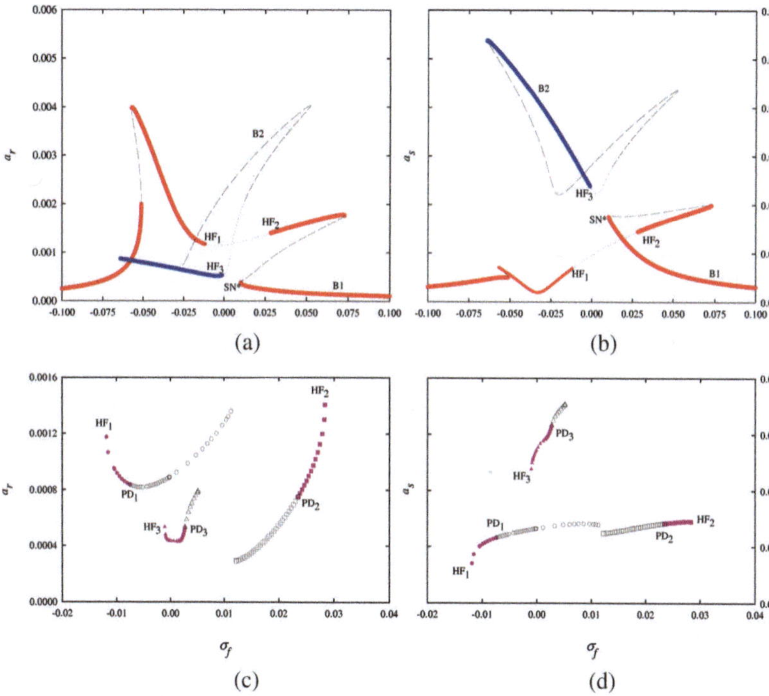

Fig. 9 Frequency-response curves (**a, b**) and dynamic solution branches after Hopf bifurcations (**c, d**) of 1:1 resonant inclined cable at first avoidance. $\theta = 45°$, $\sigma = 0.04$, $F = 0.002$

3.2 Modulated, Non-regular, and Multi-harmonic Responses

Overall, depending on the considered resonance cases, system control parameters, and initial conditions, a multiplicity of (stable/unstable) equilibrium and periodic solutions may occur, along with meaningful transitions from periodic to quasi-periodic and chaotic responses. Upon Hopf bifurcations, limit cycles may undergo cyclic-fold as well as direct period doubling bifurcations. A cornucopia of nonlinear phenomena are observed: sequences of period doubling bifurcations to chaos, funnel-shaped chaos, on-off intermittency mechanisms, sudden switching of solutions via boundary crises, with also competing effects of dynamic solutions. Multi-harmonic response features occur mainly due to contributions from higher-order non-resonant modes. The dynamic deflections occurring in chaotically resonant vibrations exhibit non-periodic multi-mode features, with time-varying amplitudes which may become significantly large.

In the sequel, for representative purposes, the non-regular responses of the 1:1 resonant inclined cable at first avoidance, established in Figs. 9c, 9d upon the various Hopf bifurcations, are reported. They have been obtained via direct numerical integration of the APMEs written in the Cartesian variables p, q [55], aimed at validating the continuation prediction and showing rich nonlinear dynamics after HF and PD bifurcations.

Along the branch starting from HF_2, a sequence of period doublings leading to chaos is highlighted by the (p_r, q_r) phase portraits in Fig. 10, where σ_f is slowly decreased with fixed initial conditions corresponding to the PD_2 solution at $\sigma_f = 0.02357$. It is seen that as the limit cycle (Fig. 10a) experiences two-cycle (Fig. 10b), four-cycle (Fig. 10c) and multiple-cycle (Fig. 10d) closed loops generated by consecutive PD bifurcations, the last trajectory undergoes a chaotically-modulated response at $\sigma_f = 0.0225$ (Fig. 10e), thereby giving rise to aperiodic time histories. Further slightly decreasing σ_f (Fig. 10f), transient chaos appears and then the orbit undergoes a likely boundary crisis, where the chaotic set is destroyed and the post-bifurcation state suddenly escapes to the bounded fixed-point B1 solution (Figs. 9a, 9b).

A route to chaos via PD is also detected when continuing the solution from HF_1 with a slow increment in σ_f. By assigning initial conditions corresponding to the PD_1 solution ($\sigma_f = -0.00735$), the (p_r, q_r) and (p_s, q_s) phase planes which exhibit a funnel-shaped chaotic attractor [45] at $\sigma_f = -0.004812$ are shown in Figs. 11a and 11b. The associated Poincaré (a_r, a_s) section reveals a collection of points spreading out over the limited region in Fig. 11c. Increasing σ_f thereafter, a crisis occurs at $\sigma_f = -0.004312$ and the resulting response ends up with a stable fixed point of the overlapped B2 branch in Figs. 9a and 9b.

Qualitative differences occur regarding the route to chaos when continuing the dynamic solution from HF_3 of B2 branch in Figs. 9c and 9d. Since neither fixed point nor periodic stable solutions take place in the region between PD_3 and the saddle node SN* of B1 branch, one would expect either quasi-periodic or aperiodic response, depending on initial conditions. By initiating with the PD_3 solution and slowly increasing σ_f towards $\sigma_f \approx 0.0035$, the ensuing nonlinear time histories

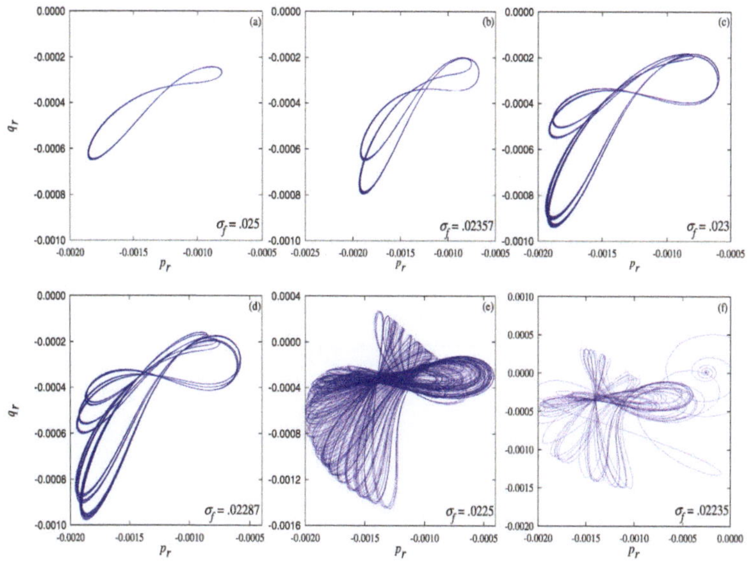

Fig. 10 Phase portraits showing the transition from periodically to chaotically modulated dynamic response and boundary crisis of inclined cable in Fig. 9 (from HF_2)

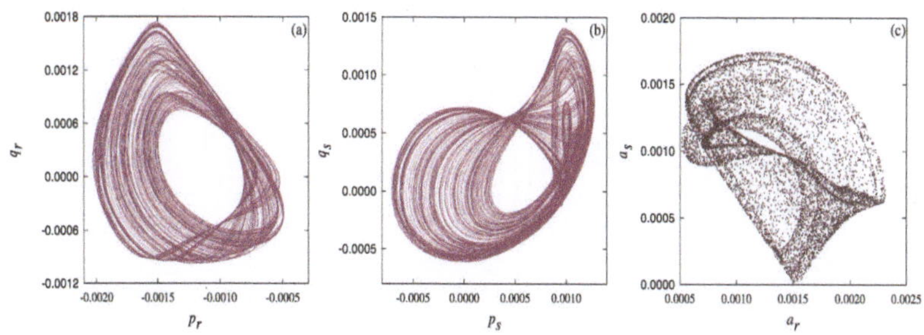

Fig. 11 Phase portraits and Poincaré map showing chaotically funnel-shaped dynamic response of inclined cable in Fig. 9 with $\sigma_f = -0.004812$ (from HF_1)

exhibit intermittently chaotic outbursts, as shown in Fig. 12. Thus, the PD_3 bifurcation may be subcritical, and the ensuing intermittency may be of type III (on-off intermittency) [45]. Applying the same initial conditions and varying σ_f further, the nonlinear dynamics may be quasi-periodic as well as chaotic. The occurrence of quasi-periodic motion is exemplified by the time histories and Poincaré map of (a_r, a_s) amplitudes in Fig. 13 for $\sigma_f = 0.006$. Figs. 13a and 13b highlight multi-frequency responses, whereas Fig. 13c exhibits a closed-loop map. Of course, when σ_f meets a critical value, a crisis finally occurs ($\sigma_f \approx 0.011$) and the response jumps onto the stable fixed point of B1 branch. Thus, the mentioned sudden changes in

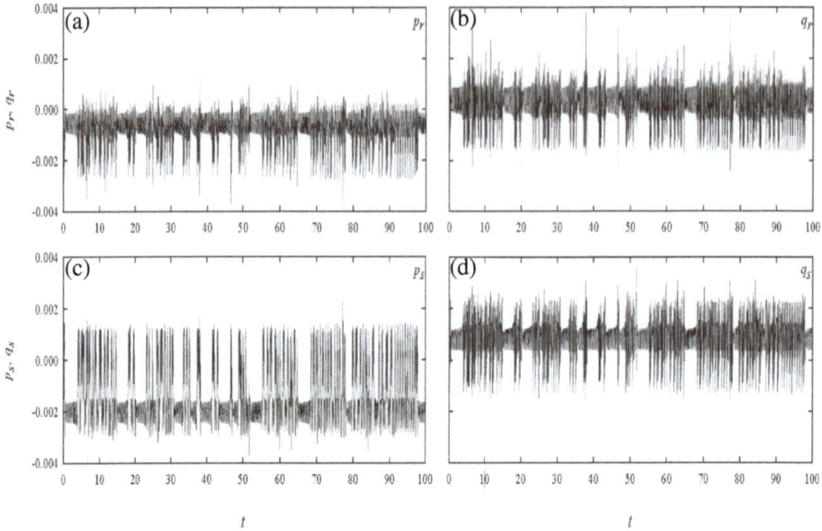

Fig. 12 Time histories of the Cartesian components p, q of the APMEs showing the intermittent chaotic response of inclined cable in Fig. 9 with $\sigma_f = 0.0035$ (from HF_3)

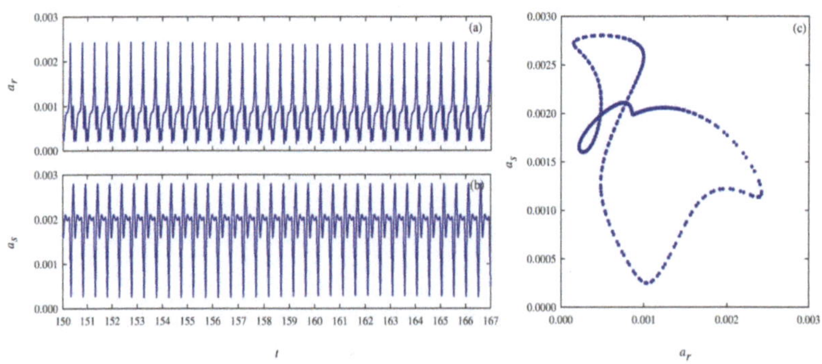

Fig. 13 Time response histories and Poincaré map showing the quasi-periodic response of inclined cable in Fig. 9 with $\sigma_f = 0.006$ (from HF_3)

nonlinear response via the crisis may lead to switching between dynamic and equilibrium solutions of the two overlapping branches.

Examples of power spectral density (PSD) corresponding to q_s responses in Figs. 10d, 10e, 11 and 13 are shown in Figs. 14a-d, respectively. The PSD of Fig. 14a illustrates how, when period doubling takes place, the dominant sharp peak and its sub-harmonics simultaneously occur along with their integer multiples, justifying the multiple closed-loop orbit in Fig. 10d. On the contrary, due to chaotically-modulated oscillations (Figs. 10e, 11), the PSDs of Figs. 14b and 14c highlight a major spike associated with the excitation frequency and many side-banding components distributed continuously over a broad frequency band with high density. Differently, the PSD of Fig. 14d exhibits many harmonics whose frequencies are not commensurate in perfectly integer ratios, which confirms how the response in Fig. 13 is quasi-periodic.

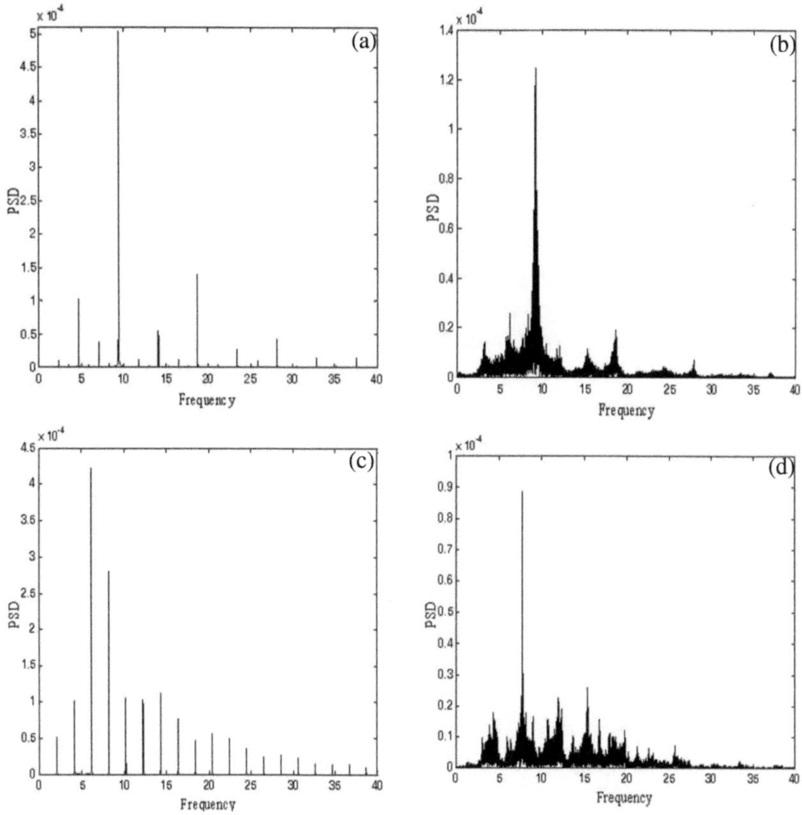

Fig. 14 PSDs of q_s-response histories corresponding to Figs. 10d **a**, 10e **b**, 11 **c** and 13 **d**

3.3 Nonlinear Dynamic Displacements and Tensions

As already anticipated in Sect. 2.4, the availability of the coupled dynamic configurations of the cable associated with the relevant nonlinear normal modes – which ensue from accomplishing the multiple scale analysis up to the second order – allows us to account for the spatial corrections, with respect to the reference linearly resonant modes, due to the quadratic nonlinearity effects of all non-resonant modes considered in a finite discretization. In turn, this is of major importance as regards the evaluation of cable nonlinear dynamic tension. Let's now address the matter for both the horizontal and the inclined cable.

The 2:1 resonant *horizontal cable* at second crossover, already addressed in Sect. 3.1.1, is considered, whose results obtained with the reference continuous model of non-condensed cable are also compared with those furnished by the condensed cable model most commonly used in the cable dynamics literature: this allows us to highlight important dynamic effects taken into account by the former (and instead, neglected by the latter) in terms of *longitudinal dynamics* and *space-dependent dynamic tension*. Based on Eq. (10), the nonlinear spatio-temporal variation of coupled vertical dynamic displacement profiles (with $\sigma = 0$, $\sigma_f = -0.2$, $F = 0.002$, see Fig. 7) is visualised in Fig. 15. The solid (dashed) lines denote time-varying spatial distributions over a half period ($t = 0 - 0.5T_f$) of forced vibrations furnished by the non-condensed (condensed) model [64]. It is evident how, due to the interactions between symmetric and antisymmetric modes at second crossover, some of the 2:1 resonant combined dynamic displacements are actually *asymmetric* with respect to cable midspan, e.g., at $t = 0$ and 0.4 T_f. These results are in qualitative agreement with the numerical finite difference results of a second-crossover cable in [70], which highlighted the 2:1 resonant normal modes and their spatial transition from nearly antisymmetric ($t = 0.2T_f$) to nearly symmetric ($t = 0.5T_f$), in the nonlinear free vibration responses. Since the corresponding (a_r, a_s) stable amplitudes of non-condensed and condensed models are quantitatively different, quantitative discrepancies occur in the time-varying second-order coupled displacements, along with some qualitative differences occurring in small portions of the superimposed shapes, e.g., at $t = 0$.

With reference to Fig. 15, the space-time varying distributions of *nonlinear dynamic tension* ($T_d = EAe_d$) of the non-condensed model (Fig. 16a) accounting for both modal coupling and contributions of longitudinal/vertical displacements are evaluated through the dynamic strain e_d given in Eq. (3), whereas the approximate, spatially-constant, nonlinear dynamic tensions of the condensed model (Fig. 16b) are evaluated based on Eq. (5). All nonlinear dynamic tension values – which can be either positive (tensile-type) or negative (compression-type) – are normalised with respect to the maximum static tension T_H, and they are displayed with time sequence following the given arrow direction. Overall, the spatially asymmetric character of the coupled (a_r, a_s) 2:1 resonant configurations associated with the non-condensed model, (see, e.g., Fig. 15 at $t = 0$), entails a remarkable *asymmetric* feature of the nonlinear dynamic tension profiles shown in Fig. 16a, whereas the condensed model, besides overlooking this essential character, provides – based on strain average – a spatially-constant dynamic tension at each time (Fig. 16b). From a dynamic design point of view, the maximum tensile or compressive T_d/T_H values of the non-condensed model in the half period $t = 0 - 0.5T_f$ are considerably larger than those of the condensed model, with percent differences up to about 80%.

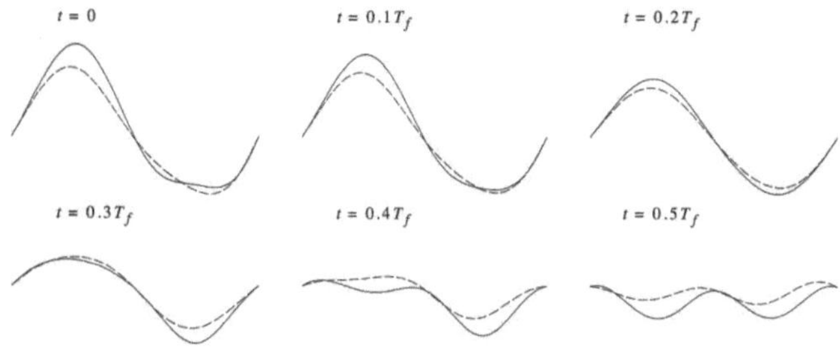

Fig. 15 Nonlinear coupled vertical dynamic displacements of 2:1 resonant second crossover cable over a half period of forced vibration, with non-condensed (solid) and condensed (dashed) models

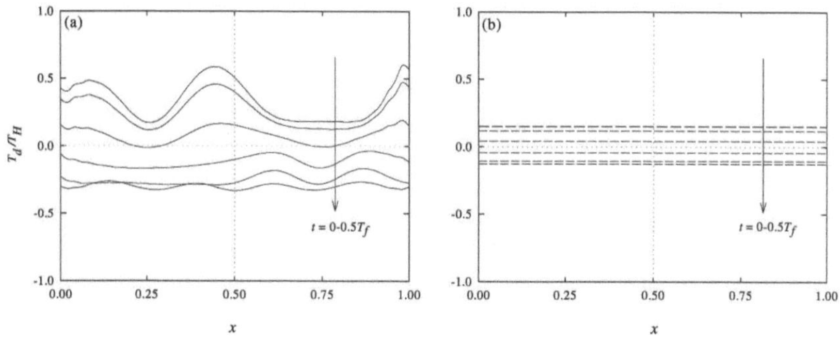

Fig. 16 Nonlinear dynamic tension distributions associated with coupled responses of 2:1 resonant second crossover cable over a half period of forced vibration: **a** non-condensed and **b** condensed models

Therefore, the latter considerably underestimates the nonlinear dynamic, both tensile and compressive, stresses in the resonantly coupled response of a suspended cable with even low extensibility and sag-to-span ratio.

Along the same line, displacement and tension results as well as model comparisons for the 1:1 resonant horizontal cable at first or second crossover and for a 2:1 resonant non-crossover cable are reported in [66].

Moving to the *inclined cable*, the space-time distribution governing *steady* and *chaotically-modulated* dynamic displacements and tensions due to 1:1 resonant interactions is highlighted [55]. Depending on the obtained amplitudes (a_r, a_s) and phases (γ_r, γ_s), the second-order (u, v) coupled dynamic deflections are constructed through Eq. (9), whereas the relevant dynamic tensions ($T_d = EAe_d$) are evaluated based on displacement gradients through the strain e_d in Eq. (3). Again, the T_d value is normalised with respect to the associated maximum static tension T_H.

The first comparison of steady responses is illustrated in Fig. 17 for first-avoidance cables having different stable fixed points. The u, v and T_d/T_H responses

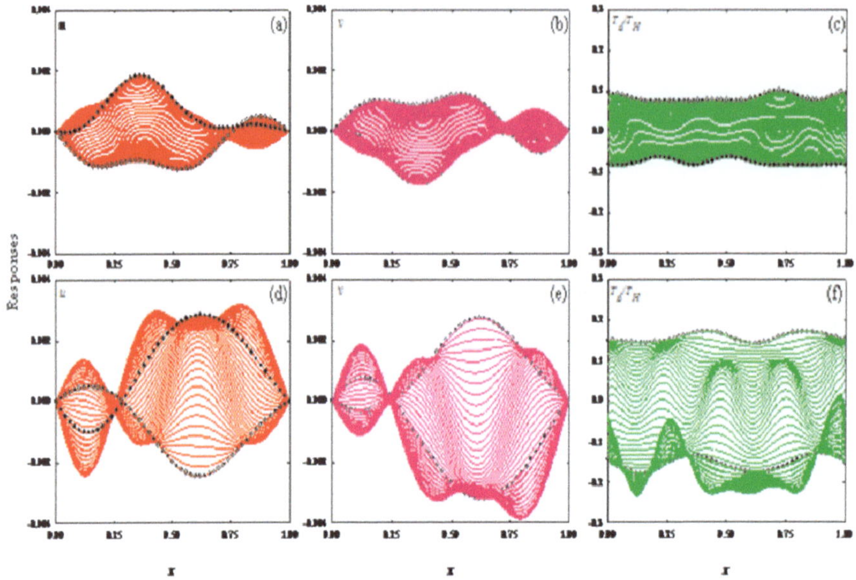

Fig. 17 Space-time varying coupled dynamic (u, v) configurations and tension distributions (T_d/T_H) of 1:1 resonant inclined cable at first avoidance ($\theta = 45°$), based on fixed-point amplitudes at $\sigma_f = -0.02$, over a half period of forced oscillation, corresponding to B1 **a–c** and B2 **d–f** solutions in Figs. 9a and 9b

at $\sigma_f = -0.02$ of the B1 and B2 solutions for the $\theta = 45°$ cable in Figs. 9a and 9b are visualised in Figs. 17a-c and 17d-f, respectively, over a half period of harmonic excitation. Here, empty (filled) circles indicate the initial (half-period) responses. With regard to the B1 solution, the u and v asymmetric profiles exhibit predominance of the indirectly-excited r mode (see Fig. 3) according to the fact that the a_r amplitude (0.00134) is significantly greater – by an order of magnitude – than the a_s amplitude (0.000475). The spatially asymmetric feature of T_d/T_H responses, due to the strain variation effect, is already visible in Fig. 17c.

With respect to B1 responses, the associated B2 (frequency-island) responses in Figs. 17d and 17e exhibit the predominance of the directly-excited s mode (see Fig. 3) because the a_s amplitude (0.00347) is now much greater than the a_r one (0.0006). Accordingly, the second-order spatial corrections from resonant and non-resonant (higher-order) modes are significant due to quadratic nonlinear effects, thereby leading to some essential multi-mode responses. Apart from the induced T_d/T_H responses being greater, both spatial and temporal variation of tensile/compressive forces are considerably important (Fig. 17f).

Next, the comparison of two coexisting periodic and chaotic responses is shown in Fig. 18. At $\sigma_f = 0.0225$, either the single-valued, steady-state amplitudes (phases) of stable B1 solution (Figs. 9a and 9b) or the multi-valued, chaotically modulated amplitudes (phases) based on numerical integrations of modulation equations (Fig. 10e) are considered. The associated periodic and non-periodic (u, v, T_d/T_H) responses are depicted in Figs. 18a-c and 18d-f, respectively, the latter being recorded over five periods of excitation.

Unlike the $\theta = 45°$ case with $\sigma_f = -0.02$ (Figs. 17a and 17b), the steady u and v responses at $\sigma_f = 0.0225$ (Figs. 18a and 18b) exhibit the predominant s-mode profiles because a_s (0.001156) is greater than a_r (0.000252). This highlights how, depending on the control parameter, one given solution (e.g., B1) may, of course, be dominated by either the directly-excited or indirectly-excited mode. On the contrary, the multi-modal asymmetric spatial responses become evident when the cable experiences chaos, as shown in Figs. 18d and 18e. Since the varying a_r, a_s amplitudes may have the same order of magnitudes, from a modal superimposition viewpoint contributions from the two resonant modes are comparatively important, with also meaningful second-order spatial corrections from all retained modes. Consequently, in Fig. 18f, the induced dynamic tensile/compressive tensions are outstandingly increased, with the associated non-periodic non-uniform spatial distribution, whereas in Fig. 18c, the small-amplitude responses show rather spatially-uniform dynamic tensions. Thus, in cable design, one has to worry about remarkable asymmetric features of spatially non-uniform, strongly time-varying, tensile/compressive dynamic tensions induced by chaotic oscillations, as shown herein for the 1:1 resonant cable at first avoidance.

4 Experimental Characterisation of Cable Nonlinear Dynamics

As exemplified in Sect. 3, flexible continuous (infinite-dimensional) systems as the sagged cables are prone to finite amplitude vibrations involving several spatial modes in many different regular or non-regular response regimes. Their reliable and possibly complete description in control parameters space is of paramount importance from both the theoretical and practical point of view, and should be accomplished by constructing systematic, yet demanding, response charts allowing us to highlight the occurrence of rich and varied classes of motion, which are also dependent on the realisation of meaningful external/internal resonance conditions. It is apparent how this is indeed a hard issue for systems with a high modal content, both if working in a substantially analytical context and via purely numerical techniques: in the former case, due to the need to preliminarily select a number of ROMs capable to reliably describe the nonlinear response of the actual underlying system in different dynamical situations, in the latter case owed to the great difficulties encountered in handling and classifying a possibly huge amount of available results.

A fundamental role is played in this respect by the nonlinear experimentation on *physical models*, which is mandatory for getting information on the actual nonlinear response of flexible systems, by properly sustaining, validating and/or driving companion theoretical and numerical investigations. As a matter of fact, confirmation of, e.g., the features of a regular response obtained in a certain parameter range, or

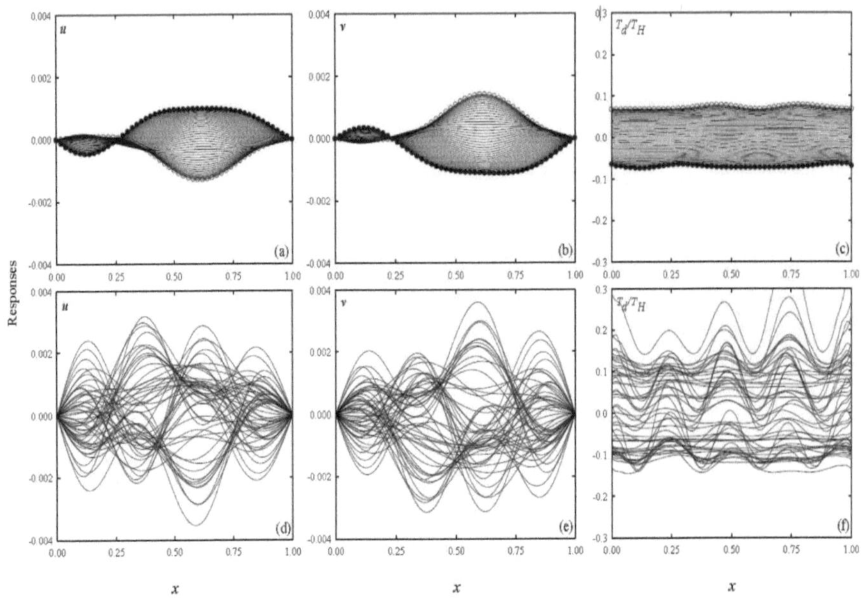

Fig. 18 A comparison of space-time varying coupled dynamic (u, v) configurations and tension distributions (T_d/T_H) of 1:1 resonant inclined cable at first avoidance ($\theta = 45°$), based on fixed-point **a–c** and time-varying **d–f** amplitudes at $\sigma_f = 0.0225$, over a half period of forced oscillation: corresponding to B1 in Figs. 9a and 9b **a–c**, and to Fig. 10e with $t \approx 211.2 - 216.2$ **d–f**, respectively

information about the possible occurrence and characterisation of new/complex phenomena associated with the governing nonlinearities but often un-modelled in the theoretical analyses, can be solely provided by careful experimental investigations.

For the sagged cable, it has been already mentioned (Sect. 3) how a large amount of meaningful nonlinear phenomena recently highlighted in the literature have been concerned with the sole, or prevailing, planar nonlinear response of the system, often addressed via models which are also rather sophisticated. Yet, the role played in the actual response scenario by the system out-of-plane behaviour, as well as the extent of the likely ensuing modifications in the highlighted phenomena, become apparent only when dealing with a physical model of sagged cable.

Apart from former experimental investigations mainly devoted to analysing the system response in regular regimes [35] or more recent ones focused on specific response features of nearly taut [12] or sagged, horizontal [18] or inclined [58], cables, a quite systematic investigation of cable nonlinear experimental dynamics has been accomplished in about the last decade by referring to a horizontal elastic cable/mass system hanging at vertically and harmonically moving supports (Fig. 19), and realising, for relatively low excitation frequencies, a fairly reliable model of bare suspended cable [15]. As mandatory in highly flexible systems, no-contact devices (optical cameras) have been used to measure the two, in-plane and out-of-plane, components of motion of a few masses located at different positions along the cable, produced by the in-phase or out-of-phase motion of the supports. System parameters have been adjusted to produce two different conditions of multiple internal resonances involving either all first four in-plane (vertical, V) and out-of-plane (horizontal, H), symmetric and antisymmetric, modes (2:2:1:2) or only three of them (2:2:1, first in-plane and out-of-plane antisymmetric, first out-of-plane symmetric), for a cable at first crossover or a slightly slacker cable, respectively. Nonlinear dynamics have been investigated in various excitation frequency ranges, which include meaningful external (primary and subharmonic) resonance conditions of either modes. Former studies were devoted to analysing and classifying the local and overall system response – in either a regular or non-regular regime – in a control parameter plane ([10], [56]). More recent studies have been dealing with the systematic characterisation of some main features of complex response, and of the relevant bifurcation mechanisms, accomplished by properly reconstructing the system nonlinear dynamics from experimental measurements ([3], [52], [53]).

As a matter of fact, based on both theoretical/numerical analyses and experimental investigations, quasi-periodic and chaotic motions are often seen to occur in the cable response, mostly in between regions of clearly dominating low-dimensional regular motions. Accordingly, one major effort has to be devoted to investigating possible *finite dimensionality* in the complex dynamics of the actual infinite-dimensional system, and to detecting *minimum numbers* and features of the *spatial configuration variables* which take meaningful part in its complex motions.

This can be done via a number of sophisticated experimental techniques, which allow us, among others, (i) to characterise the response dimensionality in terms of both time and spatial complexity, and (ii) to unveil actual low-dimensionality, if any, of both regular and non-regular response, with the ensuing identification of suitably reduced, and possibly minimal, theoretical models able to describe the system dynamics in a certain control parameter range. Experimental analyses may also provide important information about robust bifurcation features of regular responses towards complex dynamics.

In the sequel, attention is solely focused to the characterisation of complex dynamics, by shortly referring to the relevant techniques, by summarising interesting features of a number of experimentally observed scenarios of transition to chaos, which are also discussed against canonical bifurcation scenarios from dynamical systems theory, and by dwelling on the hints provided by the experimental investigation as regards formulation of proper ROMs.

4.1 System Dimensionality and Reduced-Order Models

Upon analysing and classifying local and overall system responses in regular or non-regular regimes, based on various dynamical systems tools ([42], [45]),

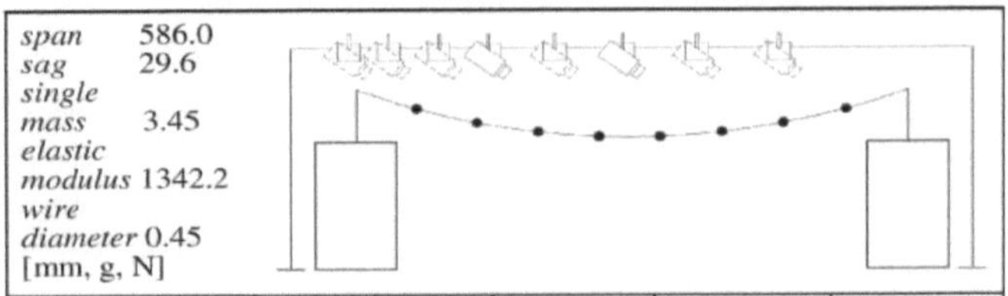

mode		H1	V2	H2	V1
shape	h-plane				
	v-plane				
frequency ω [Hz]		3.49	6.68	6.77	8.82
damping		0.0021	0.0003	0.0008	0.027

(a)

(b)

Fig. 19 Mechanical model with slacker system parameters and dynamic characteristics **a**; experimental setup: the hanged nylon wire carrying eight concentrated masses, the shakers and the two movable optical cameras **b**

attention is focused on characterising some main features of complex response, and of the relevant bifurcation mechanisms, by means of proper reconstruction techniques of the dynamics from experimental measurements. The *dimensionality* of the response is characterised in terms of both time and spatial complexity. *Time complexity* is evaluated by calculating invariant measures of the dynamics through the delay embedding procedure ([41], [72]). Besides Poincaré map inspection and power spectra analysis, information about the dimension of the quasi-periodic and chaotic attractors is obtained from correlation dimension evaluations ([21], [1]) carried out on time-delay reconstructed phase spaces. In turn, *spatial complexity* is tackled via two approaches: (i) by relating the embedding dimension of the reconstructed attractors to the dimension of the linear phase space, which gives information on the involved number of dynamical degrees of freedom; (ii) by analysing the spatial structure of the non-regular flow through the proper orthogonal decomposition [23], which allows the identification of the dominant experimental eigenfunctions (*proper orthogonal modes*, POMs) corresponding, from the mechanical viewpoint, to the configurations most visited by the system, on average, during its spatio-temporal evolution.

Classification of motions is made based on topological dimension of the manifolds where they develop (growing, in regular regime, from one to three torus) and on correlation dimension of the attractors (from one to three in regular regime, coinciding with their local topological dimension; non integer in chaotic regime). In the following, regular motions are labelled Pm-Mk, periodic, QPm-Mk, quasi-periodic, with the labels giving information on the attractor periodicity (m) and on the number of involved POMs (k), the latter coinciding with the dimension of the invariant manifold where the motion develops unless the manifold is resonant.

Though being the experimental system potentially infinite-dimensional, not only its regular response but also the complex one are shown to be *low-dimensional* in several cases. Indeed, more than 90% of the non-regular signal power can be represented by using up to three or four POMs, among those already responsible for the higher-dimensional coupled regular dynamics. In addition, an overall heuristic correspondence is seen to occur between the main POMs and the main linear physical modes of the system, so that the former are often labelled by referring directly to the corresponding in-plane (vertical), Vn, or out-of-plane (horizontal), Hn, cable modes (n =odd, symmetric; n =even, antisymmetric).

These low-dimensionality features are of notable interest to the aim of associating to each class of complex response of the experimental system, a class of *reduced order theoretical models* able to describe the observed dynamics. Within the framework of a reduction procedure, these can be built specifically in each region of control parameters space either by using just the identified POMs, or – getting hints from them – by projecting the system infinite-dimensional dynamics on the known sub-optimal basis of corresponding linear modes [4].

4.2 Bifurcation Scenarios and Complex Dynamics

Major efforts in experimental analysis of non-regular dynamics are devoted to properly characterising (i) the bifurcation paths leading to complex attractors, (ii) the extent of the relevant regions of occurrence, and (iii) the robustness of transition mechanisms in parameter space. The most challenging task consists of satisfactorily characterising the various possible *bifurcation scenarios* to complex response, as well as the response itself. Generally speaking, this can be done by realising a profitable *feedback between experiments and theory* which allows us: (i) to possibly trace preliminary experimental results back to a canonical scenario from dynamical systems theory, (ii) to exploit hints from the latter to improve and steer the experimental analysis, (iii) to systematically pursue ahead the physical investigation by detailing the most robust features of system response and by clarifying to which extent they can be referred to theoretical scenarios, (iv) to improve cable modelling, and (v) to identify and analyse a proper reduced-order cable model aimed at (partially) reproducing the highlighted experimental scenarios.

A general overview of the richness and robustness of different global bifurcation scenarios to chaos, occurring with different support motion in various regions of the excitation control parameter space, is reported in Table 2. Two main routes to chaos, possibly coexisting and competing with each other, do occur, namely a quasi-periodic, QP, scenario and a scenario involving the global bifurcation of an homoclinic, HOM, (or heteroclinic, HET) invariant set of the flow [52].

Table 2 Transition to non-regular dynamics. A summary of bifurcation scenarios to non-regular responses for different cables and support motions, in various ranges of V1-resonant excitation frequency, with the involved configuration variables (POMs)

cable	support motion							
	in-phase external resonance condition				anti-phase external resonance condition			
	primary		subharmonic-½		primary		subharmonic-½	
	scenario	POMs	scenario	POMs	scenario	POMs	scenario	POMs
slacker	QP	V1 H1 H2	HOM (HET)	V5 H5 V3 H3	HOM	V2 H2	HOM (QP)	V4 H4 (+$H1^T$H2)
crossover	no chaos		HOM	V5 H5 V3 H3	HOM	V2 H2	HOM	V4 H4

4.2.1 Quasi-periodic Transition to Chaos

Quasi-periodic transition to chaos via breakdown of regular dynamics on 3-Tori is robust for the slacker cable excited by in-phase support motion at primary resonance with the first symmetric in-plane mode (V1) (Table 2). This scenario looks rich and involved due to complicated interaction between internally resonant and non-resonant modes, motion on a 2- or 3-Torus, and phase locked three-mode motions on 3-Tori [3].

In the QP transition to chaos, a close and intertwined sequence of regular and non-regular response classes occurs, as shown by the schematic bifurcation diagram of Fig. 20, obtained by varying the excitation frequency. Therein, the additional label nT, $n = 2,3$, is used to distinguish between quasi-periodic motions with a number n of incommensurate frequencies (equal to correlation dimension) different from – due to a resonant torus –, or equal to, the dimension of the manifold where they develop, the label Mk being omitted in the latter case for the sake of simplicity. Moving from the reference period 1 response dominated by the V1 mode, the following response classes are identified: (i) two-frequency quasi-periodic motions on two-dimensional manifolds (2T-QP1); (ii) two-frequency phase-locked quasi-periodic motions on three-dimensional manifolds (2T-QP2-M3); (iii) stable three-frequency quasi-periodic motions (3T-QPm); (iv) chaotic motions (CH1, CH2); (v) phase-locked periodic motions which are invariant sets of dimension 1 on three-dimensional manifolds (PmM3). Various kinds of bifurcation (Fig. 20) mark the passage from one response class to another.

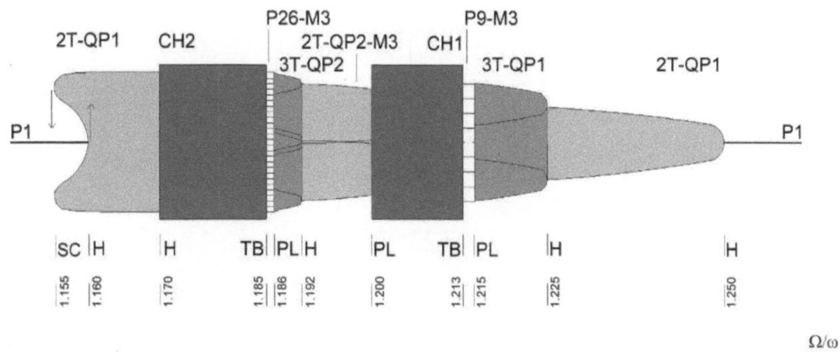

Fig. 20 Schematic bifurcation diagram (in-phase support motion, primary resonance). H: Hopf, PL: phase-locking, TB: torus breakdown, SC: saddle cycle. The highlighted motion classes refer to periodic (Pm-M3), quasi-periodic (nT-QPm-Mk) and chaotic (CH1, CH2) attractors

Though being quite rich and involved, the experimental transition mechanism is substantially traceable, from a qualitative viewpoint, to a canonical scenario of bifurcation of flows via breakdown of regular dynamics on 3-Tori known from dynamical systems theory and numerics (see the response chart in Fig. 21, where the thick vertical line corresponds to the bifurcation path followed by the experimental cable/mass system when varying the excitation frequency). Yet, not all of the experimental response features can be explained according to the torus breakdown paradigm, due to the occurrence of intermittent synchronisation on high-periodicity solutions and competing complex phase-modulated regimes which are conjectured to represent toroidal chaos. This is a case showing how a possible interplay between sophisticated experimental investigations and theoretical/numerical achievements can indeed allow for a proper modelling and understanding of the characterising nonlinear phenomena.

The various classes of motion taking part in the transition to chaos involve well identified experimental eigenfunctions (POMs). Specifically, the meaningful involvement in the system dynamics of two more modes, besides the underlying first symmetric in-plane one (V1), is associated with some closeness of the first symmetric out-of-plane mode H1 to its order ½-subharmonic resonance and of the first antisymmetric out-of-plane mode H2 to its primary resonance. The latter, in particular, is seen to play a decisive role as regards transition to chaos.

In a theoretical/numerical framework, the identified prevailing POMs constitute the optimal basis for a decomposition of the spatio-temporal flow aimed at reproducing the bifurcation features and the responses highlighted in the experimental investigation. Indeed, a three-mode mathematical model including the linear modes V1, H1 and H2 – like the one in [11], containing also a fourth mode, here unnecessary, – is sufficient to capture the important non-regular dynamics of the system in the considered frequency range [4].

Going back to the experimental investigation, it is worth noting how, with the same primary resonance excitation condition, the quasi-periodic scenario to chaos

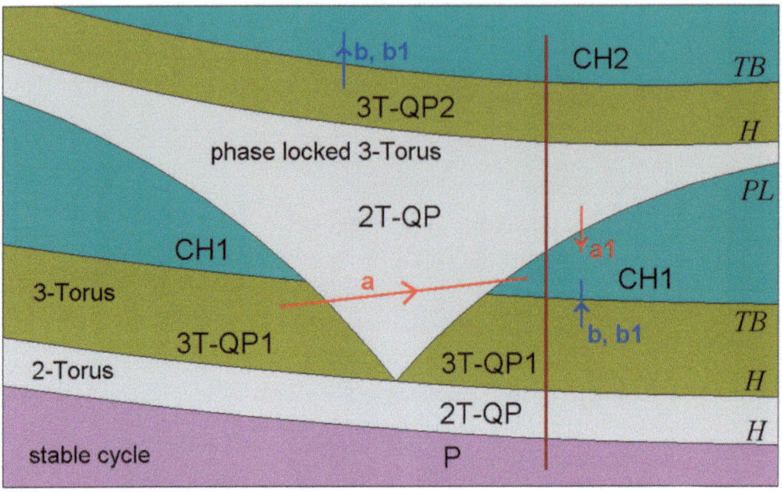

Fig. 21 Sketch of the qualitative two-parameter bifurcation chart for the breakdown of 3-Torus following [7]. (H: Hopf bifurcation, TB: torus breakdown, PL: phase locking)

does not occur for the crossover cable. Indeed, in this case, the existence of a nearly perfect 2:2:1:2 internal resonance involving all first four in-plane and out-of-plane, symmetric and antisymmetric, modes is likely to make periodic coupled responses more robust, while preventing quasi-periodicity and chaos from actual occurrence.

Finally, for the slacker cable at ½-subharmonic resonance under out-of-phase motion, the quasi-periodic scenario is seen to compete with the homoclinic bifurcation scenario. Herein, the main role in the transition to chaos seems to be played by the first torsional symmetric out-of-plane mode ($H1^T$), which somehow contributes to the overall response of the experimental system owing to the presence of the (though small) concentrated masses; of course, such a role cannot be reproduced by a companion ROM unless properly enriching the underlying continuous cable model with a rotational degree-of-freedom.

4.2.2 A Global Bifurcation Organising Regular Nonlinear Dynamics and Homoclinic Chaos

Homoclinic bifurcation involves just two main – though variable – POMs, and shows to be quite a robust scenario with respect to variations of both cable geometrical-mechanical parameters and excitation conditions (see Table 2), thus being of *general* interest [52].

To characterise in-depth the features of this seemingly paradigmatic scenario, the attention is focused on the homoclinic bifurcation of the slacker cable under anti-phase support motion at primary resonance of the antisymmetric in-plane mode V2, which ensues from the passage from the corresponding unimodal antisymmetric motion to a ballooning-type (both in-plane and out-of-plane, V2 and H2, modes) antisymmetric motion evolving from periodic to quasi-periodic up to chaotic. Yet, the systematic experimental investigation needed for in-depth characterisation of both the classes of motion and the transition scenario has been feasibly and reliably accomplished only working with a proper, thermally conditioned and controllable, experimental setup [53]. Indeed, guaranteeing a steady temperature all along the experimental investigation, this has the advantage of stabilising the response of the very flexible cable/mass system and of making it mechanically accessible without the cable loosening effects possibly entailed by too high values of the excitation amplitude, while at the same time allowing to consider the temperature as a controllable parameter in addition to the excitation amplitude and frequency. Varying the temperature as a further "external" control allows to highlight the strong role likely played by the material damping (which depends on temperature) in the unfolding of the experimental dynamics.

Based on systematic construction of bifurcation diagrams and spectra of singular values of the covariance matrix of measurements results, the various classes of motion are characterised both qualitatively and quantitatively in terms of periodicity (P), quasi-periodicity (QP) – and their degree – or chaoticity (CH), of dimension of the manifold (M) where the motion develops, of correlation dimension (D_C) of the attractor, of dimension of the possibly resonant (^{R}) invariant torus (D-T), of number (typically corresponding to manifold dimension) of contributing POMs that provide more than 90% of the experimental signal power and, finally, of mechanical meaning of the corresponding, vertical (V) or horizontal (H), linear modes (see tables in Fig. 22).

The analysis shows how the regular cable dynamics possibly ending up with homoclinic chaos is organised by a codimension-two divergence-Hopf (d-H) bifurcation point where two (a Hopf and a pitchfork) bifurcation loci of the Poincaré map of the experimental attractor cross with each other in the excitation frequency-amplitude parameter plane. Indeed, for growing frequency at temperatures $T \geq 6\,°\mathrm{C}$, two most robust experimental bifurcation paths occur in the neighbourhood of the d-H point, for forcing levels respectively higher and lower than the critical value (Fig. 22a):

$$\text{(i)} \quad \mathrm{P1M1} \xrightarrow[H]{} \mathrm{QP1M1} \xrightarrow[pf]{} \mathrm{QP1M2}^{(S)} \xrightarrow[global]{} \mathrm{QP1M2}^{(SC)} \xrightarrow[global]{} \mathrm{P1M1}$$

$$\text{(ii)} \quad \mathrm{P1M1} \xrightarrow[pf]{} \mathrm{P1M2}^{(SC)} \xrightarrow[H]{} \mathrm{QP1M2}^{(SC)} \xrightarrow[global]{} \mathrm{P1M1}$$

Focusing on the initial part of the two paths, below the critical value (path (ii)) the period 1 response P1M1, characterised by the sole antisymmetric in-plane mode V2, bifurcates through pitchfork to a symmetric couple P1M2$^{(SC)}$ of coexisting (clockwise and anti-clockwise) two-mode ballooning responses also involving the antisymmetric out-of-plane experimental eigenfunction H2. Thereafter, P1M2$^{(SC)}$ bifurcates through Hopf to a quasi-periodic symmetric couple QP1M2$^{(SC)}$. In contrast, above the critical value (path (*i*)), P1M1 is lost through a Hopf bifurcation, where a quasi-periodic response QP1M1 settles down. Then, a pitchfork bifurcation

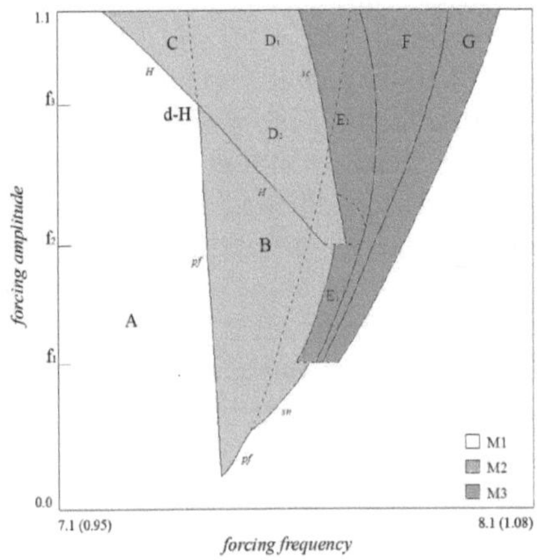

Zone		Attractor	Dimension Dc	D-T	poms	Modes
A		P1M1	1	1	1	V_2
C		QP1M1	2	2	1	V_2
B		$P1M2^{(SC)}$	1	2^R	2	$V_2 H_2$
D1		$QP2M2^{(S)}$	2	2	2	V_2H_2
		$QP2M2^{(SC)}$	2	2	2	V_2H_2
D2		$QP1M2^{(SC)}$	2	2	2	V_2H_2
E1	(a)	QP1M3	2	3^R	3	$V_2H_2H_1$
	(b)	QP2M3	2	3^R	3	$V_2H_2H_1$
		P1M1	1	1	1	V_2
E2	(a)	QP1M3	2	3^R	3	$V_2H_2H_1$
	(c)	P2M3	1	3^R	3	$V_2H_2H_1$
	(d)	P1M1	1	1	1	V_2
F	(a)	CHM3	2.8		3	$V_2H_2H_1$
	(b)	P2M3	1	3^R	3	$V_2H_2 H_1$
		P1M1	1	1	1	V_2
G	(b)	P2M3	1	3^R	3	$V_2H_2H_1$
		P1M1	1	1	1	V_2

(a)

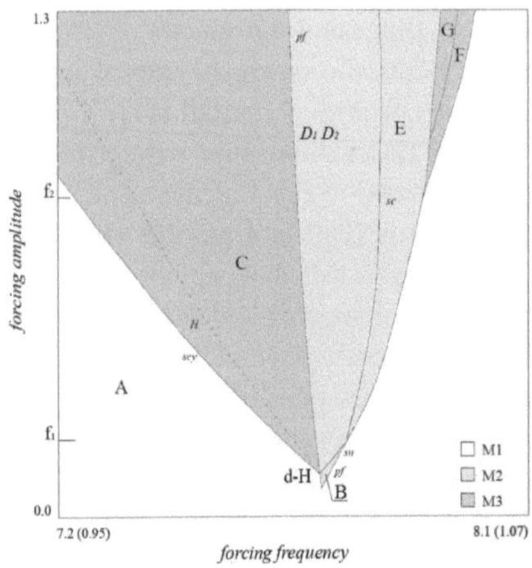

Zone	Attractor	Dimension Dc	D-T	poms	Modes
A	P1M1	1	1	1	V_2
B	P1M1	1	2^R	2	V_2
C	$QP1M1^{(+2)(S)}$	2	2	1(+2)	$V_2(H_1V_1)$
D_1	$QP1M2^{(S)}$	2	2	2	V_2H_2
D_2	$QP1M2^{(SC)}$	2	2	2	V_2H_2
E	CHM2	2.3		2	V_2H_2
G	QP5M2	2	2	2	V_2H_2
F	CHM3	3.1		3	$V_2H_2H_1$
	P1M1	1	1	1	V_2

(b)

Fig. 22 At two different temperatures, $T = 12°C$ **a** and $T = 4°C$ **b**: Response charts with zones of periodic (P), quasi-periodic (QP) and chaotic (CH) response (left), qualitative and quantitative characterisation of the relevant classes (right)

is trespassed and the out-of-plane antisymmetric experimental eigenfunction H2 is driven into the self-symmetric two-mode quasi-periodic response $QP1M2^{(S)}$.

In summary, due to the external (anti-phase primary resonance of first antisymmetric in-plane mode) and internal (1:1 resonance between antisymmetric in-plane and out-of-plane modes) resonances, the system dynamics is essentially traceable to participation of two spatial shapes closely resembling the first two antisymmetric linear modes V2 and H2. A substantially invariant bifurcation scheme persists over the whole range of temperature variation, thus allowing to refer the experimental response in the Poincaré map to the theoretical unfolding provided by the low-dimensional bifurcation system represented by the d-H normal form ([53]; see also Sect. 4.2.3 forward).

However, investigations made at different values of setup temperature highlight (i) a substantial decrease of the excitation amplitude value at which the codimension-two bifurcation occurs, along with (ii) meaningful changes in the picture of the ensuing classes of regular and/or non-regular motion extending up to regions relatively far away from the organising centre. This is clearly observable by comparing the response chart in Fig. 22a for $T = 12\,^{\circ}\mathrm{C}$ with that in Fig. 22b for $T = 4\,^{\circ}\mathrm{C}$. Each subfigure, and the corresponding table, shows the most robust classes of motion occurring in the various zones, along with the possibly competing classes (denoted by lower case labels and arrows in the second column of the table) alternatively attained for, e.g., $T = 12\,^{\circ}\mathrm{C}$ when considering different initial conditions and/or sweeping directions of the control parameter value, herein represented by the excitation frequency.

Besides lowering the critical forcing amplitude corresponding to d-H codimension-two bifurcation, lowering the temperature – which indirectly induces a material damping decrease – progressively entails a clearer scenario of regular response, along with the possibility to distinguish between the chaotic motions CHM2 and CHM3 characterised by two (the reference in-plane V2 and out-of-plane H2 antisymmetric modes) or three (also the out-of-plane symmetric mode H1) basic POMs, respectively. Overall, with decreasing temperatures, the CHM2 class of motion – into which quasi-periodic two-mode motions (QP1M2) end up at also low excitation amplitudes – becomes structurally stable and robust, and exhibits a clear evidence of low-dimensional homoclinic chaos, as shown by the results of a delay embedding reconstruction of phase space from a relevant time series. As a matter of fact, the dynamics in the second order Poincaré section of the reconstructed attractor is organised by an unstable fixed point characterised by a two-dimensional focus-stable manifold W^s and a one-dimensional saddle-unstable manifold W^u, and an invariant of the flow responsible for re-injection toward the fixed point. The fixed point on the second order Poincaré section corresponds to an unstable two-dimensional invariant of the flow resembling the formerly stable quasi-periodic solution $\mathrm{QP1M2}^{(SC)}$. In Fig. 23, besides the time series (a), two of the recorded homoclinic orbits are reported (b,c), showing the ejection along the two opposite directions of the unstable manifold (w direction), and the re-injection onto the stable manifold (the local uv plane), respectively.

Overall, the availability of the temperature as a third control parameter shows to be fundamental: (i) to qualitatively refer the *experimental unfolding* of *regular* and *non-regular* cable dynamics in the considered frequency range to the *theoretical* unfolding of the divergence-Hopf bifurcation normal form; (ii) to unfold the system dynamics not only in the strict neighbourhood of the organising d-H bifurcation but also in the ensuing post-critical regions where the dependence of material damping on temperature affects secondary bifurcations to homoclinic chaos; (iii) to show the variable involvement, in either quasi-periodic or chaotic responses, of a further POM with respect to the reference two-mode normal form scenario ending up with an *homoclinic chaos* [53].

This gives hints about the likely suitability of a two-d.o.f. ROM accounting for the sole two resonant in-plane and out-of-plane antisymmetric modes to the aim of possibly reproducing the homoclinic chaos via an analytical-numerical procedure, as well as on the likely need to include also a third d.o.f. accounting for the resonant contribution of the out-of-plane symmetric mode in order to possibly obtain, in different ranges of control parameter values, the enriched response scenario also exhibited by the experimental system in the post-critical range.

4.2.3 A Low-Dimensional Phenomenological Model of the Experimental Bifurcation Scenario

The accomplished unfolding of both the regular dynamics associated with the occurrence of a divergence-Hopf bifurcation and the ensuing non-regular dynamics exhibiting features of homoclinic chaos allows to qualitatively trace them back to a few bifurcation scenarios known from dynamical system theory and numerics.

In the context of a fruitful feedback between experiments and theory, this calls for developing a phenomenological bifurcation model of the experimental dynamics aimed at reproducing some relevant intrinsic features mostly as regards the transition scenario. Such a model relies on the normal form of the recognised bifurcation mechanism, which is well-known to the theory, while properly modifying it to account for the specific features of the underlying experimental system and for its response. The model should allow to recognise the mechanical meaning of mathematical terms playing a role in the bifurcation mechanism, and to gain insight into limitations, if any, possibly associated with any assumption made in "classical" theoretical modelling of cables. In such a case, it should also pave the way towards the independent formulation of a refined theoretical reduced-order cable model having all the necessary pre-requisites for possibly and reliably reproducing the experimentally observed phenomena.

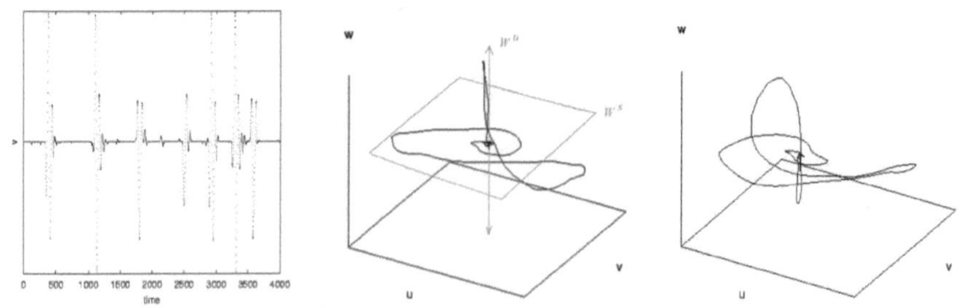

(a) (b) (c)

Fig. 23 Time delay reconstruction: **a** time series, **b** and **c** homoclinic orbits due to homoclinic tangency

Within this general perspective, a complete phenomenological model embedding the experimental bifurcation scenario but also having a discrete mathematical structure profitably comparable with the existing cable models, has been recently proposed ([5], [6]). It is based on (i) considering an equation set of order corresponding to the observed dimensionality of the experimental response, and (ii) respecting the symmetry properties of the forced experimental system, which is solely Z2 symmetric instead of being O(2) symmetric as the unforced system in the background. The model is built by reproducing the bifurcation scenario as highlighted in the experimental Poincaré section; yet, the set of equations governing the dynamics of the flow are obtained, too. The model appears suitable: (i) to analyse the interaction between the system parameters (linear frequency σ and dissipation μ) governing the unfolding of the normal form of the codimension-two bifurcation playing a role in the experimental response, and the applied excitation amplitude; (ii) to highlight the symmetry breaking effect induced in the unforced system by the asymmetric boundary conditions (sole in-plane support motion) of the experimental system.

Overall, the approach consists of building a model having not only a higher dimensionality with respect to that of the sole divergence-Hopf bifurcation but also a more convenient structure where terms and coefficients preserve a direct physical meaning.

To this aim, first an O(2) symmetric highly degenerated nilpotent double zero bifurcation set is produced. Indeed, if the cable sag-to-span ratio is small, as in the experiment, the unforced physical system has O(2) symmetry properties, and the same features are required to the unforced theoretical model in the invariant space spanned by the first two antisymmetric modes (corresponding to the V2, H2 physical POMs) where the experimental dynamics to be possibly reproduced is confined. The highly degenerated bifurcation set is produced starting from the unfolding of the O(2) symmetric nilpotent double-zero (Takens-Bogdanov) normal form [9], and considering a dependence on temperature of the dissipation terms – which include viscous and hysteretic damping – just as in the experimental setup, where the temperature has indeed been used as an essential third control parameter to detect the divergence-Hopf bifurcation and the ensuing transition to homoclinic chaos.

Then, in order to reproduce the effect of the supports motion in the experimental setup, the symmetry of the highly degenerated bifurcation set (playing the role of a map) is broken down to Z2, as in the forced physical system, in two steps. (i) By imposing non-homogeneous non-O(2) symmetric time-periodic boundary conditions (i.e., forcing) to the autonomous evolution set of equations (i.e., the flow) which stays in the background of the map. This furnishes a periodically excited highly degenerated set of evolution equations, approximate up to third order and exhibiting parametric as well as inertial and viscous terms owing to the support harmonic motion. (ii) By applying the multiple time scale method to this equation set at primary external and 1:1 internal resonance between in-plane and out-of-plane antisymmetric modes, to obtain the amplitude and phase modulation equations which govern the solution of the forced flow in the Poincaré section. They represent a perturbed normal form of the canonical O(2) symmetric nilpotent double-zero bifurcation set, with respect to which they exhibit a reduced symmetry (D4) but are indeed enriched by the contribution of parametric forcing terms. The symmetry of the perturbed normal form is further reduced to the Z2 one of the experimental case by considering solely vertical motion of the supports.

The obtained Z2 symmetric normal form constitutes a complete phenomenological model of suspended cable able, in principle, to exhibit the same bifurcation scenario as the experimental one in the post-critical range of the periodically excited highly degenerated double zero bifurcation set. A partial unfolding of the ensuing nonlinear dynamics, with the relevant variable bifurcations, has been accomplished in ([5], [6]) by analysing the stability of the trivial and non-trivial solutions of the APMEs, which depend on three control parameters as in the experiment, namely the excitation amplitude and frequency plus the temperature, whose effects are taken into account via the hysteretic damping.

Fig. 24 shows a response chart in the external frequency detuning (σ)–support motion amplitude (f) plane, with motion classes and bifurcation loci, as obtained for a fixed temperature value greater than the reference one where the effects of hysteretic damping vanish. The chart shows the occurrence of a divergence-Hopf bifurcation along a pattern qualitatively similar to the one (variable with temperature) exhibited by the experimental system. Yet, the chart also highlights precursor bifurcational events, for it is associated with a model having a higher dimensionality with respect to that of the sole divergence-Hopf bifurcation. In particular, the gray region, delimited by saddle node bifurcations, shows the existence range of the one-mode P1M1 solution, stable in a small region on the left. The Takens-Bogdanov $(\sigma, f) = (-0.02, 0.11)$ and divergence-Hopf $(\sigma, f) = (0.035, 0.05)$ codimension-two bifurcation points affect the trivial and periodic (P1M1) solutions, respectively. The divergence and Hopf loci of the trivial solution (labelled pf_1 and H_1, respectively) are reported in the behaviour chart, which also summarises the stability analysis results of the P1M1 solution by showing the corresponding divergence (pf_2) and Hopf (H_2) loci, and their intersection. With increasing frequency, the divergence involves the second mechanical mode (antisymmetric out-of-plane H2) responsible of the onset of the P1M2 motion class in the experiment, whereas the Hopf corresponds to the onset of the experimental quasi-periodic QP1M1 motion class. The

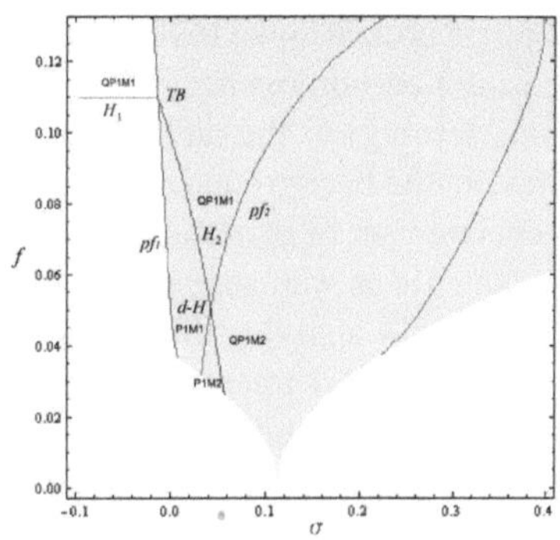

Fig. 24 Response chart in the external frequency detuning (σ)–support motion amplitude (f) plane, with motion classes and bifurcation loci

intersection of the two loci furnishes the searched codimension-two divergence-Hopf bifurcation.

Preliminary, non-systematic, results obtained for different temperature values highlight how the value of the support motion amplitude corresponding to the divergence-Hopf bifurcation occurrence decreases with decreasing hysteretic damping, i.e. with decreasing temperature. Again, this is qualitatively consistent with the decrease of the d-H bifurcation point which occurs in the experimental system with decreasing temperature (Fig. 22), up to its practical disappearance from the experimental behaviour chart.

This has important consequences from the theoretical modelling point of view, meaning that there is no practical hope to reproduce the experimentally observed phenomena and response features if considering a cable continuous model which does not include hysteretic damping. It is indeed the case of nearly all (though refined) models considered in the cable dynamics literature, including those discussed and used in the previous sections. This adds to the already made points concerned with the need to build a cable ROM which properly accounts for the cable linear modes, or other approximating functions (e.g., POMs [4] or nonlinear normal modes), taking meaningful part in the nonlinear dynamics scenario.

In summary, it is apparent how considering a further controllable parameter (material damping or, equivalently, temperature) besides the excitation amplitude and frequency allows us to interpret the response scenario of the experimental cable-mass system in the neighbourhood of a divergence-Hopf bifurcation in the framework of the symmetry breaking of a highly degenerated bifurcation set describing an O(2) symmetric Takens-Bogdanov bifurcation.

More generally, in the perspective of a fruitful cross-correlation between experimental and theoretical/numerical investigations, the construction of phenomenological models based on results of experimental nonlinear dynamics can meaningfully help to interpret them and further steer the experimental investigation, by also clarifying the extent to which any highlighted behaviour can possibly be referred to a bifurcation/response canonical scenario known from dynamical system theory. Moreover, it can pave the way for improving the cable continuous modelling and for building ROMs aimed at cross-validating and partially reproducing the observed experimental scenarios, to be implemented in a numerical procedure.

5 Further Developments and Research Topics

Notwithstanding the existing amount of available knowledge on cable nonlinear dynamics, further research developments are going on as regards several of the topics addressed in the previous sections. Just a few of them are mentioned herein, along with sample references.

(i) Concerned with modelling, attention is being paid to introducing material nonlinear constitutive laws and/or hysteretic behaviour, as well as cable bending and/or torsion capacities ([19], [40], [28]), also in view of specific analyses and applications.

(ii) Theoretical/numerical/experimental solutions for nonlinear vibration analysis of suspended cables, as obtained with different models and approaches, are being the object of further validation and comparisons ([68], [2]).
(iii) More complete descriptions of nonlinear multimodal interaction phenomena in the fully 3D cable dynamics are being accomplished, along with the analyses of the effects of further resonance conditions [76].
(iv) Improving the theoretical/numerical understanding of bifurcation scenarios to complex dynamics [14] is a topic of major interest, along with the development of properly tailored reduced-order models.

Further challenging issues are concerned with a large number of companion topics of both theoretical and practical importance. In fact, even restricting the attention to topics concerned with the solely isolated cable (i.e. leaving apart systems like, e.g., cable-stayed structures or suspension bridges, where the cable is just one, though fundamental, system component), the list of topics of interest is rather long. Comprehensively dwelling on them by giving proper credit to the research work accomplished in the last few years would require extended analyses. Herein, these further topics are just mentioned to give the reader a feeling, with no attempt of completeness, by grouping them in some main classes, and by providing sample references.

- Suspended cables in air under wind loads [39], stochastic excitations [82], or moving masses [75].
- Features of cable nonlinear dynamics in fluids ([20], [63]), with consideration of all important nonlinearities [54].
- Control of cable nonlinear vibrations through a variety of passive, active or hybrid control techniques ([81], [16], [13]).

Acknowledgements. Proper credit has to be given to Dr. Narakorn Srinil, University of Strathclyde, Glasgow, Scotland, UK and Dr. Rocco Alaggio, University of L'Aquila, Italy, who meaningfully collaborated with the writer to most of the research work in the background of the overview presented in this chapter.

References

1. Abarbanel, H.D.I., Brown, R., Sidorowich, J.J., Tsimring, L.S.: The analysis of observed chaotic data in physical systems. Rev. Mod. Phys. 65, 1331–1391 (1993)
2. Abe, A.: Validity and accuracy of solutions for nonlinear vibration analyses of suspended cables with one-to-one internal resonance. Nonlin Anal: Real World Appl. 11, 2594–2602 (2010)
3. Alaggio, R., Rega, G.: Characterizing bifurcations and classes of motion in the transition to chaos through 3d-tori of a continuous experimental system in solid mechanics. Phys. D 137, 70–93 (2000)
4. Alaggio, R., Rega, G.: Exploiting results of experimental nonlinear dynamics for reduced-order modeling of a suspended cable. In: 18th Bienn. Conf. Mechanical vibration and noise DETC01/VIB-21554 CD-Rom. ASME, New York (2001)

5. Alaggio, R., Rega, G.: Low-dimensional model of the experimental bifurcation scenario of a polymeric cable. In: Int. Conf. Nonlinear phenomena in polymer solids and low-dimensional systems, pp. 170–181. Russian Academy of Sciences, Moscow (2008)
6. Alaggio, R., Rega, G.: Unfolding of complex dynamics of sagged cables around a divergence-Hopf bifurcation: theoretical model and experimental results. In: XIX Congr. AIMETA Mecc. Teor Appl. CD-Rom. Ancona, Italy (2009)
7. Anishchenko, V.S., Safonova, M.A., Feudel, U., Kurths, J.: Bifurcations and transitions to chaos through three-dimensional tori. Int. J. Bif. Chaos. 4, 595–607 (1994)
8. Arafat, H.N., Nayfeh, A.H.: Nonlinear responses of suspended cables to primary resonance excitations. J. Sound Vib. 266, 325–354 (2003)
9. Armbruster, D., Guckenheimer, J., Kim, S.: Chaotic dynamics with square symmetry. Phys. Lett. A 140, 416–420 (1989)
10. Benedettini, F., Rega, G.: Experimental investigation of the nonlinear response of a hanging cable. Part II: Global analysis. Nonlin. Dyn. 14, 119–138 (1997)
11. Benedettini, F., Rega, G., Alaggio, R.: Nonlinear oscillations of a four-degree-of-freedom model of a suspended cable under multiple internal resonance conditions. J. Sound Vibr. 182, 775–798 (1995)
12. Berlioz, A., Lamarque, C.H.: Nonlinear vibrations of an inclined cable. J. Vib. Acoust. 127, 315–323 (2005)
13. Casciati, F., Ubertini, F.: Nonlinear vibration of shallow cables with semiactive tuned mass damper. Nonlin. Dyn. 53, 89–106 (2008)
14. Chen, H., Zhang, Z., Wang, J., Xu, Q.: Global bifurcation and chaotic dynamics in suspended cables. Int. J. Bif. Chaos. 19, 3753–3776 (2009)
15. Cheng, S.P., Perkins, N.C.: Closed-form vibration analysis of sagged cable/mass suspensions. J. Appl. Mech. 59, 923–928 (1992)
16. Gattulli, V.: Advanced control strategies in cable dynamics. In: Topping, B.H.V. (ed.) Civil Engineering Computations: Tools and Techniques, pp. 243–269. Saxe-Coburg Publications, Stirlingshire (2007)
17. Gattulli, V., Martinelli, L., Perotti, F., Vestroni, F.: Nonlinear oscillations of cables under harmonic loading using analytical and finite element models. Comput. Meth. Appl. Mech. Eng. 193, 69–85 (2004)
18. Gattulli, V., Alaggio, R., Potenza, F.: Analytical prediction and experimental validation for longitudinal control of cable oscillations. Int. J. Nonlin. Mech. 43, 36–52 (2008)
19. Goyal, S., Perkins, N.C., Lee, C.L.: Nonlinear dynamics and loop formation in Kirchoff rods with implications to the mechanics of DNA and cables. J. Comput. Phys. 209, 371–389 (2005)
20. Goyal, S., Perkins, N.C., Lee, C.L.: Nonlinear dynamic intertwining of rods with self-contact. Int. J. Nonlin. Mech. 43, 65–73 (2008)
21. Grassberger, P., Procaccia, I.: Measuring the strangeness of strange attractors. Phys. D 9, 189–208 (1983)
22. Guckenheimer, J., Holmes, P.: Nonlinear oscillations, dynamical systems and bifurcations of vector fields. Springer, New York (1983)
23. Holmes, P., Lumley, J.L., Berkooz, G.: Turbulence, coherent structures, dynamical systems and symmetry. Cambridge University Press, Cambridge (1996)
24. Ibrahim, R.A.: Nonlinear vibrations of suspended cables. Part III: random excitation and interaction with fluid flow. Appl. Mech. Rev. 57, 515–549 (2004)
25. Irvine, H.M.: Cable structures. MIT Press, Cambridge (1981)

26. Irvine, H.M., Caughey, T.K.: The linear theory of free vibrations of a suspended cable. Proc. R. Soc. London A 341, 299–315 (1974)
27. Lacarbonara, W.: Direct treatment and discretizations of non-linear spatially continuous systems. J. Sound Vibr. 221, 849–866 (1999)
28. Lacarbonara, W., Pacitti, A.: Nonlinear modeling of cables with flexural stiffness. Math. Probl. Engng., 370–767 (2008)
29. Lacarbonara, W., Rega, G.: Resonant nonlinear normal modes. Part II: Activation/Orthogonality Conditions for Shallow Structural Systems. Int. J. Nonlin. Mech. 38, 873–887 (2003)
30. Lacarbonara, W., Arafat, H.N., Nayfeh, A.H.: Non-linear interactions in imperfect beams at veering. Int. J. Nonlin. Mech. 40, 987–1003 (2005)
31. Lacarbonara, W., Paolone, A., Vestroni, F.: Elastodynamics of nonshallow suspended cables: linear modal properties. J. Vib. Acous. 129, 425–433 (2007)
32. Lacarbonara, W., Paolone, A., Vestroni, F.: Nonlinear modal properties of non-shallow cables. Int. J. Nonlin. Mech. 42, 542–554 (2007)
33. Lacarbonara, W., Rega, G., Nayfeh, A.H.: Resonant nonlinear normal modes. Part I: Analytical Treatment for One-Dimensional Structural Systems. Int. J. Nonlin. Mech. 38, 851–872 (2003)
34. Lee, C.L., Perkins, N.C.: Nonlinear oscillations of suspended cables containing a two-to-one internal resonance. Nonlin. Dyn. 3, 465–490 (1992)
35. Lee, C.L., Perkins, N.C.: Experimental investigation of isolated and simultaneous internal resonances in suspended cables. J. Vib. Acous. 117, 385–391 (1995)
36. Lee, C.L., Perkins, N.C.: Three-dimensional oscillations of suspended cables involving simultaneous internal resonances. Nonlin. Dyn. 8, 45–63 (1995)
37. Leissa, A.W.: On a curve veering aberration. J. Appl. Math. Phys. 25, 99–112 (1974)
38. Lepidi, M., Gattulli, V., Vestroni, F.: Static and dynamic response of elastic suspended cables with damage. Int. J. Solids Struct. 44, 8194–8212 (2007)
39. Luongo, A., Piccardo, G.: A continuous approach to the aeroelastic stability of suspended cables in 1:2 internal resonance. J. Vib. Control 14, 135–157 (2008)
40. Luongo, A., Zulli, D., Piccardo, G.: Analytical and numerical approaches to nonlinear galloping of internally resonant suspended cables. J. Sound Vib. 315, 375–393 (2008)
41. Mané, R.: On the dimension of the compact invariant sets of certain nonlinear maps. In: Rand, D.A., Young, L.S. (eds.) Dynamical Systems and Turbulence, vol. 898, pp. 230–242. Springer, Berlin (1981)
42. Moon, F.C.: Chaotic and fractal dynamics. Wiley, New York (1992)
43. Nayfeh, A.H.: Introduction to perturbation techniques. Wiley, New York (1981)
44. Nayfeh, A.H.: Nonlinear interactions. Wiley, New York (2000)
45. Nayfeh, A.H., Balachandran, B.: Applied nonlinear dynamics. Wiley, New York (1995)
46. Nayfeh, A.H., Arafat, H.N., Chin, C.M., Lacarbonara, W.: Multimode interactions in suspended cables. J. Vib. Control 8, 337–387 (2002)
47. Pakdemirli, M., Nayfeh, S.A., Nayfeh, A.H.: Analysis of one-to-one autoparametric resonances in cables. Discretization vs direct treatment. Nonlin. Dyn. 8, 65–83 (1995)
48. Perkins, N.C.: Modal interactions in the non-linear response of elastic cables under parametric/external excitation. Int. J. Nonlinear Mech. 27, 233–250 (1992)
49. Perkins, N.C., Mote Jr., C.D.: Comments on curve veering in eigenvalue problems. J. Sound Vib. 106, 451–463 (1986)
50. Rega, G.: Nonlinear vibrations of suspended cables. Part I: Modeling and Analysis. Appl. Mech. Rev. 57, 443–478 (2004)
51. Rega, G.: Nonlinear vibrations of suspended cables. Part II: Deterministic Phenomena. Appl. Mech. Rev. 57, 479–514 (2004)

52. Rega, G., Alaggio, R.: Spatio-temporal dimensionality in the overall complex dynamics of an experimental cable/mass system. Int. J. Solids Struct. 38, 2049–2068 (2001)
53. Rega, G., Alaggio, R.: Experimental unfolding of the nonlinear dynamics of a cable-mass suspended system around a divergence-Hopf bifurcation. J. Sound. Vib. 322, 581–611 (2009)
54. Rega, G., Sorokin, S.: Asymptotic analysis of linear/nonlinear vibrations of suspended cables under heavy fluid loading. In: Kreuzer, E. (ed.) IUTAM Symp. Fluid-Structure Interaction in Ocean Engineering, pp. 217–228. Springer, Berlin (2008)
55. Rega, G., Srinil, N.: Nonlinear hybrid-mode resonant forced oscillations of sagged inclined cables at avoidances. J. Comput. Nonlin. Dyn. 2, 324–336 (2007)
56. Rega, G., Alaggio, R., Benedettini, F.: Experimental investigation of the nonlinear response of a hanging cable. Part I: Local Analysis. Nonlin. Dyn. 14, 89–117 (1997)
57. Rega, G., Lacarbonara, W., Nayfeh, A.H., Chin, C.M.: Multiple resonances in suspended cables: direct versus reduced-order models. Int. J. Nonlin. Mech. 34, 901–924 (1999)
58. Rega, G., Srinil, N., Alaggio, R.: Experimental and numerical studies of inclined cables: free and parametrically-forced vibrations. J. Theor. Appl. Mech. 46, 621–640 (2008)
59. Ricciardi, G., Saitta, F.: A continuous vibration analysis model for cables with sag and bending stiffness. Engng. Struct. 30, 1459–1472 (2008)
60. Russell, J.C., Lardner, T.J.: Experimental determination of frequencies and tension for elastic cables. J. Engng. Mech. 124, 1067–1072 (1998)
61. Seydel, R.: Practical bifurcation and stability analysis. Springer, New York (1994)
62. Sorokin, S., Rega, G.: On modeling and linear vibrations of arbitrarily sagged inclined cables in a quiescent viscous fluid. J. Fluids Structures 23, 1077–1092 (2007)
63. Srinil, N.: Multi-mode interactions in vortex-induced vibrations of flexible curved/straight structures with geometric nonlinearities. J. Fluids Structures 26, 1098–1122 (2010)
64. Srinil, N., Rega, G.: Resonant non-linear dynamic responses of horizontal cables via kinematically non-condensed/condensed modelling. In: Mota Soares, C.A., et al. (eds.) III Eur. Conf. Comput. Mech. Solids, Structures Coupl. Probl. Engng. CD-Rom, Lisbon, Portugal (2006)
65. Srinil, N., Rega, G.: Two-to-one resonant multi-modal dynamics of horizontal/inclined cables. Part II: Internal Resonance Activation, Reduced-Order Models and Nonlinear Normal Modes. Nonlin. Dyn. 48, 253–274 (2007)
66. Srinil, N., Rega, G.: The effects of kinematic condensation on internally resonant forced vibrations of shallow horizontal cables. Int. J. Nonlin. Mech. 42, 180–195 (2007)
67. Srinil, N., Rega, G.: Nonlinear longitudinal/transversal modal interactions in highly-extensible suspended cables. J. Sound Vib. 310, 230–242 (2008)
68. Srinil, N., Rega, G.: Space-time numerical simulation and validation of analytical predictions for nonlinear forced dynamics of suspended cables. J Sound Vib. 315, 394–413 (2008)
69. Srinil, N., Rega, G., Chucheepsakul, S.: Large amplitude three-dimensional free vibrations of inclined sagged elastic cables. Nonlin. Dyn. 33, 129–154 (2003)
70. Srinil, N., Rega, G., Chucheepsakul, S.: Three-dimensional nonlinear coupling and dynamic tension in the large amplitude free vibrations of arbitrarily sagged cables. J. Sound Vib. 269, 823–852 (2004)
71. Srinil, N., Rega, G., Chucheepsakul, S.: Two-to-one resonant multi-modal dynamics of horizontal/inclined cables. Part I: Theoretical Formulation and Model Validation. Nonlin. Dyn. 48, 231–252 (2007)
72. Takens, F.: Detecting strange attractors in turbulence. In: Rand, D.A., Young, L.S. (eds.) Dynamical systems and turbulence, vol. 898, pp. 366–381. Springer, Berlin (1981)

73. Triantafyllou, M.S.: The dynamics of taut inclined cables. Q. J. Mech. Appl. Math. 37, 421–440 (1984)
74. Triantafyllou, M.S., Grinfogel, L.: Natural frequencies and modes of inclined cables. J. Struct. Engng. 112, 139–148 (1986)
75. Wang, L., Rega, G.: Modelling and transient planar dynamics of suspended cables with moving mass. Int. J. Solids Struct. 47, 2733–2744 (2010)
76. Wang, L., Zhao, Y., Rega, G.: Multimode dynamics and out-of-plane drift in suspended cable using the kinematically condensed model. J. Vib. Acous. 131, 061008 (2009)
77. Wu, Q., Takahashi, K., Nakamura, S.: Formulae for frequencies and modes of in-plane vibrations of small-sag inclined cables. J. Sound Vib. 279, 1155–1169 (2005)
78. Zhao, Y., Wang, L.: On the symmetric modal interaction of the suspended cable: three-to-one internal resonance. J. Sound Vib. 294, 1073–1093 (2006)
79. Zhao, Y., Wang, L., Cheng, D., Jiang, L.: Non-linear dynamic analysis of the two-dimensional simplified model of an elastic cable. J. Sound Vib. 255, 43–59 (2002)
80. Zheng, G., Ko, J.M., Ni, Y.Q.: Super-harmonic and internal resonances of a suspended cable with nearly commensurable natural frequencies. Nonlin. Dyn. 30, 55–70 (2002)
81. Zhou, Q., Nielsen, S.R.K., Qu, W.L.: Semi-active control of three-dimensional vibrations of an inclined sag cable with magnetorheological dampers. J. Sound Vib. 296, 1–22 (2006)
82. Zhou, Q., Larsen, J.W., Nielsen, S.R.K., Qu, W.L.: Nonlinear stochastic analysis of sub-harmonic response of a shallow cable. Nonlin. Dyn. 48, 97–114 (2007)

Engineering Applications of Non-smooth Dynamics

Marian Wiercigroch and Ekaterina Pavlovskaia

Abstract. This chapter introduces and discusses practically important concept of non-smooth dynamical systems, which are very common in engineering applications. Mathematically, such systems can be considered as piecewise smooth and therefore their global solutions are obtained by stitching local solutions, which are easy to develop by standard methods. If a dynamical system is piecewise linear then an implicit global analytical solution can be given, however the times when non-smoothness occurs have to be determined first. This leads to a set of nonlinear algebraic equations. To illustrate the non-smooth dynamical systems and the methodology of solving them, three mechanical engineering problems were studied. Firstly, a vibro-impact system in a form of moling device was modelled and analysed to understand how the progression rates can be maximised. For this system, periodic trajectories can be reconstructed as they go through three linear subspaces (no contact, contact with progression and contact without progression), and using combination of analytical and numerical methods the optimal range of the system parameters can be identified. In the second application the influence of opening and closing of a fatigue crack on the system dynamics was investigated. Specifically, a novel apparatus to induce aperiodic loading to a specimen with a fatigue crack was studied. It was shown experimentally that fatigue life can be reduced few times if the sample is loaded aperiodically. The analysis of the developed mathematical model shown that as a crack grows linearly before reaching its critical value, the response of the system remains periodic. When its size exceeds the critical value, the system behaviour becomes chaotic and then the crack growth increases exponentially. This phenomenon can be used in structural health monitoring. The last problem comes from rotordynamics, where nonlinear interactions between the rotor and the snubber

---

Marian Wiercigroch · Ekaterina Pavlovskaia
Centre for Applied Dynamics Research, School of Engineering,
Aberdeen University, King's College, AB24 3UE, Aberdeen, Scotland, UK
e-mail: {m.wiercigroch,e.pavlovskaia}@abdn.ac.uk

J. Warminski et al. (Eds.): Nonlinear Dynamic Phenomena in Mechanics, SMIA 181, pp. 211–273.
springerlink.com

ring were studied. The influence of the preloading of the snubber ring on the system behaviour was investigated and the range of the system parameters where chaotic vibrations occur was identified. The results obtained from the developed mathematical model confronted with the experiments shown a good degree of correlation.

Keywords: Non-smooth dynamics, vibrations, mechanical systems, vibro-impact moling, fatigue, rotor systems.

1 Non-smooth Dynamical Systems in Engineering Dynamics [56, 60, 44, 58, 59]

Most of engineering systems are inherently nonlinear and their nonlinearities manifest themselves in many different forms, where one of the most common is the non-smoothness. One may think of the noise of a squeaking chalk on a blackboard, or more pleasantly of a violin concert. Mechanical engineering examples include squeal in brakes, impact print hammers, percussion drilling machines or chattering of machine tools. These effects are due to the non-smooth characteristics such as clearances, impacts, intermittent contacts, dry friction, or combinations of them [60].

Non-smooth dynamical systems have been extensively studied for nearly four decades showing a huge complexity of dynamical responses even for a simple impact oscillator or Chua's circuit. The theory of non-smooth dynamical systems has been rapidly developing and now we are in much better position to understand those complexities occurring in the non-smooth vector fields and caused by generally discontinuous bifurcations. There are numerous practical applications, where the theoretical findings on nonlinear dynamics of non-smooth systems have been applied in order to verify the theory and optimize the engineering performance. However, from a mathematical point of view, problems with non-smooth characteristics are not easy to handle as the resulting models are dynamical systems whose right-hand sides are discontinuous [14], and therefore they require a special mathematical treatment and robust numerical algorithms to produce reliable solutions. Therefore, we apply a combination of numerical, analytical and semi-analytical methods to analyse such systems and this particular aspect will be a focus in this chapter.

Many characteristics of mechanical systems can be non-smooth, and such systems mathematically can be described as so-called piecewise smooth dynamical systems. Well-known examples include an impact oscillator (e.g. [49]), piecewise linear oscillators (e.g. [46, 51, 41, 63]), mechanical "bouncing ball" system [64], Jeffcott rotor with bearing clearances [33, 45, 22], systems with Coulomb friction (e.g. [13, 55]), gear-box systems [29, 38, 12] and metal cutting processes [23, 57]. A general methodology proposed in [56, 44, 58] of describing and solving non-smooth dynamical system which originates from the Fillipov's approach [14] can be found for example in [60] and [37]. It includes modelling of discontinuous systems by discontinuous functions and by smooth functions. In the latter case extra care is required as smoothing discontinuities can produce an artifact responses [32]. Here the first approach will be adopted, where a dynamical system is defined in global hyperspace Ω as continuous but not necessarily smooth autonomous system [56]

$$\dot{\mathbf{x}} = \mathbf{f}(\mathbf{x}, \mathbf{p}), \tag{1}$$

where $\mathbf{x} = [x_1, x_2, \ldots, x_n]^T$ is the state space vector (x_n is used to represent time), $\mathbf{p} = [p_1, p_2, \ldots, p_m]^T$ is a vector of the system parameters, and $\mathbf{f}(\mathbf{x}, \mathbf{p}) = [f_1, f_2, \ldots, f_n]^T$ is the vector function which is dependent upon the system structure or the process being modelled. Then we assume that the dynamical system (1) is smooth but only within subspace $\mathbf{X}_i$ of the global hyperspace Ω (see Fig. 1). Therefore, for each subspace $\mathbf{X}_i$ ($\mathbf{x} \in \mathbf{X}_i$), the right hand side of equation (1) may be described by a different function, $\mathbf{f}_i(\mathbf{x}, \mathbf{p})$ where $i \in [1, N]$. Such system description will be used throughout this chapter where for each individual problem we will specify global and local subspaces which will define its regimes of operations.

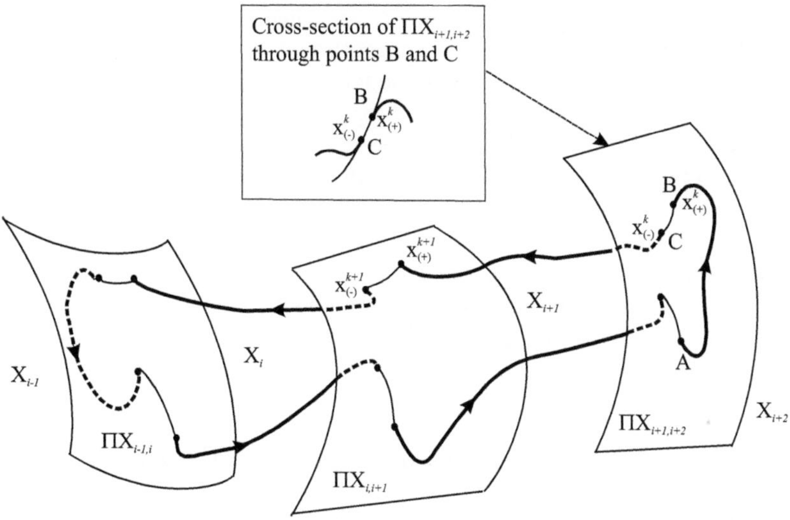

Fig. 1 Conceptual model of a piecewise smooth dynamical system, where thick lines (solid and dash) denote trajectories within subspaces and thin lines – trajectories sliding on the hypersurfaces [44]

The main aim of this chapter is to outline a general methodology of modelling and analysing of non-smooth dynamical systems. The methodology will be illustrated through three case studies. Firstly, the dynamics of a drifting vibro-impact system will be investigated through a novel semi-analytical method developed in [46, 42, 48, 61, 43, 47, 44], which allows to determine the favourable operating conditions. The model accounts for visco-elastic impacts and is capable to mimic dynamics of progressive motion (a drift). Secondly, we will study the evolution of a fatigue crack in a specimen subjected to aperiodic loading [18, 15, 16, 19, 17]. A unique experimental rig developed to investigate the dynamic crack growth under aperiodic excitation will be briefly discussed and then the modelling of the crack growth embedded into the mechanical model of the entire system will be shown. A comparison between the theoretical and experimental results will verify the model. Finally, we will examine the complex dynamics of a rotor system with bearing clearances [33, 32, 30, 31, 45, 34], where a contact incident is modelled as a piecewise smooth effect. Influence of the preloading of the snubber ring on the dynamic responses and the comparison with the experiments will be discussed.

2 Drifting Oscillator as an Effective Model of Vibro-impact Moling [48]

Vibro-impact systems are strongly nonlinear and have been widely used in civil and mechanical engineering applications. These include ground moling machines, percussive drilling, ultrasonic machining and mechanical processing (cold and hot forging). Let us focus here on a vibro-impact system driving a pile into the ground, where during its operation the driving module moves downwards, and its motion can be viewed as a superposition of a progressive motion and bounded oscillations. The simplest physical model exhibiting a such behaviour is comprised of a mass loaded by a force having static and harmonic components, and a dry friction slider, as shown in Fig. 2(a). This model was introduced and preliminary analysed in [35, 36]. Despite its simple structure, a very complex dynamics was revealed. The main result from that work was the finding that the best progression occurs during period one responses. A more realistic model including viscoelastic properties of the ground and its optimal periodic regimes were studied extensively in [46, 42, 48, 61, 43, 47, 44].

2.1 Mathematical Modelling and Experimental Study

A simple model of a vibro-impact moling system may be represented as an oscillating mass with a frictional visco-elastic slider, as shown in Fig. 2(b). This model defines the motion of the moling device (mass) and the soil (slider) which allows to make a distinction between the motion of the mole head and the front face of the hole.

The model of the soil represented by a frictional slider is depicted in Fig. 3(a). The dependence of the soil resistance on the penetration of the tool for this model is marked by a solid line in Fig. 3(b). This dependence is similar to one obtained by Spektor [54], who approximated it by a piece-wise linear relationship. The other approach used in the previous studies [50] considered the overall soil resistance as a superposition of the frontal elasto-plastic force (dash line in Fig. 3(b)) and the viscous damping force created by the fluidized soil.

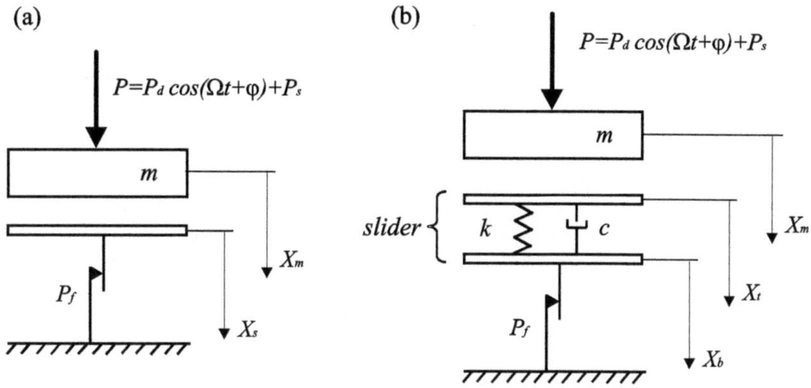

Fig. 2 Physical models of progressive impact systems; **a** simplest models [35, 36], **b** more realistic model [46]

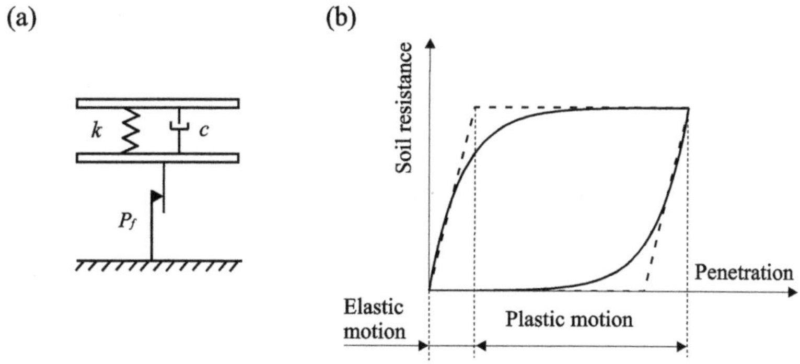

Fig. 3 Model of soil [48]; **a** physical model as a frictional slider with elasticity and viscous damping, **b** soil resistance versus penetration characteristics

In order to gain some insight into the dynamics of vibro-impact moling systems, a brief experimental study was carried out on the rig depicted in Fig. 4, which was originally designed by Lok *et al.* [39]. It comprises a steel tube hosting an oscillatory mass. The vibro-impact mechanism is comprised of a vibratory mass and a hammer mass both supported by coil springs. The hammer mass impacts upon the nose cone. The springs are precompressed in the tube and the excitation is provided by a cam driven system with a variable rotational speed.

As can be seen from Fig. 4 the moling system is set to operate in a sand box. The relative displacement of the mole with respect to the sand box was measured by an

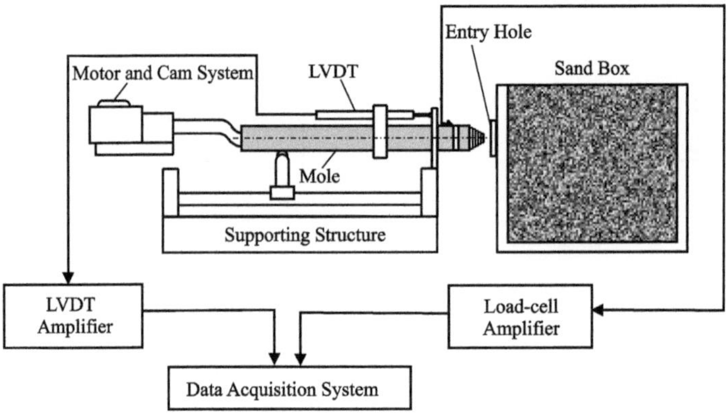

Fig. 4 Schematic of the experimental set-up to study the dynamics of vibro-impact ground moling system [48]

LVDT. The recorded signal was passed through an amplifier and monitored by an oscilloscope before being captured on a data storage system. A circular strain gauge load-cell was fitted between the impacting mass and the nose cone to measure the impact forces.

In spite of its very simple structure the model confirms the main phenomena occurring during the soil-penetrator interactions. In Fig. 5 the experimental (a, c) and the numerical (b, d) results are compared. As can be seen, there is a good qualitative agreement between the time histories. This similarity allows us to draw a conclusion that this simple model can be used to study the soil-mole dynamic interactions.

In the considered model shown in Fig. 2(b) a mass m is excited by an external force comprised of a harmonic component of amplitude P_d, frequency Ω and phase shift φ; and a static component, P_s. The slider has weightless top and bottom plates connected to each other by a linear spring with stiffness k, and a viscous damper with damping coefficient c. The progressive motion of the mass occurs when the force acting on the slider exceeds the threshold of the dry friction force, P_f. X_m, X_t and X_b represent the absolute displacements of the mass, slider top and slider bottom respectively.

For the case when the mass and the slider move separately the dynamics of the system is described by one second and two first order differential equations

$$\begin{cases} m\ddot{X}_m = P_s + P_d\cos(\Omega t + \varphi), \\ c\left(\dot{X}_t - \dot{X}_b\right) + k\left(X_t - X_b\right) = 0, \\ \dot{X}_b = 0. \end{cases} \tag{2}$$

The case when the mass and the slider are in contact, is described by one second order and one first order differential equations, which can be either oscillatory

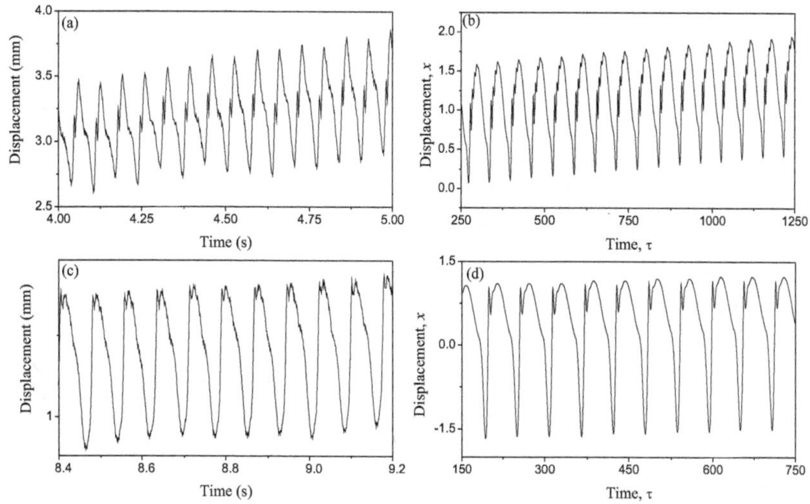

Fig. 5 **a, c** Experimental data [48] obtained from the rig with a double cam of 4 mm lift rotating at 7.75 Hz and effectively inducing a forcing frequency of **a** 15.5 Hz and **c** 13.2 Hz; **b, d** Results of numerical simulation for **b** $a = 0.5,\ \omega = 0.11,\ \xi = 0.35,\ b = 0.456$, and for **d** $a = 0.5,\ \omega = 0.1,\ \xi = 0.11,\ b = 0.48$

$$\begin{cases} m\ddot{X}_m + c\left(\dot{X}_t - \dot{X}_b\right) + k\left(X_t - X_b\right) = P_s + P_d\cos(\Omega t + \varphi), \\ \dot{X}_b = 0, \end{cases} \tag{3}$$

or progressive

$$\begin{cases} m\ddot{X}_m = -P_f + P_s + P_d\cos(\Omega t + \varphi), \\ c\left(\dot{X}_t - \dot{X}_b\right) + k\left(X_t - X_b\right) = P_f. \end{cases} \tag{4}$$

Note that for Eqs. (3) and (4) the displacement of the slider top, X_t is in phase with the displacement of the mass, X_m, but differs by a gap, G (G is the distance between the mass and the slider top at the initial point $t = 0$)

$$X_t = X_m - G.$$

The equations of motion (2) – (4) are transformed to a system of first order differential equations by using the following non-dimensional variables

$$\tau = \Omega_0 t,\ x = \frac{k}{P_{max}} X_m,\ y = \frac{dx}{d\tau} = \frac{k}{\Omega_0 P_{max}} \dot{X}_m,$$

$$z = \frac{k}{P_{max}} X_t,\ v = \frac{k}{P_{max}} X_b,$$

and parameters

$$\omega = \frac{\Omega}{\Omega_0}, \ \Omega_0 = \sqrt{\frac{k}{m}}, \ a = \frac{P_d}{P_{max}}, \ b = \frac{P_s}{P_{max}},$$

$$d = \frac{P_f}{P_{max}}, \ \xi = \frac{c}{2m\Omega_0}, \ g = \frac{k}{P_{max}}G.$$

The considered system can operate at any time in one of the following modes:

- No contact,
- Contact without progression,
- Contact with progression.

For the simplicity of further analysis, the dimensionless friction threshold force, d is set to 1. We also assume the parameters $g = 0.02$ and $\varphi = \pi/2$ to be constant in this study. Consequently, Eqs. (2) – (4) can be formulated for these specified modes in the first order ODEs.

If the displacement of the mass is smaller than the displacement of the slider top plus the gap,

$$x < z + g, \tag{5}$$

then the mass and the slider top move separately. The motion of the mass can be determined from the following set of equations

$$\begin{cases} x' = y, \\ y' = a\cos(\omega\tau + \varphi) + b, \end{cases} \tag{6}$$

where $'$ denotes $d/d\tau$. The equations of motion for the top and the bottom of the slider are

$$z' = -\frac{1}{2\xi}(z - v), \tag{7}$$

$$v' = 0. \tag{8}$$

Contact without progression occurs when the relative displacement of the mass exceeds the displacement of the slider top plus the gap, i.e.

$$x \geq z + g, \tag{9}$$

and the force acting on the mass from the slider is greater than zero but smaller than the threshold of the dry friction force, which can be expressed as

$$0 < 2\xi z' + (z - v) < 1. \tag{10}$$

In this case the mass and the slider top move together but without progression, and the second equation of (6) has additional elastic and viscous terms:

$$\begin{cases} x' = y, \\ y' = -2\xi z' - (z - v) + a\cos(\omega\tau + \varphi) + b. \end{cases} \tag{11}$$

The velocity of the slider top is equal to the velocity of the mass, and the displacement of the slider top is in phase with the mass displacement but differs by g in position

$$z' = x', \tag{12}$$

$$x = z + g. \tag{13}$$

When there is no progression, the bottom of the slider remains stationary, hence its velocity is equal to zero

$$v' = 0. \tag{14}$$

When the displacement of the mass is equal or greater than the displacement of the slider top plus the gap (see Eq.(9)), and the force acting on the mass is greater than the threshold of dry friction force

$$2\xi z' + (z - v) \geq 1, \tag{15}$$

then the mass and the top and the bottom of the slider are moving together, and progression takes place. The equations of motion for mass are

$$\begin{cases} x' = y, \\ y' = a\cos(\omega\tau + \varphi) + b - 1. \end{cases} \tag{16}$$

The displacement and the velocity of the slider top are described as before (see Eqs. (12) and (13)). The velocity of the slider bottom motion can be calculated from the expression below

$$v' = z' + \frac{1}{2\xi}(z - v - 1). \tag{17}$$

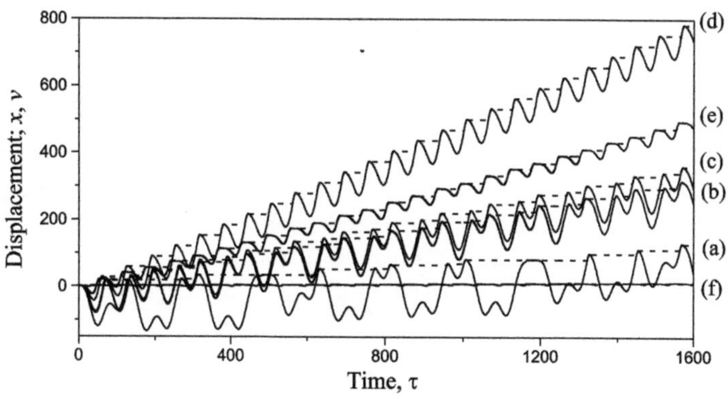

Fig. 6 Time histories [46] of the mass, x (solid curves) and slider bottom, v (dash curves) calculated for $a = 0.3$, $\omega = 0.1$, $\xi = 0.05$ and (a) $b = 0.05$; (b) $b = 0.095$; (c) $b = 0.1$; (d) $b = 0.15$; (e) $b = 0.19$ and (f) $b = 0.27$

Equations (5)–(17) were used to conduct an extensive nonlinear dynamic studies by means of numerical simulation using *Dynamics* software [65], and also to develop an algorithm to determine periodic responses [46, 42].

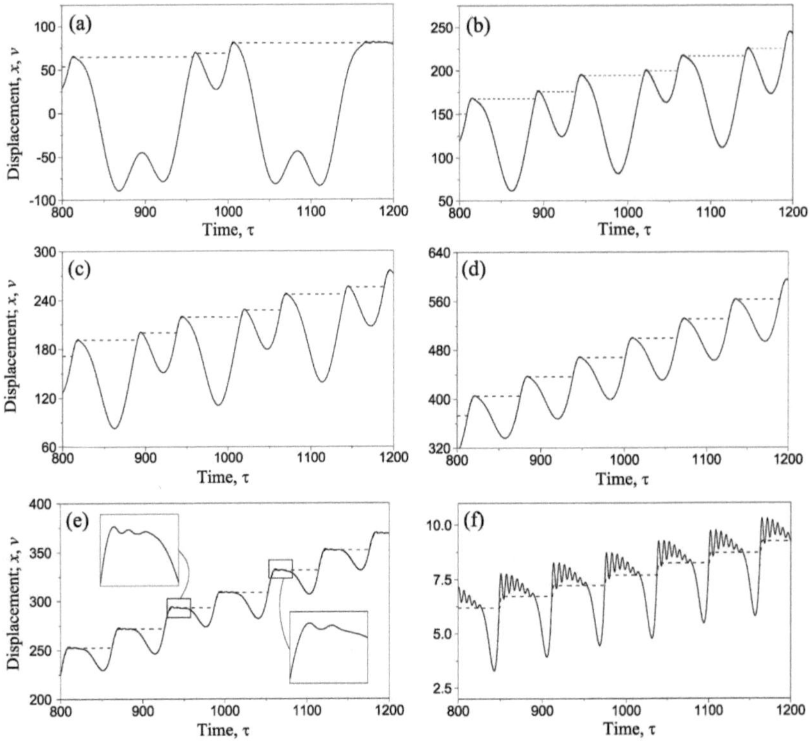

Fig. 7 Time histories of steady-state responses [46] of the mass, x (solid curves) and slider bottom, v (dash curves) for $a = 0.3$, $\omega = 0.1$, $\xi = 0.05$ and **a** $b = 0.05$; **b** $b = 0.095$; **c** $b = 0.1$; **d** $b = 0.15$; **e** $b = 0.19$ and **f** $b = 0.27$

Our study has revealed that the best progression (the largest drift) is achieved for the period one motion, which can be clearly seen by examining displacement of the slider bottom (dash lines) in Fig. 6. These steady-state responses are also depicted in Fig. 7(a)-(e) for $\tau \in (800,\ 1200)$. Figure 7 shows an important sequence of subcritical bifurcations, where the system bifurcates from a period four (Fig. 7(b)) to a period two (Fig. 7(c)), then from a period two to a period one (Fig. 7(d)). A transition from a period one to a chaotic motion with a high frequency component (Fig. 7(e)) determines the interval of static force, b, for which the best progression exists. The system can also exhibit chaos for a wide range of the system parameters (e.g. see Fig. 8(a)).

2.2 Determination of the Best Progression

Figure 8 demonstrates a number of bifurcation diagrams showing the velocity and progression of the system as function of the static force for different values of the dynamic amplitude. As can be seen in Fig. 8(a)-(c) the system responds aperiodically for small values of static force, b with some narrow windows of periodic motion. This is followed by a large window of periodic motion marked by dash lines, which is increasing for the larger values of dynamic force, a. Then a series of aperiodic and periodic windows appear. It should be noted that if the dynamic force, a, is large the system responds periodically for the most values of the static force, b (Fig. 8(c)).

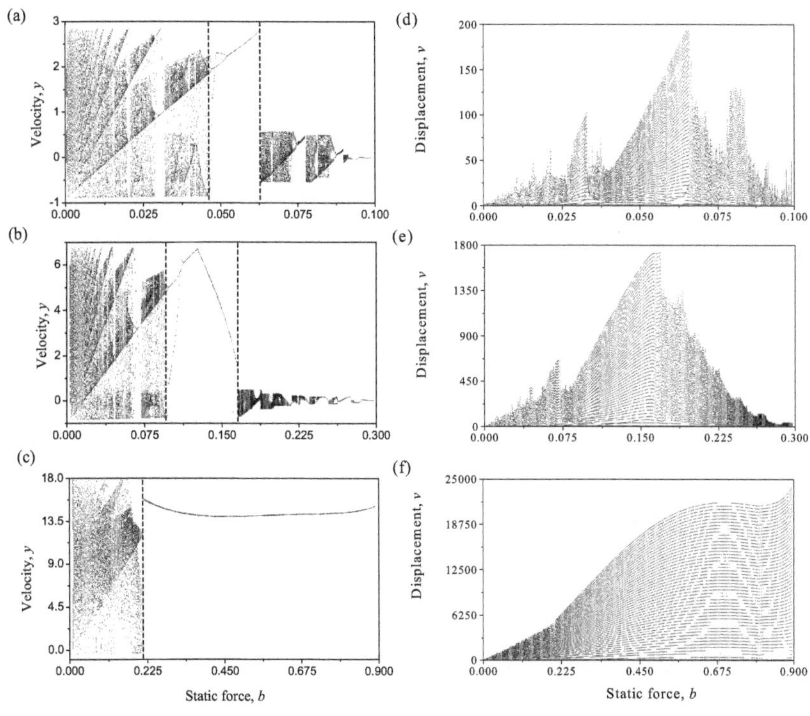

Fig. 8 Bifurcation diagrams [48] $y = y(b)$ and progression during the first 50 periods respectively for $\xi = 0.05$, $\omega = 0.1$ and **a**, **d** $a = 0.1$; **b**, **e** $a = 0.3$; and **c**, **f** $a = 0.9$

The analysis of the bifurcation diagrams has also unveiled some insight regarding the progression rates achieved by the system. It was obtained that the best progression is reached when system responds periodically with the period of external excitation. As can be clearly seen from Fig. 8(d)-(f), the maximum penetration rate coincides with the point where periodic regime becomes aperiodic. For a special case of large dynamic force, $a = 0.9$, shown in Fig. 8(f), we have periodic regimes for all values of static force, b starting from $b \approx 0.225$, and the progression rate increases starting from this value of static force, b. In general, as the system is essentially nonlinear, better progression rates are not necessarily achieved for larger static forces.

The finding regarding the position of the maximum penetration rate (the end of periodic regime) was used to develop a semi-analytical algorithm for determining this point and details can be found in [42]. This method constructs a periodic response assuming that each period is comprised of a sequence of distinct phases for which analytical solutions are explicitly known. For example, a period may consist of the following sequential phases: Contact with progression, Contact without progression, No contact and Contact without progression. Using this information a system of four piecewise linear first order differential equations is transformed to a system of nonlinear algebraic equations. Then an accurate prediction of the range of control parameters for which the best progression rates are obtained, can be made without laborious numerical simulation.

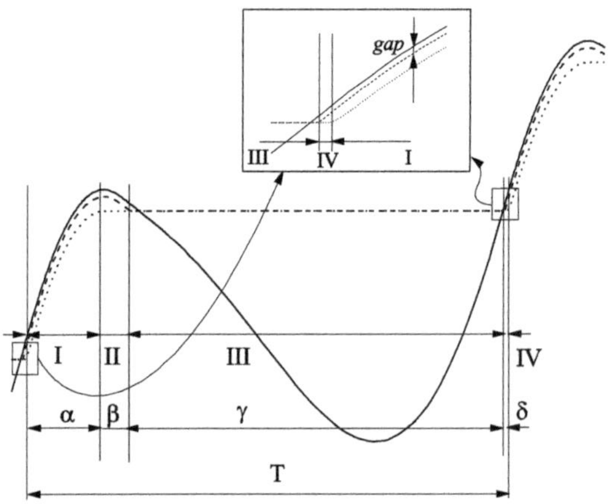

Fig. 9 Four stages of a periodic response [46]; solid line – displacement of the mass, dash line – displacement of the slider top, dotted line – displacement of the slider bottom. The blow-up window shows the displacements of the mass, slider top and bottom at the beginning and at the end of the period

As our system is piecewise linear, the periodic solutions can be constructed by stitching linear solutions obtained in the subspaces at points of discontinuities. Initially, it is assumed that the displacement and velocity of the mass have certain (as yet unknown) values. Starting from these values, the system operates in one of the phases described earlier. For the period one motion it goes through a sequence of four phases as presented in Fig. 9. As mentioned earlier a typical pattern is comprised of Contact with progression (Phase I), Contact without progression (Phase II), No contact (Phase III) and again Contact without progression (Phase IV). The solutions for each phase are constructed by allowing the final displacements and velocities of the preceding phase to be the initial conditions for the next phase. The initial conditions of the first phase are determined from the periodicity condition.

In order to simplify our consideration, the beginning of progression was chosen as an initial point. Based on Eqs. (12), (13) and (15), the following relation between the initial displacement and velocity can be written

$$2\xi y_0 + (x_0 - g - v_0) = 1. \tag{18}$$

As the initial displacement of the slider bottom, v_0 does not influence the motion of the mass, it is set to zero. Then we have

$$x_0 = 1 + g - 2\xi y_0. \tag{19}$$

The other unknown is a phase shift, ψ_0, between the external force and the system response at the beginning , $\tau = \tau_0$

$$\psi_0 = \varphi + \omega\tau_0. \tag{20}$$

Two periodic conditions for the mass displacement and velocity can be established

$$\begin{cases} x(\tau+T) = x(\tau) + \Delta, \\ y(\tau+T) = y(\tau), \end{cases} \tag{21}$$

where T is the period equal to the period of external forcing

$$T = \frac{2\pi}{\omega}, \tag{22}$$

and Δ is progression of the slider per period.

Thus three unknown functions ψ_0, x_0 and y_0 can be determined from the equations (21) and (22). However an arbitrary solution of these equations cannot guarantee that x_0 and y_0 satisfy Eq.(19). For this reason we first substitute x_0 by the function of y_0 (expression Eq.(19)), and then construct a special function F to monitor a difference between the exact periodic solution and the one calculated for these arbitrary initial conditions

$$F = \sqrt{(x(\tau+T) - x(\tau) - \Delta)^2 + (y(\tau+T) - y(\tau))^2}. \tag{23}$$

If the minimum of this function is equal to zero, then the periodic regime exists, and the durations of all four stages can be determined. More details can be found in [42].

The results obtained using this method are presented in Fig. 10 showing the influence of the frequency, the dynamic force and the static force on the progression per period. It can be deduced from Fig. 10(a) that a better progression can be achieved for a smaller excitation frequency. Here a number of values of the dynamic amplitude was considered, and the monotonically decreasing curves indicate higher penetration rates for larger amplitudes. From Fig. 10(b), it is seen that the larger the dynamic and static force are, the larger the progressions per period are obtained. Figure 10(c) shows that the progression reaches a maximum value at some certain values of static force and close to this maximum the periodic solution breaks down. The parts of the curves represented by dash lines are calculated using the same algorithm, however for these values of static force the motion detected is not periodic.

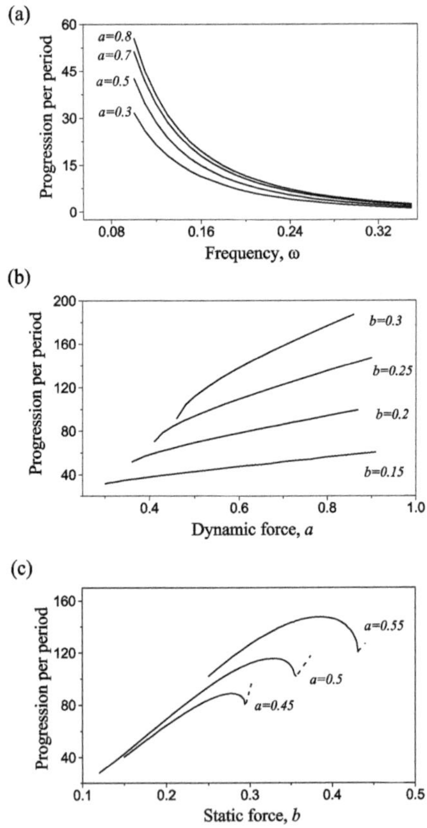

Fig. 10 Progression per period as a function of **a** frequency ω for $\xi = 0.01$, $b = 0.15$; **b** dynamic force a for $\omega = 0.1$, $\xi = 0.01$; **c** static force b for for $\xi = 0.01$, $\omega = 0.1$. Adopted from [61]

2.3 Separation of Bounded Oscillatory Motion from Drift [43]

The investigated system exhibits a non-stationary motion, which means that the displacements of the mass and the slider are unbounded. In addition, the dynamics of this system is very complex ranging from different types of periodic motion to chaos [46]. These facts rise some difficulties in analysing the system dynamics in a standard way. A simple co-ordinates transformation was proposed in [43], which resolved the problem, and it is described below.

We introduce a new system of co-ordinates (p,q,v) instead of (x,z,v):

$$\begin{aligned} p &= x - v, \\ q &= z - v. \end{aligned} \tag{24}$$

The main aim of this transformation is to separate the oscillatory motion of the system from the drift. In the new co-ordinates system, p and q are displacements of the mass and the slider top relative to the current position of the slider bottom v. We will demonstrate that the introduction of the new co-ordinates allows to study a non-stationary motion shown in Fig. 11(a) as independent bounded oscillations depicted in Fig. 11(b) and a dependent on them drift shown in Fig. 11(c).

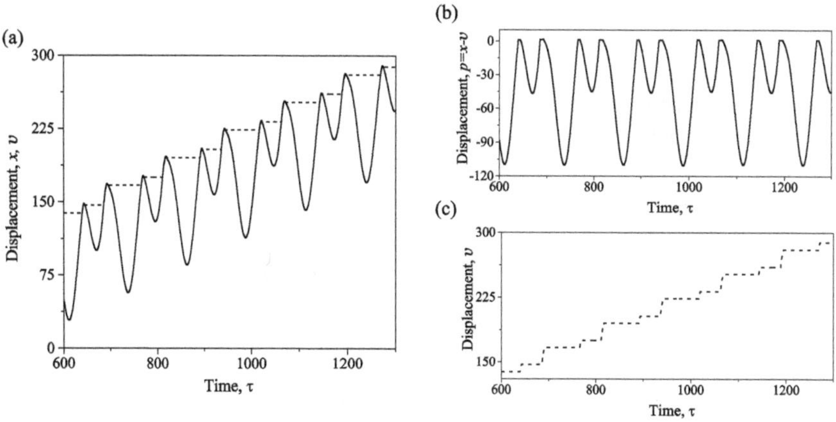

Fig. 11 Time histories of **a** drifting displacement of the mass (solid line) and the slider bottom (dash line); **b** bounded displacement of the mass $p = x - v$; and **c** progressive displacement of the slider bottom v. Adopted from [43]

The equations of motion for each phase can be rewritten as follow:

No contact

$$
\begin{aligned}
p' &= y, \\
y' &= a\cos(\omega\tau+\varphi)+b, \qquad \text{for } p<q+g, \\
q' &= -\frac{1}{2\xi}q, \\
v' &= 0.
\end{aligned}
\tag{25}
$$

Contact without progression

$$
\begin{aligned}
p' &= y, \\
y' &= -2\xi y-q+a\cos(\omega\tau+\varphi)+b, \quad \text{for } p=q+g \text{ and } 0<2\xi y+q<1, \\
q' &= y, \\
v' &= 0.
\end{aligned}
\tag{26}
$$

Contact with progression

$$
\begin{aligned}
p' &= -\frac{1}{2\xi}(q-1), \\
y' &= a\cos(\omega\tau+\varphi)+b-1, \qquad \text{for } p=q+g \text{ and } 2\xi y+q\geq 1, \\
q' &= -\frac{1}{2\xi}(q-1), \\
v' &= y+\frac{1}{2\xi}(q-1).
\end{aligned}
\tag{27}
$$

As can be seen by closer examination of Eqs. (25), (26) and (27), the first three equations of each set describing the mass and slider top motions do not depend on the displacement of the slider bottom, v. Therefore, there is no influence of the progression v on the bounded system dynamics. On contrary the motion of the slider bottom depends on the mass velocity and the displacement of the slider top (see the last equation of Eqs. (27)), hence the progression (drift) can be determined once the oscillatory mass and the slider top motions are known.

The equations of motion describing bonded oscillations are linear for each phase, and therefore the global solution can be constructed by stitching the local solutions for each phase at the points of discontinuities (non-smoothness) in the same way as it was done for the unbounded system. The set of initial values $(\tau_0; p_0, y_0, q_0)$ defines in which phase the system will operate. If $p_0 < q_0 + g$, it will be *No contact* phase. For $p_0 = q_0 + g$, it will be *Contact without progression* phase if $0 < 2\xi y_0 + q_0 < 1$ or *Contact with progression* phase if $2\xi y_0 + q_0 \geq 1$. When the conditions in the current phase fail, the next phase begins, and the final displacements and velocity for the preceding phase define the initial conditions for the next one.

The progression $v(\tau)$ can be calculated separately if the dynamics of the bounded system (p, y, q) is known (i.e. the sequence of the phases and the initial conditions for them). By solving the forth part of Eqs. (25), (26) and (27), we learnt that during the *No contact* and *Contact without progression* phases the progression does not change its value and is equal to

$$v(\tau) = v_0. \tag{28}$$

The progression during the *Contact with progression* phase can be calculated as

$$v(\tau) = v_0 + p_0 - g - 1 + (p_0 - g - 1)\exp\left(-\frac{\tau-\tau_0}{2\xi}\right) + y_0(\tau-\tau_0) + \frac{b-1}{2}(\tau-\tau_0)^2$$
$$-\frac{a}{\omega^2}\left[\cos(\omega\tau+\varphi) - \cos(\omega\tau_0+\varphi) + \omega(\tau-\tau_0)\sin(\omega\tau_0+\varphi)\right]. \tag{29}$$

The introduced co-ordinate transformation allows to study dynamics of the systems with drift in the same way as systems exhibiting bounded oscillations. Once the drift is subtracted from the oscillatory motion the standard nonlinear dynamics tools including the cell-to-cell mapping method [26] can be deployed to find co-existing attractors and their basins. For example, an evolution of the strange attractor and co-existing periodic orbits under increasing frequency ω is shown in Fig. 12 for $a = 0.3,\ b = 0.1,\ \xi = 0.1$. The presented basins of attractions were constructed using *Dynamics* software [65] adopting the following colour coding. The strange chaotic attractor and its basin are marked in orange and yellow respectively. The period four motion and its basin are coloured in black and purple. Red colour marks the attractor for blue basin, and green colour marks the attractor for pink basin. All presented cases have fractal boundaries of attractions. As can be seen from Fig. 12 co-existence of two (Fig. 12(a) and 12(d)) and three (Fig. 12(b), 12(c), 12(e) and 12(f)) attractors were found for this set of parameters.

2.4 Conclusions

In this section we presented the study of a drifting oscillator as an effective model for a vibro-impact moling system. The detailed mathematical modelling of the drifting oscillator was presented and the developed model was calibrated by the experiments.

The undertaken nonlinear dynamics analysis suggests the best progression rates are achieved for a period one motion, which means the period of response is equal to the period of excitation. Based on this finding a semi-analytical method to determine the best progression rates was developed. The parametric studies unveiled the best progression can be obtained for: (i) the ratio between the dynamic amplitude and static force around 2 and (ii) low excitation frequencies when compared with the natural frequency of the penetrated media.

To simplify the analysis further, a special coordinate transformation was proposed in order to separate the bounded oscillations of the impacting mass from the drifting motion. This transformation allows to use the standard nonlinear dynamics tools to analyse the dynamics of the bounded motion first and then reconstruct the drift (progression). After applying this transformation, the basins of attractions were calculated for the bounded system, and a number of co-existing solutions were observed for the higher excitation frequency.

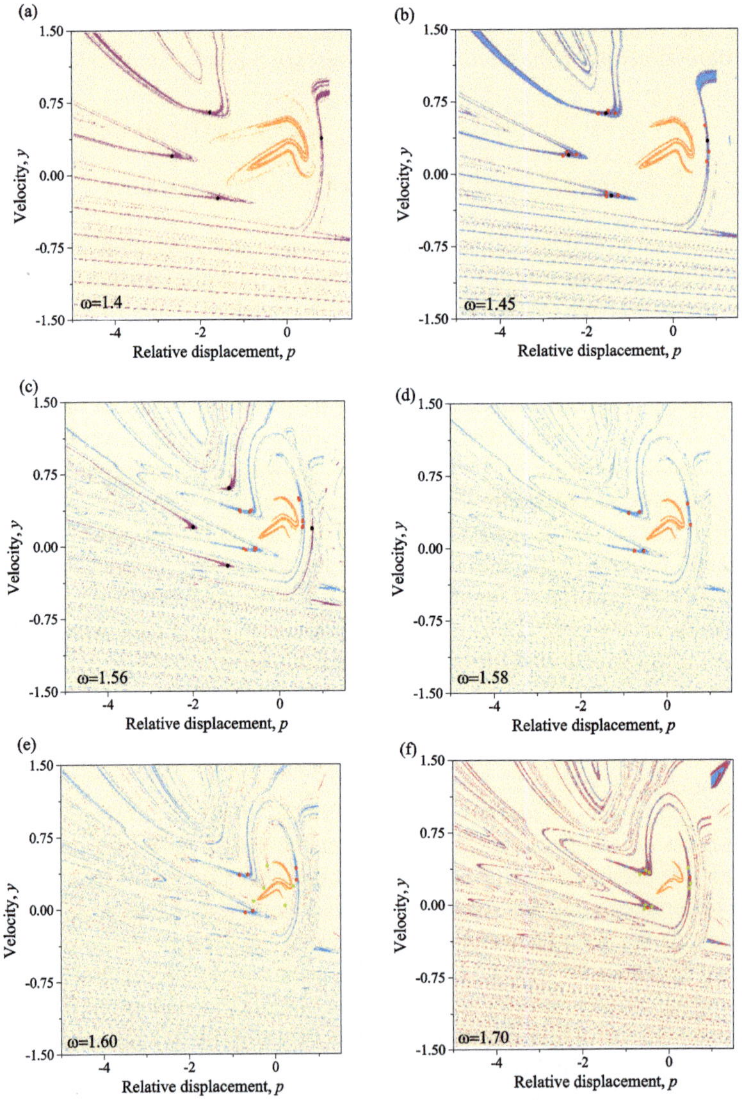

Fig. 12 Evolution of basins of attraction [43] for $a = 0.3$, $b = 0.1$, $\xi = 0.1$; **a** $\omega = 1.4$, **b** $\omega = 1.45$, **c** $\omega = 1.56$, **d** $\omega = 1.58$, **e** $\omega = 1.60$ and **f** $\omega = 1.70$

3 Nonlinear Dynamics Caused by Fatigue Crack Growth [18, 15, 16, 19, 17]

The main aim of this section is to study the effects of fatigue crack growth on the dynamic responses of engineering components and structures [18, 15, 16, 19, 17]. Due to the fact that the dynamic responses of a standard cracked specimen are often constrained by the kinematics of the forcing mechanisms in the conventional fatigue-testing machines, the natural response of the cracked specimen cannot be easily obtained. Therefore, a novel fatigue-testing rig was designed and built as detailed in [15, 16]. This novel rig consists of two base-excited oscillators, one positioned above and the other below a Single-Edge-Notched Beam (SENB) specimen and being excited by an electro-dynamic shaker. The main operating principle of the rig is that inertial forces generated by the oscillators act on the specimen, in which crack opens and closes. The rig was modelled mathematically as a two mass and one mass system, and the non-smoothness in these systems comes from the opening and closing of a fatigue crack.

3.1 Fatigue-Testing Rig and Experimental Set-Up [16, 19]

The fatigue-testing rig as shown in Fig. 13 consists of two base-excited oscillators, which are positioned above and below a SENB specimen. Each mass sandwiches a pair of leaf springs, which are also sandwiched and bolted on tower 1. These two

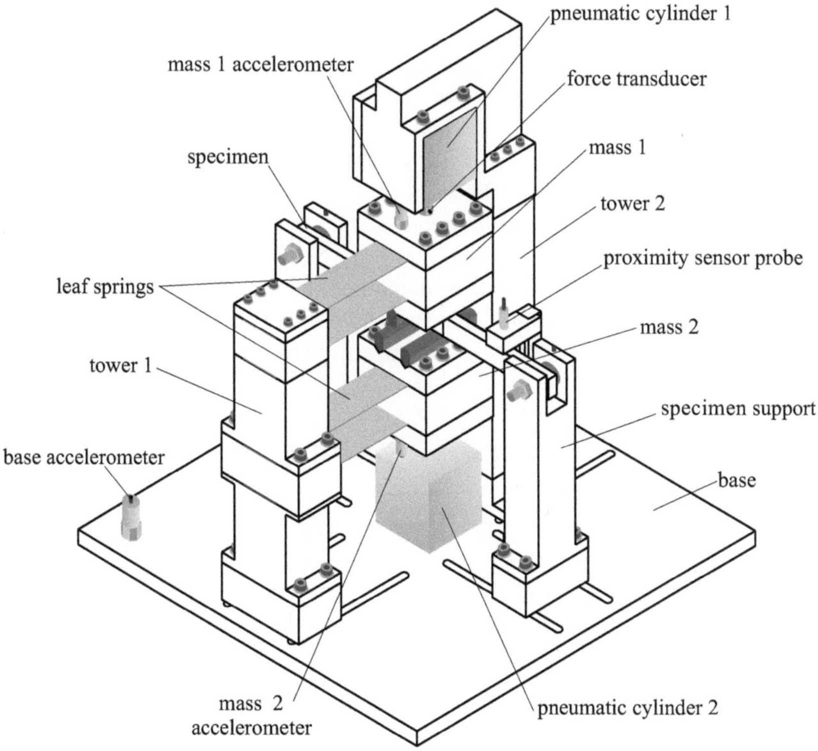

Fig. 13 Fatigue-testing rig [19] developed at the Centre for Applied Dynamics Research of the University of Aberdeen

pairs of leaf springs prevent the masses from rotating during oscillations, hence, the line of action of the applied force is perpendicular to the neutral axis of the specimen. The stiffness of the leaf springs can be adjusted by sliding tower 1 along the slot on the base. The specimen is held by supports at both ends, which are adjustable (by sliding along the slot on the base) to accommodate a range of specimen lengths. The specimen is then in turn held in place on the supports by a means of loading pins.

During experiments, both oscillators are kept in contact with the specimen by the aid of the pre-loads from the pneumatic cylinders. In addition, the pneumatic cylinders are used to set the prescribed mean load on the specimen by appropriately adjusted pressures at the top and the bottom. With this loading arrangement fatigue testing of positive (tension), negative (compression) or zero mean stresses can be carried out.

The test rig has been mounted on an electro-dynamic shaker which provides the base excitation causing the inertial forces of both oscillators to be generated and to act on the specimen. During the downward motion, the inertia of mass 1 exerts a load on the specimen causing the crack to open and the inertia of mass 2 is responsible for closing the crack during the upward motion. The amount of inertial force induced on a specimen is controlled by the adjustment of amplitude and frequency of the base excitation. The excitation waveform used to drive the shaker can be periodic or aperiodic (quasi periodic, chaotic or stochastic).

In the present studies, samples were made out of aluminium alloy 2024-T351 with the mechanical properties and chemical composition given in Tables 1 and 2, respectively. For all experiments, the same sample type, SENB, was used as shown in Fig. 14 having a width and thickness of $W = 20$mm and $B = 10$mm, respectively. The size of the notch was 5mm long and 1.5mm wide, and the loading span was 270mm.

The rig was mounted on the electro-dynamic shaker, and the base excitation was controlled by the data acquisition unit using *Labview*. The accelerations of the base ($\ddot{x}_b$), mass 1 ($\ddot{x}_{m_1}$), and mass 2 ($\ddot{x}_{m_2}$), were measured by calibrated accelerometers. A

Table 1 Material properties for aluminium alloy 2024-T351 [19]

Property	Value
Tensile strength	454MPa
Yield strength	317MPa
Young's modulus	72.4GPa
Density	2780kg/m^3

Table 2 Chemical composition for aluminium alloy 2024-T351 [19]

AL	Si	Fe	Cu	Mn	Mg	Cr	Zn	Ti
93.63	0.09	0.21	4.06	0.47	1.37	0.01	0.14	0.02

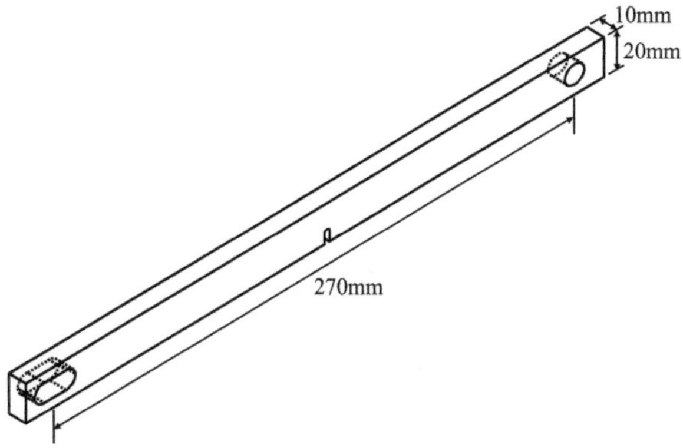

Fig. 14 Geometry of a Single-Edge-Notched Beam (SENB) specimen [19]

signal from each accelerometer and an output of each force transducer (f_{c_1} and f_{c_2}) were passed through a charge amplifier and monitored on an oscilloscope before being captured on the data acquisition unit. A schematic of the measurement block diagram is shown in Fig. 15.

The signal of the proximity sensor which represents the relative displacement (z_s) between the specimen and the base was passed through a 1kHz low pass filter to remove the high frequency noise. The output of the low pass filter was then connected to a power supply before being registered on the data acquisition unit. Finally, the signal from the ACPD crack growth monitor was fed to the data acquisition unit via a power amplifier and an oscilloscope.

In addition to measuring the relative displacement between the test specimen and the base, the proximity sensor was also used to set the dynamic load amplitude of the specimen. For these reasons, the proximity sensor was calibrated to obtain the displacement versus the sensor output voltage and the load versus the sensor output voltage relationships. During the calibration procedure the central position of the specimen was displaced by a known distance and, at each increment, the output voltage of the proximity sensor was recorded.

To complete the calibration procedure, the compliance of the cracked specimen was evaluated. The total compliance, C_{tot}, as a function of crack length of a through-thickness cracked beam can be calculated as [3]

$$C_{tot} = C_{nc} + C_c, \tag{30}$$

where C_{nc} is the compliance in the absence of a crack and C_c is the additional compliance due to the crack. For the case of a three-point loaded crack-free beam, the compliance, C_{nc}, is given as,

$$C_{nc} = \frac{L_{span}^3}{48EI}, \tag{31}$$

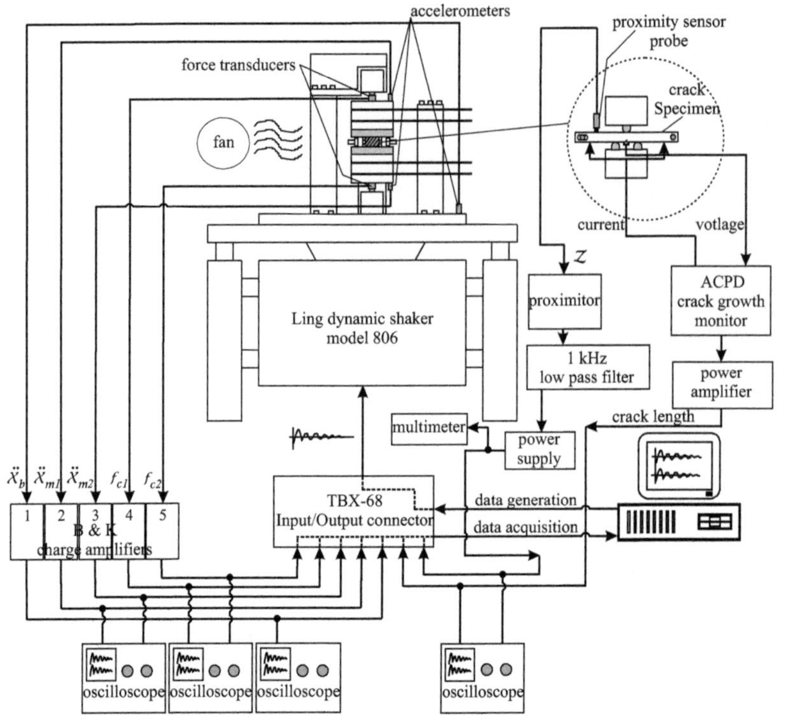

Fig. 15 Experimental layout of the fatigue-testing device for inducing dynamic loading [19]

where L_{span} is the loading span, E is Young's modulus, I is the second moment of area $\left(BW^3/12\right)$, B is the thickness and W is the width of the beam. The compliance, C_c, of the three-point loaded SENB for any value of span-to-width ratio (e.g. $\beta = L_{span}/W$) larger than 2.5 has been developed by Guinea *et al.* [25] in the following form

$$C_c = \frac{c_1(\alpha) + \beta c_2(\alpha) + \beta^2 c_3(\alpha)}{EB}, \tag{32}$$

where

$$c_1(\alpha) = -0.378\alpha^3 \ln(1-\alpha) + \alpha^2 \frac{0.29 + 1.39\alpha - 1.6\alpha^2}{1 + 0.54\alpha - 0.84\alpha^2},$$

$$c_2(\alpha) = 1.1\alpha^3 \ln(1-\alpha) + \alpha^2 \frac{-3.22 - 16.4\alpha + 28.1\alpha^2 - 11.4\alpha^3}{(1-\alpha)(1+4.7\alpha-4\alpha^2)},$$

$$c_3(\alpha) = -0.176\alpha^3 \ln(1-\alpha) + \alpha^2 \frac{8.91 - 4.88\alpha - 0.435\alpha^2 + 0.26\alpha^3}{(1-\alpha)^2(1+2.9\alpha)},$$

$\alpha = a/W$ is the crack ratio and a is the crack length. Substituting Eqs. (31) and (32) into Eq. (30), and rearranging, the load versus sensor output voltage relationship as a function of crack ratio is obtained:

$$P = \frac{\frac{G_f}{1000} \times V_{so}}{\frac{L_{span}^3}{48EI} + \frac{c_1(\alpha) + \beta c_2(\alpha) + \beta^2 c_3(\alpha)}{EB}}, \tag{33}$$

where G_f is the gain factor (displacement versus sensor output voltage relationship as obtained from above), V_{so} is the sensor output voltage and P is the load. It is worth noting that $\frac{G_f}{1000} \times V_{so}$ represents the deflection of the specimen.

A typical test was conducted in the following manner. Before a specimen was pre-loaded, an initial value of the proximity sensor was offset to zero. The amplitude of the proximity sensor output voltage corresponding to the load amplitudes acting on the bending specimen was calculated using Eq. (33). The excitation frequency was set to a value of interest and the base amplitude was varied gradually until the amplitude of the output voltage from the proximity sensor coincided with the calculated value.

3.2 *Experimental Results [16]*

To investigate the interactions between vibrations and crack growth by experimental means an extensive experimental fatigue test programme was conducted. Fatigue tests were carried out on SENB specimen shown in Fig. 14 and described earlier.

An example result of the measured time histories with the dynamic shaker driven by a sinusoidal waveform is shown in Fig. 16, where the base acceleration (Fig. 16(a)), absolute acceleration of mass 1 (Fig. 16(b)), relative displacement of the SENB specimen (Fig. 16(c)) and dynamic force between pneumatic cylinder 1 and mass 1 (Fig. 16(d)) are presented. For this particular test, the sample was subjected to a mean load of 100N and an amplitude of 100N. Examining time histories of Figs. 16(b) and 16(c), it is evident that the existence of a growing fatigue crack affects the dynamic response of the system, by breaking the symmetry of both acceleration and displacement with respect to their equilibrium positions. Fig. 17 shows the experimental time histories with the shaker driven by a chaotic waveform. The chaotic signal was obtained from the response of a forced Duffing oscillator. The test sample was subjected to zero mean load, in which the fatigue crack was opened and closed (depending on the phase of oscillations) during vibrations. A close scrutiny of the time histories (Fig. 17) shows that the pattern of the responses (Fig. 17(b)) are very similar to the base excitation (Fig. 17(a)).

The load parameters given in Table 3 were selected to compare the fatigue life of the specimen for the effects of mean load and load amplitude under constant amplitude loading.

The excitation frequency of all tests was fixed at 20Hz. Since in our tests aperiodic loading will be used (e.g. Fig. 17), instead of having a typical abscissa of

Table 3 Loading parameters for harmonic excitation fatigue test

Test name	Mean load (N)	Load amplitude (N)	R ratio	Time to fracture (secs)
hl0(78.5)	0	78.5	-1	39801
hl0(100)	0	100	-1	9191
hl0(273.5)	0	273.5	-1	1319
hl100(100)	100	100	0	1628
hl150(50)	150	50	0.5	2288
hl150(100)	150	100	0.2	547
hl200(50)	200	50	0.6	2037
hl200(100)	200	100	0.33	331

number of cycles, excitation time was used. The crack growth curves for the effect of mean load are depicted in Fig. 18, and for the effect of load amplitude are shown in Figs 19 and 20.

For the same load amplitude of 100N on each curve (Fig. 18), the fatigue life decreases as the mean load increases. In Fig. 19, when the load amplitude increases

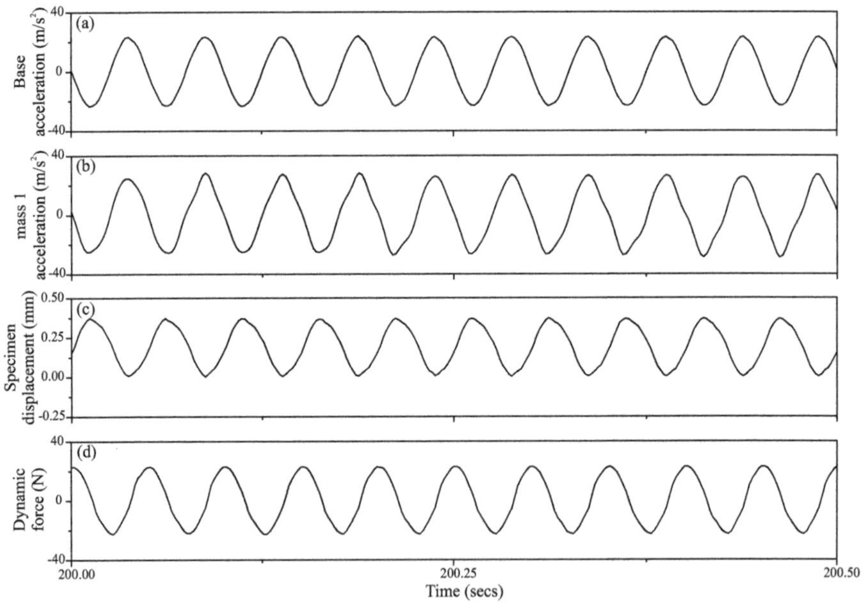

Fig. 16 Time histories [16] under harmonic excitation: (a) base acceleration, (b) mass 1 acceleration, (c) specimen displacement and (d) dynamic force between pneumatic cylinder 1 and mass 1.

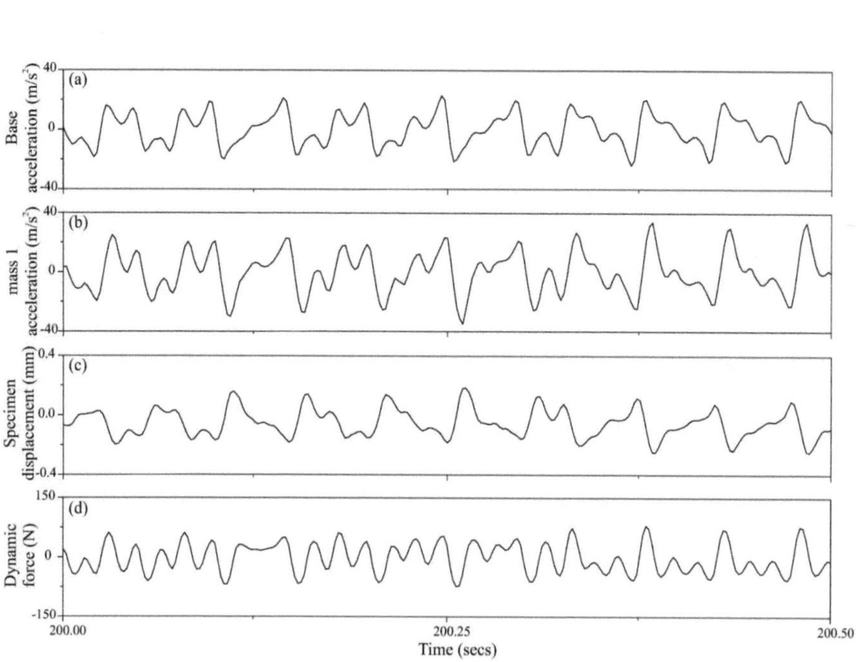

Fig. 17 Time histories [16] under chaotic excitation: **a** base acceleration, **b** mass 1 acceleration, **c** specimen displacement and **d** dynamic force between pneumatic cylinder 1 and mass 1

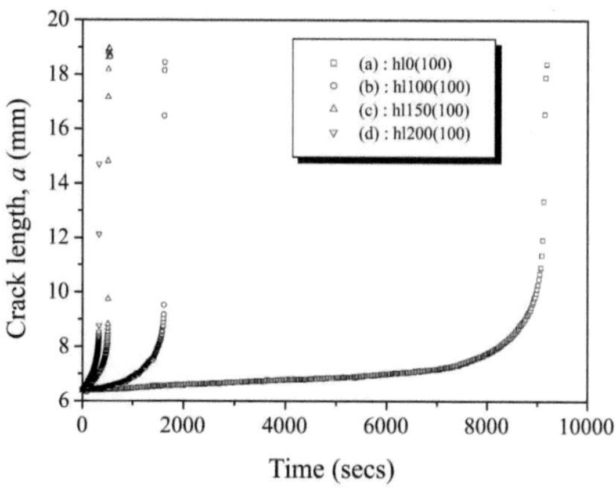

Fig. 18 Fatigue crack length [16] versus time for fixed load amplitude of 100N and with mean load of (a) 0N, (b) 100N, (c) 150N and (d) 200N

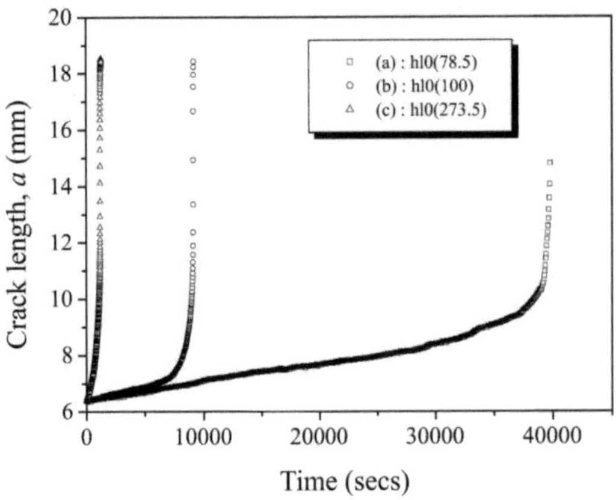

Fig. 19 Fatigue crack length [16] versus time for fixed mean load of 0N and with load amplitude of (a) 78.5N, (b) 100N and (c) 273.5N

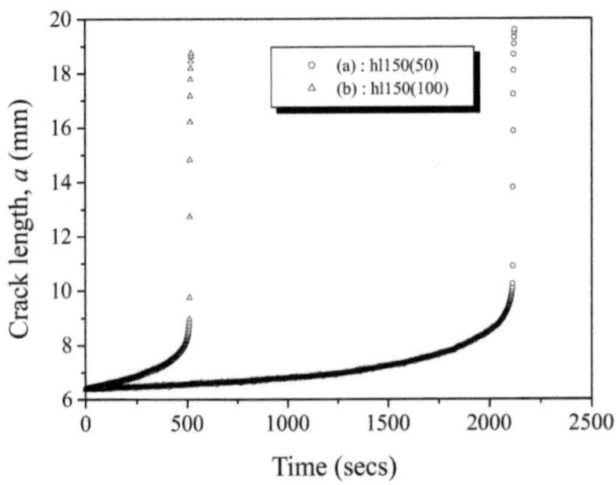

Fig. 20 Fatigue crack length [16] versus time: fixed mean load of 150N for load amplitude of (a) 50N and (b) 100N

the fatigue life decreases for the same mean load of 0N applied on each curve. A similar trend can be found for a fixed mean load of 150N in Fig. 20. The time to fracture starting from a pre-cracked length of 6.4mm for each test are given in the last column of Table 3. Time to fracture is longer for test hl150(50), which has a lower load amplitude but a higher mean load when compared to test hl100(100) (in which both tests have identical maximum load of 200N). A similar trend has been found for tests hl150(100) and hl200(50) where both tests have the same maximum load of 250N. Hence, it is not sensible to compare fatigue life by the maximum load, but rather by the mean load and load amplitude.

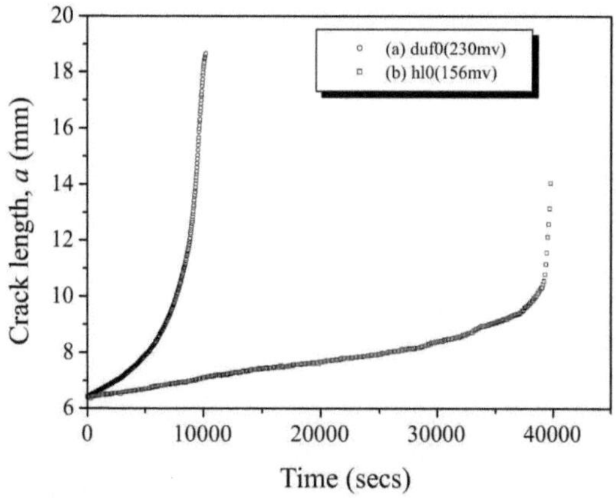

Fig. 21 Fatigue crack length [16] versus time for mean load of 0N: (a) Duffing excitation and (b) harmonic excitation

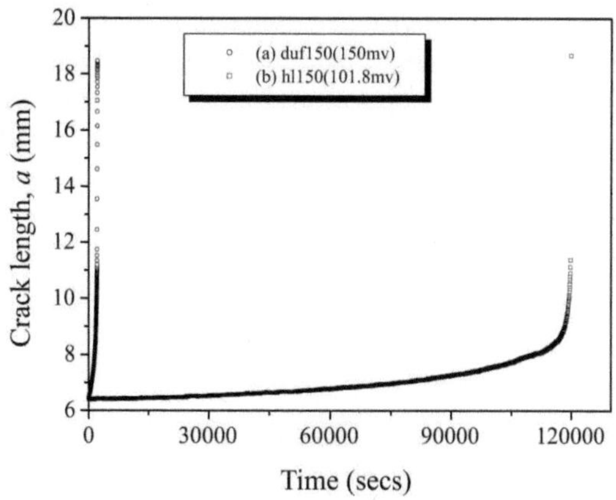

Fig. 22 Fatigue crack length [16] versus time for mean load of 150N: (a) Duffing excitation and (b) harmonic excitation

The fatigue life of SENB specimen is compared between harmonic and chaotic excitations for the same amount of elastic energy generated in the specimen. Practically, an equal amount of power was used to drive the dynamic shaker for both excitations (harmonic and chaotic). The dominant frequency for the Duffing wave is 20Hz (an example of the Duffing wave is shown in Fig. 17(a)). The excitation frequency for harmonic wave is also 20Hz. Figs 21 and 22 show the crack growth curves for the specimen subjected to mean load of 0N and 150N, respectively. On both figures, the crack curves for harmonic loading are plotted using squares and for chaotic loading uses circles. Referring to Fig. 21, for the same amount of energy the time to fracture for harmonic excitation requires 4 times more than chaotic excitation. A similar trend can be seen in Fig. 22 when the specimen is subjected to a mean load of 150N, where the fatigue life of the specimen under harmonic excitation is 56 times longer than chaotic excitation. These results indicate that for the same amount of energy pumped in into a specimen chaotic excitation is much more damaging than harmonic. The main reason for a much shorter fatigue life with specimen undergoing chaotic excitation is that the loading contains high frequency oscillations (see Fig. 17a).

3.3 *Two Mass Model [18, 19]*

The experimental rig shown in Fig. 13 can be described by a physical model depicted in Fig. 23, where the mass m_1 is attached to two springs and two dashpot dampers, in which k_{Ls1} and c_{Ls1} represent stiffness and damping of the top leaf spring, respectively, and k_{p1} and c_{p1} are stiffness and damping of the top pneumatic cylinder. The mass m_2 is attached to the base in the same way as mass m_1. Here, k_{Ls2} and c_{Ls2} represent the stiffness and damping of the bottom leaf spring, k_{p2} and c_{p2} are the stiffness and damping of the bottom pneumatic cylinder. All springs are assumed to be linear and the dampers are assumed to be linearly viscous. Masses of all springs and dampers are neglected. To ensure that both masses are kept in contact with the specimen during operation, the pneumatic forces P_1 and P_2 act on masses m_1 and m_2, respectively. The system is excited harmonically from the base with amplitude A_b and frequency Ω.

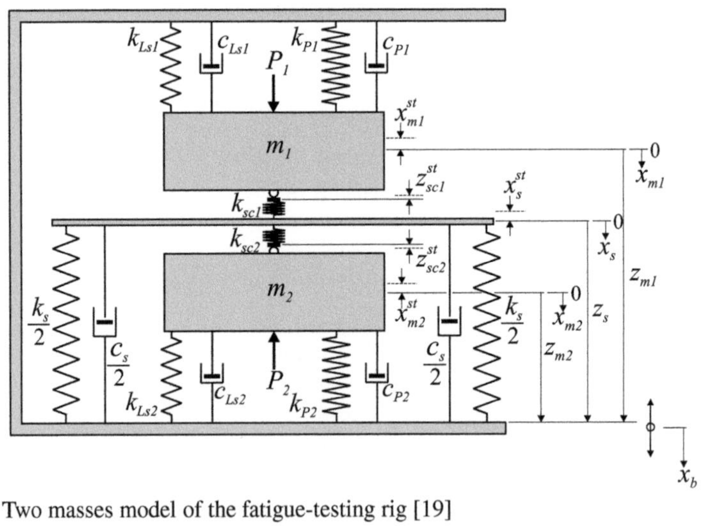

Fig. 23 Two masses model of the fatigue-testing rig [19]

Due to the fact that the operating frequency range is much lower than the fundamental frequency of the crack-free specimen, its inertial effects affecting the dynamics of the entire rig are considered to be negligible. As a consequence, the crack-free sample is modelled as a discrete linear spring k_s with a viscous damping coefficient c_s. As can be seen in Fig. 23, the model of the specimen comprises of a rigid massless beam attached to two massless springs of $0.5k_s$ stiffness each together with two massless dashpot dampers of $0.5c_s$ viscous damping each.

Elastic deformations at the point of contacts between the pneumatic cylinders and the surface of the specimen were assumed. The introduced contact stiffnesses k_{sc1} and k_{sc2} as shown in Fig. 23, are much larger than the stiffness of the specimen k_s. These springs are needed to determine the change from *Contact* and *No contact* phases.

Due to the pneumatic forces (P_1 and P_2) and also the gravitational forces ($m_1 g$ and $m_2 g$), the specimen is loaded from the top and the bottom by $k_{sc1} z_{sc1}^{st}$ and $-k_{sc2} z_{sc2}^{st}$, respectively, where

$$z_{sc1}^{st} = x_{m1}^{st} - x_s^{st}, \quad z_{sc2}^{st} = x_{m2}^{st} - x_s^{st}$$

and, x_{m1}^{st}, x_{m2}^{st} and x_s^{st} are the static displacements of masses m_1, m_2 and the specimen, respectively.

Depending on the pre-loading forces ($P_1 + m_1 g$ and $-P_2 + m_2 g$), the fatigue rig may operate in one of the following phases: (i) Full contact of both masses with the specimen, (ii) Partial contact where the mass m_1 lost contact while the mass m_2 is in contact with the specimen, (iii) Partial contact where the mass m_2 lost contact while the mass m_1 is in contact with the specimen, and finally (iv) No contact where both masses lost contact with the specimen. When both masses are in contact with the specimen (Phase (i)), the relative displacements of the masses, z_{m1} and z_{m2}, and the specimen, z_s, are oscillating in phase. In addition, if k_{sc1} and k_{sc2} are equal and also if the springs and the dampers that are attached to mass m_1 are identical to those on m_2, then for the considered system the displacements of masses m_1 and m_2 are equal, $z_{m1} = z_{m2}$. For Phase (i) the equations of motion are

$$\begin{aligned}
&m_1 \ddot{z}_{m1} + (c_{Ls1} + c_{p1})\dot{z}_{m1} + (k_{Ls1} + k_{p1} + k_{sc1}) z_{m1} - k_{sc1} z_s = m_1 A_b \Omega^2 \sin(\Omega t),\\
&c_s \dot{z}_s + (k_s + k_{sc1} + k_{sc2}) z_s - k_{sc1} z_{m1} - k_{sc2} z_{m2} = 0, \qquad (34)\\
&m_2 \ddot{z}_{m2} + (c_{Ls2} + c_{p2})\dot{z}_{m2} + (k_{Ls2} + k_{p2} + k_{sc2}) z_{m2} - k_{sc2} z_s = m_2 A_b \Omega^2 \sin(\Omega t).
\end{aligned}$$

For Phase (ii) to occur, the following inequality $z_{m1} < (z_s - z_{sc1}^{st})$, has to be satisfied and the force acting between m_1 and the specimen needs to vanish; $k_{sc1}(z_{m1} - z_s + z_{sc1}^{st}) = 0$. Hence the equations of motion are as follows

$$\begin{aligned}
&m_1 \ddot{z}_{m1} + (c_{Ls1} + c_{p1})\dot{z}_{m1} + (k_{Ls1} + k_{p1}) z_{m1} + k_{Ls1} x_{m1}^{st} = P_1 + m_1 g\\
&\qquad\qquad + m_1 A_b \Omega^2 \sin(\Omega t),\\
&c_s \dot{z}_s + (k_s + k_{sc2}) z_s - k_{sc2} z_{m2} + k_s x_s^{st} - k_{sc2} z_{sc2}^{st} = 0, \qquad (35)\\
&m_2 \ddot{z}_{m2} + (c_{Ls2} + c_{p2})\dot{z}_{m2} + (k_{Ls2} + k_{p2} + k_{sc2}) z_{m2} - k_{sc2} z_s = m_2 A_b \Omega^2 \sin(\Omega t).
\end{aligned}$$

For Phase (iii) to occur, $z_{m2} > (z_s - z_{sc2}^{st})$ has to be satisfied and the forces acting between m_2 and the specimen needs to vanish; $k_{sc2}(z_{m2} - z_s + z_{sc2}^{st}) = 0$. The equations of motion are

$$m_1\ddot{z}_{m1} + (c_{Ls1} + c_{p1})\dot{z}_{m1} + (k_{Ls1} + k_{p1} + k_{sc1})z_{m1} - k_{sc1}z_s = m_1 A_b \Omega^2 \sin(\Omega t),$$

$$c_s\dot{z}_s + (k_s + k_{sc1})z_s - k_{sc1}z_{m1} + k_s x_s^{st} - k_{sc1}z_{sc1}^{st} = 0, \tag{36}$$

$$m_2\ddot{z}_{m2} + (c_{Ls2} + c_{p2})\dot{z}_{m2} + (k_{Ls2} + k_{p2})z_{m2} + k_{Ls2}x_{m2}^{st} = -P_2 + m_2 g + m_2 A_b \Omega^2 \sin(\Omega t).$$

Finally, for Phase (iv) to occur, $z_{m1} < (z_s - z_{sc1}^{st})$ and $z_{m2} > (z_s - z_{sc2}^{st})$ have to be satisfied simultaneously. The equations of motion are

$$m_1\ddot{z}_{m1} + (c_{Ls1} + c_{p1})\dot{z}_{m1} + (k_{Ls1} + k_{p1})z_{m1} + k_{Ls1}x_{m1}^{st} = P_1 + m_1 g + m_1 A_b \Omega^2 \sin(\Omega t),$$

$$c_s\dot{z}_s + k_s z_s + k_s x_s^{st} = 0, \tag{37}$$

$$m_2\ddot{z}_{m2} + (c_{Ls2} + c_{p2})\dot{z}_{m2} + (k_{Ls2} + k_{p2})z_{m2} + k_{Ls2}x_{m2}^{st} = -P_2 + m_2 g + m_2 A_b \Omega^2 \sin(\Omega t).$$

The following simplifying assumptions were made

$$m_1 \approx m_2 = m, \quad c_{Ls1} \approx c_{Ls2} = c_{Ls}, \quad k_{Ls1} \approx k_{Ls2} = k_{Ls}, \quad k_{sc1} \approx k_{sc2} = k_{sc}.$$

The equations of motion, Eqs. (34)-(37), are now non-dimensionalised by introducing the following non-dimensional variables

$$\tau = \omega_{Ls} t, \quad X_1 = \frac{z_{m1}}{A_b}, \quad X_2 = X_1' = \frac{\dot{z}_{m1}}{A_b \omega_{Ls}}, \quad X_3 = \frac{z_s}{A_b}, \quad X_4 = \frac{z_{m2}}{A_b},$$

$$X_5 = X_4' = \frac{\dot{z}_{m2}}{A_b \omega_{Ls}}, \quad X_b = \frac{x_b}{A_b} = \sin(\eta_{Ls}\tau),$$

and parameters

$$\omega_{Ls} = \sqrt{\frac{k_{Ls}}{m}}, \quad \xi_{Ls} = \frac{c_{Ls}}{2m\omega_{Ls}}, \quad \omega_s = \sqrt{\frac{k_s}{m}}, \quad \xi_s = \frac{c_s}{2m\omega_s}, \quad \omega_{p1} = \sqrt{\frac{k_{p1}}{m}},$$

$$\xi_{p1} = \frac{c_{p1}}{2m\omega_{p1}}, \quad \omega_{p2} = \sqrt{\frac{k_{p2}}{m}}, \quad \xi_{p2} = \frac{c_{p2}}{2m\omega_{p2}}, \quad \lambda_1 = \sqrt{\frac{k_{p1}}{k_{Ls}}}, \quad \lambda_2 = \sqrt{\frac{k_{p2}}{k_{Ls}}},$$

$$\vartheta = \sqrt{\frac{k_s}{k_{Ls}}}, \quad \varepsilon = \frac{k_{Ls}}{k_{sc}}, \quad \delta_1 = \frac{x_{m1}^{st}}{A_b}, \quad \delta_2 = \frac{x_{m2}^{st}}{A_b}, \quad \Delta_1 = \frac{z_{sc1}^{st}}{A_b}, \quad \Delta_2 = \frac{z_{sc2}^{st}}{A_b},$$

$$\gamma_s = \frac{x_s^{st}}{A_b}, \ \eta_{Ls} = \frac{\Omega}{\omega_{Ls}}, \ f_1 = \frac{P_1 + mg}{mA_b\omega_{Ls}^2}, \ f_2 = \frac{P_2 - mg}{mA_b\omega_{Ls}^2},$$

where $'$ denotes $d/d\tau$.

Eqs. (34) to (37) are transformed to a set of the first order differential equations which can be written for each phase as

Phase (i)

$$\begin{aligned}
X_1' &= X_2, \\
X_2' &= -\left(1 + \lambda_1^2 + \frac{1}{\varepsilon}\right) X_1 - (2\xi_{Ls} + 2\xi_{p1}\lambda_1) X_2 + \frac{1}{\varepsilon} X_3 + \eta_{Ls}^2 \sin(\eta_{Ls}\tau), \\
X_3' &= \frac{1}{2\varepsilon\vartheta\xi_s} X_1 - \left(\frac{1}{\varepsilon\vartheta\xi_s} + \frac{\vartheta}{2\xi_s}\right) X_3 + \frac{1}{2\varepsilon\vartheta\xi_s} X_4, \\
X_4' &= X_5, \\
X_5' &= \frac{1}{\varepsilon} X_3 - \left(1 + \lambda_2^2 + \frac{1}{\varepsilon}\right) X_4 - (2\xi_{Ls} + 2\xi_{p2}\lambda_2) X_5 + \eta_{Ls}^2 \sin(\eta_{Ls}\tau).
\end{aligned} \tag{38}$$

Phase (ii)

$$\begin{aligned}
X_1' &= X_2, \\
X_2' &= -(1 + \lambda_1^2) X_1 - (2\xi_{Ls} + 2\xi_{p1}\lambda_1) X_2 - \delta_1 + f_1 + \eta_{Ls}^2 \sin(\eta_{Ls}\tau), \\
X_3' &= -\left(\frac{1}{2\varepsilon\vartheta\xi_s} + \frac{\vartheta}{2\xi_s}\right) X_3 + \frac{1}{2\varepsilon\vartheta\xi_s} X_4 + \frac{\Delta_2}{2\varepsilon\vartheta\xi_s} - \frac{\vartheta\gamma_s}{2\xi_s}, \\
X_4' &= X_5, \\
X_5' &= \frac{1}{\varepsilon} X_3 - \left(1 + \lambda_2^2 + \frac{1}{\varepsilon}\right) X_4 - (2\xi_{Ls} + 2\xi_{p2}\lambda_2) X_5 + \eta_{Ls}^2 \sin(\eta_{Ls}\tau).
\end{aligned} \tag{39}$$

Phase (iii)

$$\begin{aligned}
X_1' &= X_2, \\
X_2' &= -\left(1 + \lambda_1^2 + \frac{1}{\varepsilon}\right) X_1 - (2\xi_{Ls} + 2\xi_{p1}\lambda_1) X_2 + \frac{1}{\varepsilon} X_3 + \eta_{Ls}^2 \sin(\eta_{Ls}\tau), \\
X_3' &= \frac{1}{2\varepsilon\vartheta\xi_s} X_1 - \left(\frac{1}{2\varepsilon\vartheta\xi_s} + \frac{\vartheta}{2\xi_s}\right) X_3 + \frac{\Delta_1}{2\varepsilon\vartheta\xi_s} - \frac{\vartheta\gamma_s}{2\xi_s}, \\
X_4' &= X_5, \\
X_5' &= -(1 + \lambda_2^2) X_4 - (2\xi_{Ls} + 2\xi_{p2}\lambda_2) X_5 - \delta_2 - f_2 + \eta_{Ls}^2 \sin(\eta_{Ls}\tau).
\end{aligned} \tag{40}$$

Phase (iv)

$$
\begin{aligned}
X_1' &= X_2, \\
X_2' &= -(1+\lambda_1^2)X_1 - (2\xi_{Ls} + 2\xi_{p1}\lambda_1)X_2 - \delta_1 + f_1 + \eta_{Ls}^2 \sin(\eta_{Ls}\tau), \\
X_3' &= -\frac{\vartheta}{2\xi_s}X_3 - \frac{\vartheta\gamma_s}{2\xi_s}, \\
X_4' &= X_5, \\
X_5' &= -(1+\lambda_2^2)X_4 - (2\xi_{Ls} + 2\xi_{p2}\lambda_2)X_5 - \delta_2 - f_2 + \eta_{Ls}^2 \sin(\eta_{Ls}\tau).
\end{aligned}
\tag{41}
$$

To obtain a more elegant and compact formulation, we used the Heaviside step functions to describe the piecewise linear nature of the considered system by defining a set of switch functions G_3 and G_4,

$$
\begin{aligned}
G_3 &= G_3(X_1, X_3) = H(X_1 - (X_3 - \Delta_1)), \\
G_4 &= G_4(X_3, X_4) = H(-X_4 + (X_3 - \Delta_2)).
\end{aligned}
\tag{42}
$$

In Eq. (42), when m_1 loses contact with the specimen, $X_1 < (X_3 - \Delta_1)$, the function G_3 is equal to 0. When m_2 loses contact with the specimen, $X_4 > (X_3 - \Delta_2)$, the function G_4 is equal to 0, otherwise G_3 and G_4 are equal to 1. The equations of motion that describe all the possible phases are:

$$
\begin{aligned}
X_1' &= X_2, \\
X_2' &= -(1+\lambda_1^2)X_1 - (2\xi_{Ls} + 2\xi_{p1}\lambda_1)X_2 - \frac{G_3}{\varepsilon}(X_1 - X_3 + \Delta_1) - \delta_1 + f_1 \\
&\quad + \eta_{Ls}^2 \sin(\eta_{Ls}\tau), \\
X_3' &= \frac{G_3}{2\varepsilon\vartheta\xi_s}(X_1 - X_3 + \Delta_1) + \frac{G_4}{2\varepsilon\vartheta\xi_s}(-X_3 + X_4 + \Delta_2) \\
&\quad - \frac{\vartheta}{2\xi_s}X_3 - \frac{\vartheta\gamma_s}{2\xi_s}, \\
X_4' &= X_5, \\
X_5' &= -(1+\lambda_2^2)X_4 - (2\xi_{Ls} + 2\xi_{p2}\lambda_2)X_5 - \frac{G_4}{\varepsilon}(-X_3 + X_4 + \Delta_2) - \delta_2 - f_2 \\
&\quad + \eta_{Ls}^2 \sin(\eta_{Ls}\tau).
\end{aligned}
\tag{43}
$$

Equation (43) was used to compute the dynamic responses including the one shown in Fig. 24 where four phases of motion can be clearly observed for the system with a specimen without a crack. The displacements of mass 1, X_1, and mass 2, X_4, are plotted by thick solid and dashed lines, respectively. The displacement of the specimen, X_3, is plotted in dotted line.

Referring to Fig. 24(b), at the instant when m_1 hits m_2 both masses are kept in contact with the specimen for a very short time (labelled as Phase(i)). After this short period of light impact, due to the energy transfer between m_1 and m_2, m_2 starts to separate while m_1 still follows the trajectory of the specimen (labelled as Phase (iii)).

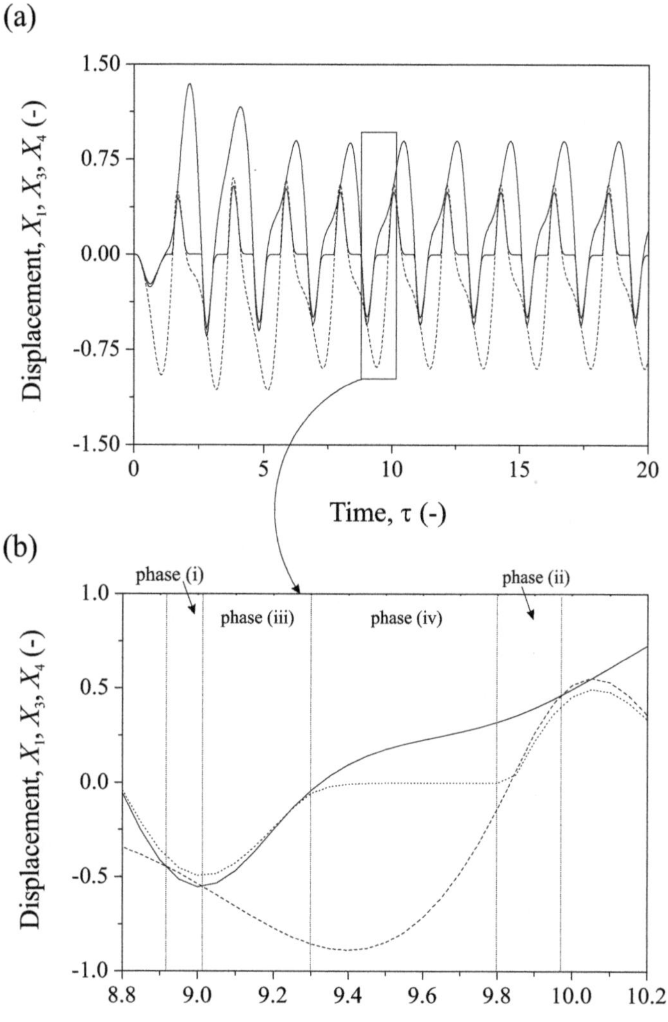

Fig. 24 Time histories [19] computed for $\eta_{Ls} = 3$, $\xi_{Ls} = 0.00255$, $\xi_s = 0.002$, $\xi_{p1} = 0.4$, $\xi_{p2} = 0.4$, $\lambda_1 = 2$, $\lambda_2 = 2$, $\vartheta = 7.828$, $\varepsilon = 0.002012$, $f_1 = 3.75$, $f_2 = 3.75$. Thick solid line represents X_1, dotted line represents X_3 and dash line represents X_4

When m_1 moves upwards to the vicinity of the equilibrium point, Phase (iv) begins. In this phase, the displacement of the specimen decays and remains stationary as at the equilibrium position. When the position of m_2 coincides with the position of the specimen Phase (ii) occurs, in which m_2 moves in phase with the specimen while m_1 is still away from the specimen.

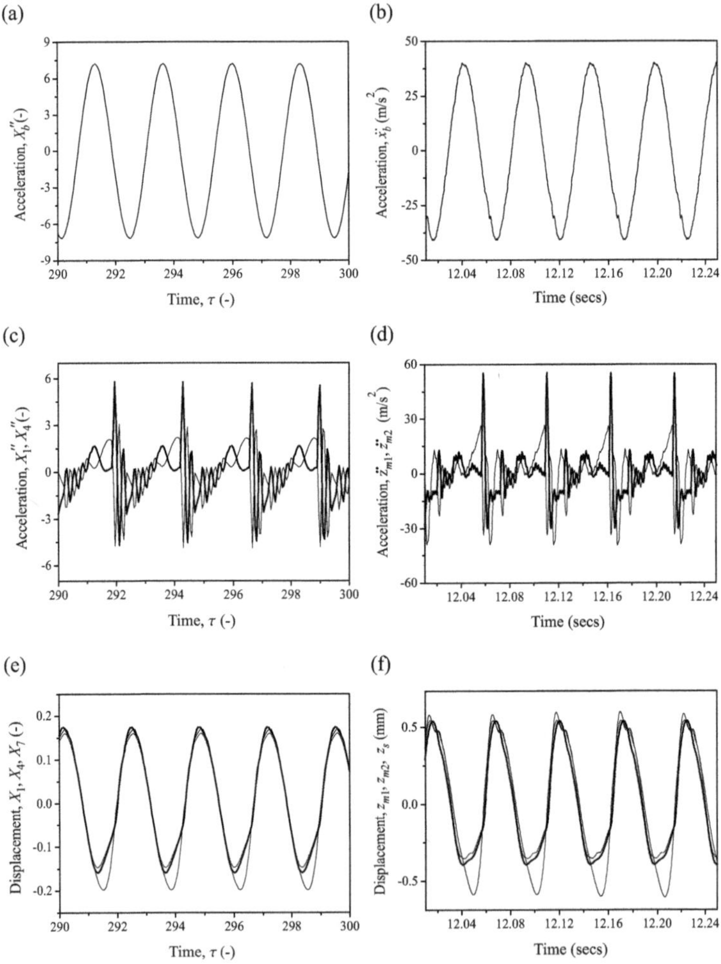

Fig. 25 **a**, **c**, **e** Theoretical time histories for $\eta_{Ls} = 2.5$, $\xi_{Ls} = 0.00255$, $\xi_s = 0.002$, $\xi_{p1} = 0.4$, $\xi_{p2} = 0.4$, $\lambda_1 = 4.36$, $\lambda_2 = 4.36$, $\vartheta = 7.828$, $\varepsilon = 0.002012$, $f_1 = 5.44$, $f_2 = 3.52$; **b**, **d**, **f** Experimental time histories for excitation frequency of 19.15Hz and an amplitude of 2.8mm. Thick solid line represents X_1'', $\ddot{z}_{m1}$, X_1 and z_{m1} and thin solid line represents X_4'', $\ddot{z}_{m2}$, X_4 and z_{m2} in Figs. 25 **c**, **d**, **e** and **f**. Dotted line represents X_3 and z_s in Figs. 25 **e** and **f**. Adopted from [19]

A comparison between numerical and experimental results is shown in Fig. 25. The theoretical time histories for the base acceleration X_b'' is shown in Fig. 25(a); the accelerations of mass 1 X_1'' (thick solid line) and mass 2 X_4'' (thin solid line) are depicted in Fig. 25(c). The displacements of mass 1 X_1 (thick solid line), mass 2 X_4 (thin solid line) and of the specimen X_3 (dotted line) were plotted on the same graph as shown in Fig. 25(e). The numerical integrations were performed with zero initial conditions using the following values of the system parameters: $\eta_{Ls} = 2.5$, $\xi_{Ls} = 0.00255$, $\xi_s = 0.002$, $\xi_{p1} = 0.4$, $\xi_{p2} = 0.4$, $\lambda_1 = 4.36$, $\lambda_2 = 4.36$, $\vartheta = 7.828$, $\varepsilon = 0.002012$, $f_1 = 5.44$, $f_2 = 3.52$. The experimental time histories acquired are depicted in Figs. 25(b), (d) and (f). The excitation frequency and amplitude used for this experiment were 19.15Hz and 2.8mm, respectively. In these figures, Fig. 25(b) is the base acceleration $\ddot{x}_b$, Fig. 25(d) depicts the relative accelerations $\ddot{z}_{m1}$ (thick solid line) and $\ddot{z}_{m2}$ (thin solid line), and Fig. 25(f) shows the relative displacements z_{m1} (thick solid line), z_{m2} (thin solid line) and z_s (dotted line). The relative displacements of m_1 and m_2 were obtained by a double numerical integration of the relative accelerations, $\ddot{z}_{m1}$ and $\ddot{z}_{m2}$, respectively of Fig. 25(d). It can be concluded from Fig. 25 that a good qualitative and quantitative agreement between the theoretical and the experimental results was obtained.

3.4 Reduction of Two Mass Model to a Single Degree-of-Freedom System [18, 19]

The pre-loading forces from the pneumatic cylinders can be set to keep both masses and the specimen always in contact. In this case the two masses system shown in Fig. 23 can be described by Eq. (38). Consequently the model can be reduced to a single degree-of-freedom by introducing the following variables,

$$y_1 = \frac{X_1 + X_4}{2},\ \ y_2 = \frac{X_2 + X_5}{2},\ \ y_3 = X_3,\ \ z_1 = \frac{X_1 - X_4}{2},\ \ z_2 = \frac{X_2 - X_5}{2},$$

and assuming that

$$\lambda_1 = \lambda_2 = \lambda,\ \ \xi_{p1} = \xi_{p2} = \xi_p.$$

Using the new variables Eq. (38) can be re-written as follows

$$\begin{aligned}
y_1' &= y_2,\\
y_2' &= -\left(1 + \lambda^2 + \frac{1}{\varepsilon}\right) y_1 - (2\xi_{Ls} + 2\xi_p\lambda) y_2 + \frac{1}{\varepsilon} y_3 + \eta_{Ls}^2 \sin(\eta_{Ls}\tau),\\
y_3' &= \frac{1}{\varepsilon\vartheta\xi_s} y_1 - \left(\frac{1}{\varepsilon\vartheta\xi_s} + \frac{\vartheta}{2\xi_s}\right) y_3,\\
z_1' &= z_2\\
z_2' &= -\left(1 + \lambda^2 + \frac{1}{\varepsilon}\right) z_1 - (2\xi_{Ls} + 2\xi_p\lambda) z_2.
\end{aligned} \tag{44}$$

As can be seen from Eq. (44), the first three equations do not depend on variable z, while the last two are independent of variable y. Hence, two independent systems have been obtained and they can be solved separately. Assuming that $X_1 = X_4$ and $X_2 = X_5$, we have $z_1 = z_2 \equiv 0$, which means that the displacements and the velocities of both masses are equal. Furthermore, for the present system, $k_{sc} \gg k_{Ls}$ and thus when $\varepsilon \to 0$, it has been assumed that the relative displacements and velocities of the springs k_{sc1} and k_{sc2} are negligible. For this case, the displacements of both masses and the specimen will be equal. Therefore, the two mass model given in Fig. 23 can be simplified to a single degree-of-freedom model as shown in Fig. 26.

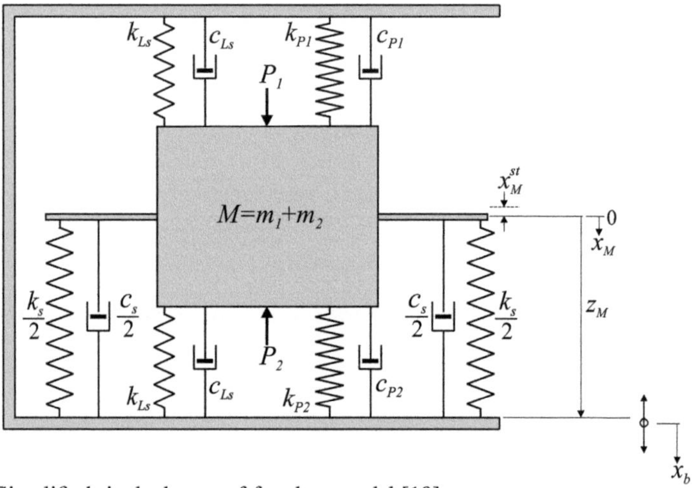

Fig. 26 Simplified single degree-of-freedom model [19]

Consequently, the dynamics of both masses can be described by the following equation of motion

$$M\ddot{z}_M + c_M\dot{z}_M + (k_M + k_s)z_M = MA_b\Omega^2\sin(\Omega t), \tag{45}$$

where,

$$z_M = \frac{z_{m1} + z_{m2}}{2} = z_{m1} = z_{m2}, \quad M = m_1 + m_2,$$

$$c_M = (2c_{Ls} + c_{p1} + c_{p2} + c_s), \quad k_M = (2k_{Ls} + k_{p1} + k_{p2}).$$

Transforming Eq. (45) into the non-dimensional first order differential equations, we have

$$\begin{aligned} y_1' &= y_2, \\ y_2' &= -(1+\kappa_M)y_1 - 2\xi_M y_2 + \eta_M^2\sin(\eta_M\tau), \end{aligned} \tag{46}$$

whereby the non-dimensional parameters are defined as,

$$\tau = \omega_M t, \ \ \omega_M = \sqrt{\frac{k_M}{M}}, \ \ \xi_M = \frac{c_M}{2M\omega_M}, \ \ \kappa_M = \frac{k_s}{k_M}, \ \ \eta_M = \frac{\Omega}{\omega_M},$$

where $'$ denotes $d/d\tau$.
As was demonstrated above, when both masses are in contact with the specimen, the set of equations that represent the model in Fig. 23 can be simplified from five first order equations (Eq. (44)) to just two (Eq. (46)).

3.5 *Stiffness of a Cracked Beam [19]*

Many investigations have been conducted to study the nonlinear effects due to discontinuous stiffness characteristics. For example externally forced piecewise linear oscillators were studied theoretically by Shaw and Holmes [51], and experimentally by Wiercigroch *et al.* [62] and Sin and Wiercigroch [53]. Obtained results in [51, 62, 53] revealed complex dynamics including existence of periodic, subharmonic and chaotic motion. The bilinear or piecewise oscillators have also been used to model the dynamic behaviour of cracked structures [66, 27, 2, 28, 9, 10, 20, 52, 6, 1]. However, the work carried out in [66, 27, 2, 28, 9, 10, 20, 52, 6, 1] assumed a stationary fatigue crack, in which, the restoring force of the cracked structure has a stiffness characteristic as shown in Fig. 27(a). In the present work, taking into account a growing fatigue crack, the stiffness of the bending specimen decreases as a function of crack length and time when crack opens. For a completely closed crack, the stiffness of a crack-free specimen is assumed. The bilinear piecewise smooth restoring force of this behaviour is shown in Fig. 27(b) for different lengthes of the crack. The main rationale behind the assumption is that, unless the remaining material at the front of the crack tip starts to yield, the decrease of the bending specimen stiffness is small even though a significant fatigue crack has propagated. Furthermore, Gudmundson [24] experimentally showed that the effect of the closing crack has a small influence on the natural frequencies, and therefore, making the assumption justified.

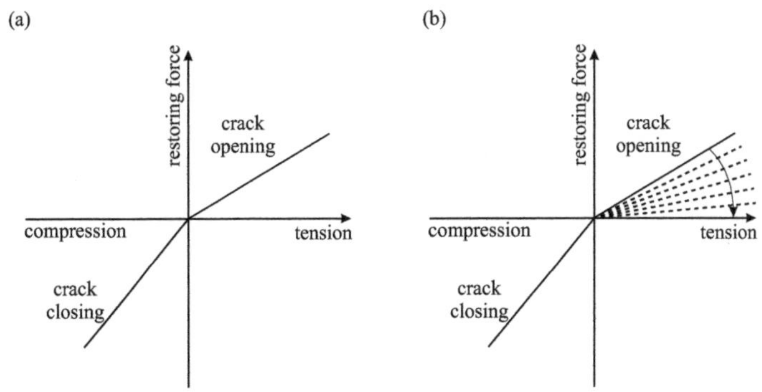

Fig. 27 Stiffness characteristics of the specimen [19]

The stiffness of the specimen with a closed crack assumed as a simply supported beam can be calculated from the formula

$$k_s^{cl} = \frac{48EI}{L_{span}^3}. \tag{47}$$

When a crack opens, the stiffness k_s^{op} can be computed from Eq. (33), which after a simple rearrangement can take the following form

$$k_s^{op} = \frac{1}{\frac{L_{span}^3}{48EI} + \frac{c_1(\alpha)+\beta c_2(\alpha)+\beta^2 c_3(\alpha)}{EB}}, \tag{48}$$

where all parameters are as defined for Eq. (33). k_s^{op} decreases in term of crack length, a, which is modelled by an exponential function of time t, [18]

$$a = a_o + a_1 \exp\left(\frac{t-a_2}{a_3}\right), \tag{49}$$

where the constants a_o, a_1, a_2 and a_3 are obtained from experiments. Eq. (49) is used to calculate the crack-depth ratio $\alpha = a/W$ in the functions $c_1(\alpha)$, $c_2(\alpha)$ and $c_3(\alpha)$ of Eq. (48).

During our fatigue tests, the two oscillating masses of the fatigue rig were kept in contact with the specimen at all time. A model that represents the full contact case was already presented in Fig. 26 and its equation of motion is given by Eq. (45). Referring to Fig. 26, for crack opening, the kinematic condition, $(z_M + x_M^{st}) > 0$, must be satisfied, and for crack closing, $(z_M + x_M^{st}) \leq 0$. After including the static forces into Eq. (45), the resulting equation was transformed to a set of two non-dimensional first order differential equations,

$$\begin{aligned} y_1' &= y_2, \\ y_2' &= -(1+\kappa_M)y_1 - 2\xi_M y_2 - (\rho+\kappa_M)\gamma_M + f_M + \eta_M^2 \sin(\eta_M \tau), \end{aligned} \tag{50}$$

where

$$\rho = \frac{2k_{Ls}}{k_M}, \quad f_M = \frac{P_1 + Mg - P_2}{MA_b\omega_M^2},$$

and $\gamma_M = x_M^{st}/A_b$. A Heaviside function $H(y_1+\gamma_M)$, which is equal to 1 for crack opening and 0 for crack closing is used to model the piecewise linear stiffness characteristics κ_M in Eq. (50), in which

$$\kappa_M = H(y_1+\gamma_M)\kappa_M^{op} + (1 - H(y_1+\gamma_M))\kappa_M^{cl}, \tag{51}$$

where

$$\kappa_M^{op} = \frac{k_s^{op}}{k_M}, \quad \kappa_M^{cl} = \frac{k_s^{cl}}{k_M}.$$

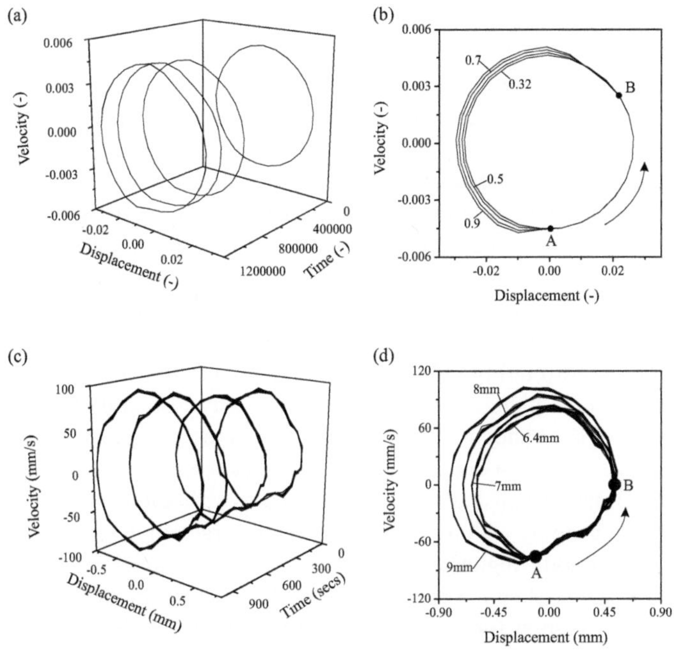

Fig. 28 Phase portraits obtained for 0N mean load and load amplitude of 274N [19]; **a** theoretical 3-dimensional, **b** theoretical 2-dimensional, **c** experimental 3-dimensional and **d** experimental 2-dimensional

A three dimensional and a two dimensional phase portraits generated from Eq. (50) representing the dynamic responses of a specimen under a growing breathing crack (crack opens and closes) are depicted in Figs. 28(a) and (b), respectively. The smallest to the largest orbits correspond to the growing crack ratio, α of 0.32, 0.5, 0.7 and 0.9, respectively. All orbits form close loops indicating period one motion with clearly visible piecewise linear nature of the dynamic responses (see Fig. 28(b)). This is due to the fact that when fatigue crack propagates, the stiffness of the specimen decreases when the crack opens, and hence, causes a larger amplitude of oscillations. In addition, the trajectories at different crack length follows the same path as the crack closes (point A to B in Fig. 28(b)), which confirms that the stiffness during the crack closure remains constant. These theoretical results were validated by the experimental phase portraits as shown in Figs. 28(c) and d.

3.6 Strange Attractor [18]

The behaviour of the system was simulated using the one mass model and the results are shown in Figs. 29 and 30. Fig. 29 presents a crack growth curve obtained for

$\xi_M = 0.014$, $\eta_M = 1.1475$, $k_s^{cl} = 14.327$, $k_s^{op} = 4.245 \times 10^{-3}(20 - a)^3$ [18]. As can be seen from Fig. 29 it takes a significant time for a crack to be initiated and to start propagating. To monitor the behaviour of the system, Poincaré maps for the system with the specified crack length were constructed and they are superimposed on the crack growth curve in Fig.29. As can be seen from this figure, while the crack length remains around 5.0 mm, the system response is stable period one motion. As the crack growths, the system response is changing and at a significantly larger crack length it becomes chaotic (for example, shown strange attractors have been identified at crack length of 12.09 and 13.03 mm). As the crack increases further, a period four response was obtained at a crack length of 13.4 mm.

To observe the behaviour of the system due to the growing fatigue crack transient Poincaré maps shown in Fig. 30 were also constructed. In contrast to the standard Poincaré maps, these pictures do not reflect the steady response of the system, but give a number of snap-shots showing different dynamic responses under a growing crack. As can be seen from Fig. 30(a) when the crack length is smaller than 10.7 mm ($\tau < 900{,}514$), the response of the system remains period one and the location of the attractor is changing with time and the growing crack in the direction shown by the arrow. This periodic regime eventually changes into a chaotic regime for a rather short time and then, when the crack length is about 12.0 mm, the system responds with period two motion for some time as shown in Fig.30(b). Again the arrows on the plot show how the location of the attractor is changing with the growing time. Later for $\tau \approx 915{,}662$ we can distinguish period three motion (Fig.30(c)), for $\tau \approx 928{,}100$ a period two motion (Fig.30(d)) and for $\tau \approx 936{,}903$ a period two motion again (Fig.30(e)). All these regimes are separated by a chaotic behaviour of the system and possibly other periodic regimes which the system experiences for rather short periods of time. Thus the system with constant crack length allows us to obtain the same responses (one for each particular crack length) as was observed for the system with a growing crack.

3.7 Conclusions

In this section we presented a study on the nonlinear dynamics caused by a fatigue crack growth in a beam specimen. Specifically we investigated the behaviour of the system with a cracked specimen under periodic and chaotic loading, where the stiffness of the specimen during crack opening and closing was modelled as piecewise nonlinear function. The dynamic interactions in this system can be strongly nonlinear resulting in aperiodic responses which have an influence on the fatigue crack growth.

As conventional fatigue testing machine cannot easily generate a flexible aperiodic loading, we designed a new fatigue testing device, which was tested and modelled. The conducted experimental study revealed that chaotic excitation is more damaging for the system than the harmonic one as for the same amount of energy pumped into a specimen, the fatigue life of the specimen subjected to the aperiodic loading was significantly reduced. We developed two mathematical models of

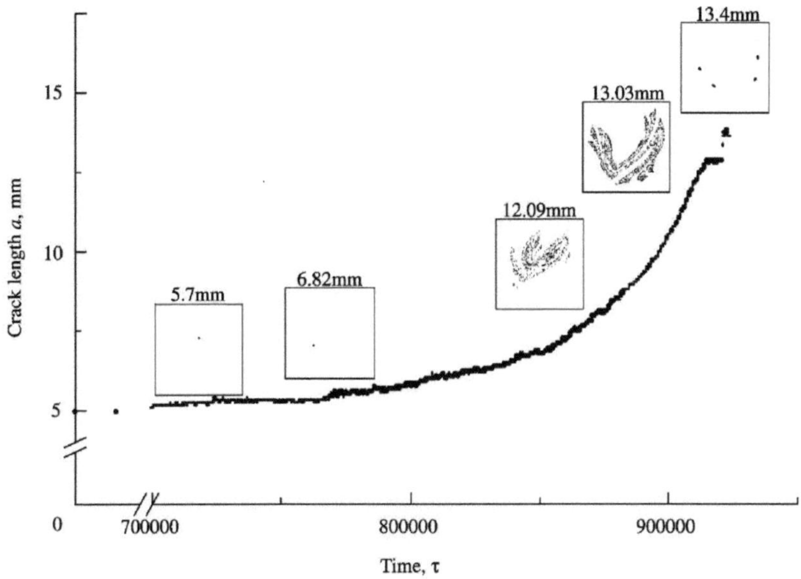

Fig. 29 Crack growth curve with superimposed Poincaré maps calculated for the system with the specified crack length obtained for $\xi_M = 0.014$, $\eta_M = 1.1475$, $k_s^{cl} = 14.327$, $k_s^{op} = 4.245 \times 10^{-3}(20-a)^3$. Adopted from [18]

the testing device to forecast a fatigue crack growth in beam samples. Specifically the experimental rig was modelled mathematically as a two mass and one mass systems. Studying one mass system, we found that when the crack size reaches a critical value, a strange attractor is born and this phenomenon can be used in structural health monitoring. The numerical results compared with the experiments show a good correspondence.

4 Regular and Chaotic Dynamics of a Rotor System with a Bearing Clearance [33, 32, 30, 31, 45, 34]

In rotor systems a non-smoothness may appear due to bearing clearances. Physically speaking this results in piecewise nonlinear stiffness characteristics, which can consequently lead to complex nonlinear behaviour including chaotic motion. The appearance of such phenomena implies a possibility of an intermittent contact between the components of the rotor system, which is difficult to predict. Rotor systems with bearing clearances have been studied in the past, and some investigations concentrated primarily on the Jeffcott rotors. For example, Choy and Padovan [5], Muszynska and Goldman [40], Childs [4] and Chu and Zhang [7, 8] examined rubbing in rotating machinery. Ehrich [11] investigated spontaneous sidebanding,

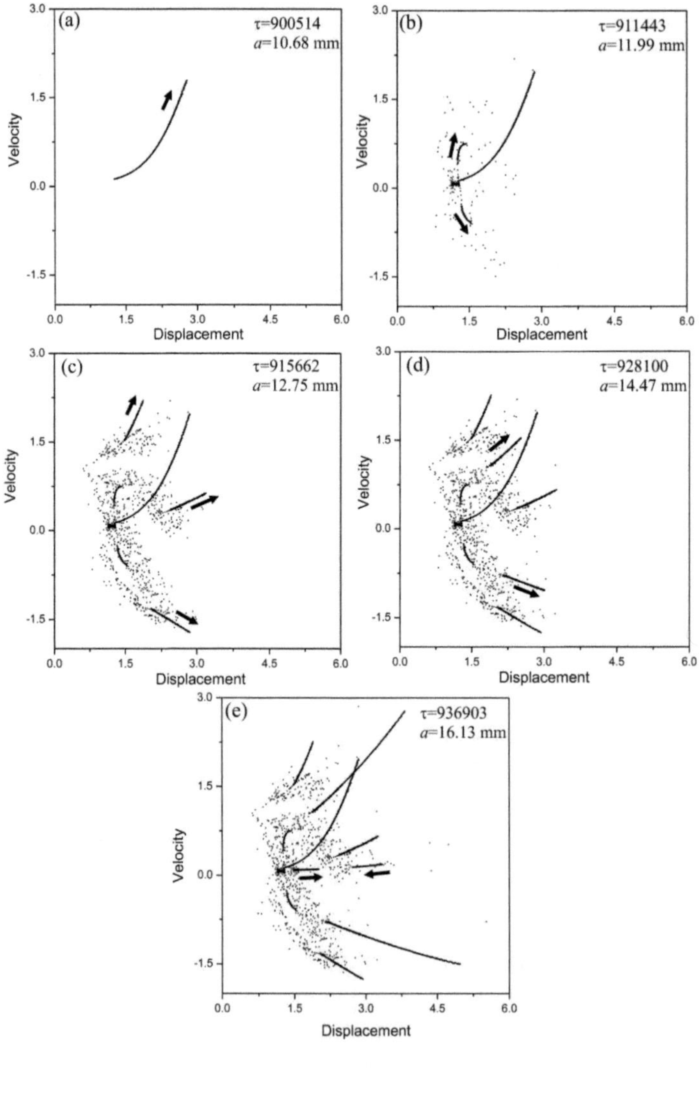

Fig. 30 Transient Poincaré maps [18] calculated for $\xi_M = 0.014$, $\eta_M = 1.1475$, $k_s^{cl} = 14.327$, $k_s^{op} = 4.245 \times 10^{-3}(20 - a)^3$, $a = 5 + 10^{-3}\exp(\tau/54734 - 7.8)$. The arrows indicate the direction of appearance of new points with the growing time

while Ganesan [21] looked at the stability analysis. Numerical investigations of the model of the Jeffcott rotor with a snubber ring presented in [33] shown the existence of multiple attractors and fractal basins of attraction. Influence of the preloading and viscous damping of the snubber ring was investigated in [31, 45], where it was shown how the preloading of the snubber ring could stabilize the dynamic responses. In this section we will discuss the work undertaken in the Centre for Applied Dynamics Research (CADR) at the University of Aberdeen. Specifically we will focuss on the finding published in [33, 32, 30, 31, 45, 34].

4.1 Physical Model and Equations of Motion [45]

The most up to date two-degrees-of-freedom model of the rotor system with a preloaded snubber ring developed by the CADR is shown in Fig. 31(a). During operation the rotor of mass M makes intermittent contact with the preloaded snubber ring and the excitation is provided by an out-of-balance rotating mass $m\rho$. It is assumed that contact is non-impulsive and that the friction between the snubber ring and the rotor is neglected. Since the mass ratio between the snubber ring and the mass of the rotor is small (for existing experimental rig it is equal to $\approx 1/17$) and the ratio between the stiffnesses of the snubber ring and the rotor is large, it is assumed that the snubber ring itself is massless. The stiffness and the viscous damping of the snubber ring are equal to k_s and c_s. The stiffness and the damping of the rotor are respectively k_r and c_r. The springs supporting the snubber ring are preloaded by Δ_x in horizontal and Δ_y in vertical directions respectively. There is a gap γ between the rotor and the snubber ring. Also in the initial position, the centre of the rotor is displaced from the centre of the snubber ring which is characterized by the eccentricity vector ε.

The presented model can operate in one of two following regimes: (a) no contact and (b) contact between the rotor and the snubber ring. In the latter regime, a preloading makes the dynamics of the system more complicated as the force acting from the snubber ring on the rotor depends on whether the displacement of the snubber ring exceeds the preloadings (in one or both directions) or not. Thus the following unique regimes can be distinguished as given in [45]:

I. No contact between the rotor and the snubber ring.
II. Contact between the rotor and the snubber ring, where the both displacements of the snubber ring are smaller than the preloadings, i.e. $|x_s| \leqslant \Delta_x$ and $|y_s| \leqslant \Delta_y$.
III. Contact between the rotor and the snubber ring, where the displacement of the snubber ring in the horizontal direction is larger than the preloading, $|x_s| > \Delta_x$, and in the vertical direction is smaller than preloading, $|y_s| \leqslant \Delta_y$.
IV. Contact between the rotor and the snubber ring, where the displacement of the snubber ring in the horizontal direction is smaller than the preloading, $|x_s| \leqslant \Delta_x$, and in the vertical direction is larger than preloading, $|y_s| > \Delta_y$.
V. Contact between the rotor and the snubber ring, where the displacements of the snubber ring are larger than the preloadings, i.e. $|x_s| > \Delta_x$ and $|y_s| > \Delta_y$.

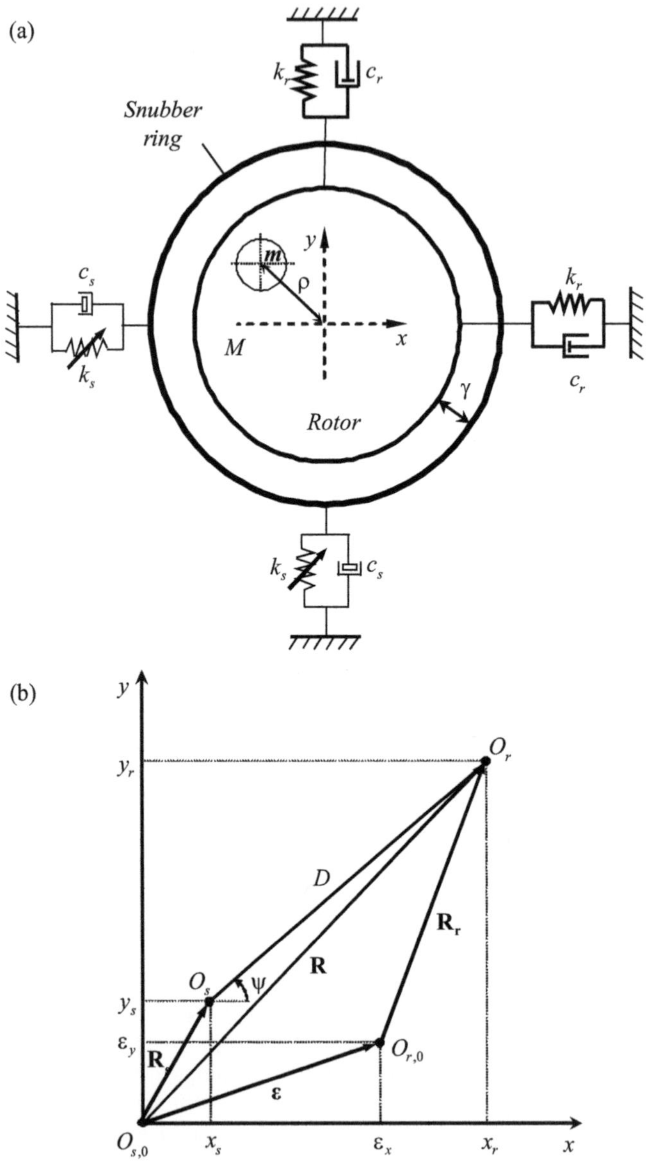

Fig. 31 **a** Physical model of the Jeffcott rotor with bearing clearance and **b** adopted co-ordinate system [45]

These four contact regimes were nesessary to be introduced as the results of our earlier theoretical work [32, 31] not correlating well with the experiments [30].

The co-ordinate system adopted in this section is presented in Fig. 31(b). The initial position of the rotor $O_{r,0}$ differs from the initial position of the snubber ring $O_{s,0}$ by the eccentricity vector $\varepsilon = (\varepsilon_x, \varepsilon_y)$. The vectors $\mathbf{R}_r = (x_r, y_r)$ and $\mathbf{R}_s = (x_s, y_s)$ show the current positions of the rotor and the snubber ring, and $D = \sqrt{(x_r - x_s)^2 + (y_r - y_s)^2}$ is the distance between the centres of the rotor and the snubber ring at any given time. $R = \sqrt{x_r^2 + y_r^2}$ is the radial displacement of the rotor.

For No contact regime the distance between the centres of the rotor and the snubber ring is smaller than the gap, γ, that is $R \leqslant \gamma$, and the equations of motion for the rotor and the snubber ring are

$$
\begin{aligned}
&M\ddot{x}_r + c_r\dot{x}_r + k_r(x_r - \varepsilon_x) = m\rho\,\Omega^2\cos(\varphi_0 + \Omega t),\\
&M\ddot{y}_r + c_r\dot{y}_r + k_r(y_r - \varepsilon_y) = m\rho\,\Omega^2\sin(\varphi_0 + \Omega t),\\
&k_s x_s + c_s\dot{x}_s = 0, \quad k_s y_s + c_s\dot{y}_s = 0,
\end{aligned}
\tag{52}
$$

where φ_0 is an initial phase shift and Ω is the shaft rotational velocity.

When $D = \gamma$, the rotor contacts the snubber ring and one of the specified contact regimes occurs, for which the equations of motion can be written as

$$
\begin{aligned}
&M\ddot{x}_r + c_r\dot{x}_r + k_r(x_r - \varepsilon_x) + F_{s_x} = m\rho\,\Omega^2\cos(\varphi_0 + \Omega t),\\
&M\ddot{y}_r + c_r\dot{y}_r + k_r(y_r - \varepsilon_y) + F_{s_y} = m\rho\,\Omega^2\sin(\varphi_0 + \Omega t),\\
&x_s = x_s(x_r, y_r), \quad y_s = y_s(x_r, y_r).
\end{aligned}
\tag{53}
$$

Here the restraining force in the snubber ring $\mathbf{F}_s = \left(F_{s_x}, F_{s_y}\right)$, shown in Fig. 32, varies for different contact regimes. The unknowns $x_s(x_r, y_r)$ and $y_s = y_s(x_r, y_r)$ give the current location of the snubber ring as a function of the current location of the rotor. To determine these functions, the principle of minimum elastic energy of the snubber ring is used.

It is worth noting that during any contact regime the distance between the centres of the rotor and the snubber ring remains constant, $D = \gamma$, despite of the fact that the force in the snubber ring, $\mathbf{F}_s$ may vary. In order to determine the moment when the contact is lost the force $\mathbf{F}_s$ should be monitored. If the projection of this force $\mathbf{F}_s$ on the vector $\mathbf{n}$ is positive (see Fig. 32), it is assumed that the rotor and the snubber ring are still in contact. Thus the contact is lost when

$$
\mathbf{n}\cdot\mathbf{F}_s \leqslant 0 \quad \text{or} \quad \cos(\varphi - \psi) \leqslant 0,
\tag{54}
$$

where

$$
\psi = \arccos\left(\frac{x_r - x_s}{\sqrt{(x_r - x_s)^2 + (y_r - y_s)^2}}\right), \qquad \varphi = \arctan\left(F_{s_y}/F_{s_x}\right).
$$

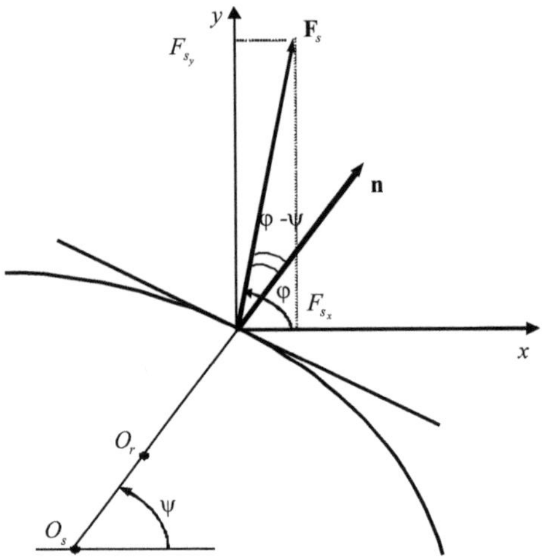

Fig. 32 Position of the force in the snubber ring $\mathbf{F}_s$ relative to the normal vector to the surface of contact [45]

Let us assume that the rotor and the snubber ring are in contact and the rotor moves the snubber ring in the direction as indicated by an arrow in Fig. 33. The forces $\mathbf{F}_1$, $\mathbf{F}_2$, $\mathbf{F}_3$ and $\mathbf{F}_4$ generated in the snubber ring as a result of the rotor and the snubber ring contact can be described in vector form as

$$\mathbf{F}_1 = \begin{cases} -\mathbf{j}\left[k_s\left(\Delta_y + y_s\right) + c_s\dot{y}_s\right], & y_s > -\Delta_y \\ 0 & y_s \leq -\Delta_y \end{cases} \tag{55}$$

$$\mathbf{F}_2 = \begin{cases} \mathbf{j}\left[k_s\left(\Delta_y - y_s\right) - c_s\dot{y}_s\right], & y_s < \Delta_y \\ 0 & y_s \geq \Delta_y \end{cases} \tag{56}$$

$$\mathbf{F}_3 = \begin{cases} -\mathbf{i}\left[k_s\left(\Delta_x + x_s\right) + c_s\dot{x}_s\right], & x_s > -\Delta_x \\ 0 & x_s \leq -\Delta_x \end{cases} \tag{57}$$

$$\mathbf{F}_4 = \begin{cases} \mathbf{i}\left[k_s\left(\Delta_x - x_s\right) - c_s\dot{x}_s\right], & x_s < \Delta_x \\ 0 & x_s \geq \Delta_x \end{cases} \tag{58}$$

Now the force in the snubber ring $\mathbf{F}_s$ can be conveniently defined as the resultant force taken with the opposite sign

$$\mathbf{F}_s = -\left(\mathbf{F}_1 + \mathbf{F}_2 + \mathbf{F}_3 + \mathbf{F}_4\right) \tag{59}$$

The formulae of this force for different regimes of operation are given in Table 4. These expressions for $\mathbf{F}_s$ should be substituted to Eq.(53) to obtain equations of motion for different contact regimes and to Eq. (54) to determine the moments when the contact is made or lost.

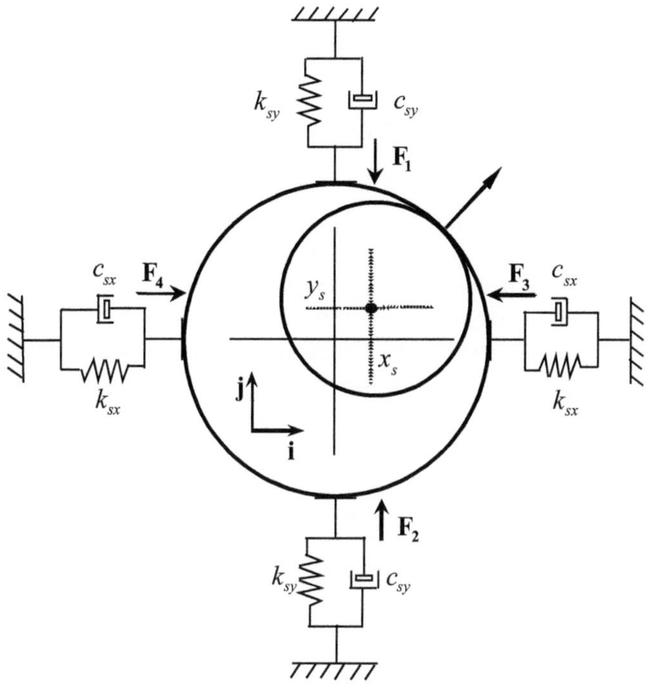

Fig. 33 Forces generated in the snubber ring [45]

Table 4 The force in the snubber ring $\mathbf{F}_s$ for various regimes

$\lvert x_s\rvert < \Delta_x$ $\lvert y_s\rvert < \Delta_y$	$\mathbf{i}\,[2k_s x_s + 2c_s\dot{x}_s] + \mathbf{j}\,[2k_s y_s + 2c_s\dot{y}_s]$
$\lvert x_s\rvert < \Delta_x$ $\lvert y_s\rvert \geq \Delta_y$	$\mathbf{i}\,[2k_s x_s + 2c_s\dot{x}_s] + \mathbf{j}\,[\mathrm{sign}\,(y_s)\,k_s\,(\Delta_y + \lvert y_s\rvert) + c_s\dot{y}_s]$
$\lvert x_s\rvert \geq \Delta_x$ $\lvert y_s\rvert < \Delta_y$	$\mathbf{i}\,[\mathrm{sign}\,(x_s)\,k_s\,(\Delta_x + \lvert x_s\rvert) + c_s\dot{x}_s] + \mathbf{j}\,[2k_s y_s + 2c_s\dot{y}_s]$
$\lvert x_s\rvert \geq \Delta_x$ $\lvert y_s\rvert \geq \Delta_y$	$\mathbf{i}\,[\mathrm{sign}\,(x_s)\,k_s\,(\Delta_x + \lvert x_s\rvert) + c_s\dot{x}_s] + \mathbf{j}\,[\mathrm{sign}\,(y_s)\,k_s\,(\Delta_y + \lvert y_s\rvert) + c_s\dot{y}_s]$

4.2 Location of the Snubber Ring and Contact Regimes

If the rotor and the snubber ring are in contact, the distance between their centres remains constant and equal to the gap, so $(x_r - x_s)^2 + (y_r - y_s)^2 = \gamma^2$. In order to find the location of the snubber ring centre when it moves and is in contact with the rotor, the following approach was adopted. It was assumed that the snubber ring being in contact with the rotor finds its position through minimum energy principle.

The potential (elastic) energy accumulated in the snubber ring at the position (x_s, y_s) is equal to the work, which is spent to bring the snubber ring to this position:

$$E = \int_{(s)} \mathbf{F}_s \cdot d\mathbf{s} = -\int_0^{x_s} (\mathbf{F_3} + \mathbf{F_4}) \cdot \mathbf{i} dx_s - \int_0^{y_s} (\mathbf{F_1} + \mathbf{F_2}) \cdot \mathbf{j}\, dy_s \tag{60}$$

Assuming that the dissipation of energy is negligible in comparison with the work of elastic forces, the expressions for the energy of the snubber ring take forms listed in Table 5. Consequently, the problem of finding the location of the snubber ring can be reduced to determining the minimum of the energy E with the constraint condition $D = \gamma$. This can be done using the Lagrange multipliers method by constructing the Lagrange function $L = E + \lambda\delta$, where λ is Lagrange multiplier, E is the elastic energy of the snubber ring, δ is the constraint function $\delta = (x_r - x_s)^2 + (y_r - y_s)^2 - \gamma^2$. As E and δ are the continuous and differentiable functions, the current position of the snubber ring (x_s and y_s) as a function of the of the current rotor position (x_r and y_r) can be determined from the conditions of the extremum existence:

$$\frac{\partial L}{\partial x_s} = 0, \qquad \frac{\partial L}{\partial y_s} = 0, \qquad \frac{\partial L}{\partial \lambda} = \delta = 0 \tag{61}$$

where $L = E + \lambda\left((x_r - x_s)^2 + (y_r - y_s)^2 - \gamma^2\right)$.

Then by minimising the energy E with respect to the constraint $(x_r - x_s)^2 + (y_r - y_s)^2 = \gamma^2$, the functions $x_s(x_r, y_r)$ and $y_s(x_r, y_r)$ can be obtained (see Table 6).

Table 5 The elastic energy of the snubber ring E for various regimes

$\lvert x_s\rvert < \Delta_x$ $\lvert y_s\rvert < \Delta_y$	$k_s x_s^2 + k_s y_s^2$
$\lvert x_s\rvert < \Delta_x$ $\lvert y_s\rvert \geq \Delta_y$	$0.5k_s\left((\lvert x_s\rvert + \Delta_x)^2 - 2\Delta_x^2\right) + k_s y_s^2$
$\lvert x_s\rvert \geq \Delta_x$ $\lvert y_s\rvert < \Delta_y$	$k_s x_s^2 + 0.5k_s\left((\lvert y_s\rvert + \Delta_y)^2 - 2\Delta_y^2\right)$
$\lvert x_s\rvert \geq \Delta_x$ $\lvert y_s\rvert \geq \Delta_y$	$0.5k_s\left((\lvert x_s\rvert + \Delta_x)^2 - 2\Delta_x^2\right) + 0.5k_s\left((\lvert y_s\rvert + \Delta_x)^2 - 2\Delta_y^2\right)$

As explained earlier the rotor can move either in or out of contact with the snubber ring. When in contact, the force acting between the rotor and the snubber ring depends on the strength of contact and four different regimes can occur. This can be clearly explained using (x_r, y_r) plane, where each regime is mapped into an associated region as shown in Fig. 34. The boundaries between regions I, II, III, IV and V are determined from the conditions listed at the top right quadrant in Fig. 34. The equations describing these boundaries were developed and are graphically depicted

Table 6 The functions $x_s(x_r, y_r)$ and $y_s(x_r, y_r)$ for various regimes of operation

$\lvert x_s\rvert < \Delta_x$ $\lvert y_s\rvert < \Delta_y$	$x_s = x_r\left(\sqrt{x_r^2+y_r^2}-\gamma\right)/\sqrt{x_r^2+y_r^2}$	$y_s = y_r\left(\sqrt{x_r^2+y_r^2}-\gamma\right)/\sqrt{x_r^2+y_r^2}$
$\lvert x_s\rvert < \Delta_x$ $\lvert y_s\rvert \geq \Delta_y$	$x_s = \operatorname{sign}(x_r)\left[\frac{2y_s(\lvert x_r\rvert+\Delta_x)}{y_r+y_s}-\Delta_x\right]$	$(y_r-y_s)^2\left((\lvert x_r\rvert+\Delta_x)^2+(y_r+y_s)^2\right)=\gamma^2(y_r+y_s)^2$
$\lvert x_s\rvert \geq \Delta_x$ $\lvert y_s\rvert < \Delta_y$	$(x_r-x_s)^2\left((\lvert y_r\rvert+\Delta_y)^2+(x_r+x_s)^2\right)=\gamma^2(x_r+x_s)^2$	$y_s = \operatorname{sign}(y_r)\left[\frac{2x_s(\lvert y_r\rvert+\Delta_y)}{x_r+x_s}-\Delta_y\right]$
$\lvert x_s\rvert \geq \Delta_x$ $\lvert y_s\rvert \geq \Delta_y$	$x_s = \operatorname{sign}(x_r)\left[\frac{(\lvert x_r\rvert+\Delta_x)(\tilde{R}-\gamma)}{\tilde{R}}-\Delta_x\right]$, $\tilde{R}=\sqrt{(\lvert x_r\rvert+\Delta_x)^2+(\lvert y_r\rvert+\Delta_y)^2}$	$y_s = \operatorname{sign}(y_r)\left[\frac{(\lvert y_r\rvert+\Delta_y)(\tilde{R}-\gamma)}{\tilde{R}}-\Delta_y\right]$

in Fig. 35, which shows one quadrant of (x_r, y_r) plane since the problem is symmetric. A detailed explanation how all these regions were determined is given below.

Region I, or No contact region is realised inside the circle

$$x_r^2+y_r^2=\gamma^2 \tag{62}$$

When the rotor makes a contact with the snubber ring the contact regime II begins. The boundaries of the corresponding region on (x_r, y_r) plane can be determined as follows. The inner boundary is described by Eq. (62). The outer boundaries are described by the conditions $|x_s| = \Delta_x$ and $|y_s| = \Delta_y$. Substituting x_s and y_s as the functions of x_r and y_r given in Table 6 for $|x_s| \leq \Delta_x$ and $|y_s| \leq \Delta_y$, in the first quadrant of (x_r, y_r) plane the outer boundaries are given by

$$y_r = \frac{x_r}{x_r-\Delta_x}\sqrt{\gamma^2-(x_r-\Delta_x)^2}, \tag{63}$$

$$x_r = \frac{y_r}{y_r-\Delta_y}\sqrt{\gamma^2-(y_r-\Delta_y)^2}. \tag{64}$$

If the rotor pushes the snubber ring strongly enough in the horizontal direction, i.e. its displacement, x_s becomes larger than the preloading Δ_x, the regime III begins. The inner (left) border for the region III is described by Eq. (63) and outer (upper) border again is governed by the conditions $|y_s| = \Delta_y$. As this border is simultaneously the inner (lower) border for the region V, the explicit expression for y_s as function of x_r and y_r (Table 6) for $|x_s| > \Delta_x$ and $|y_s| > \Delta_y$ is used

$$x_r = -\Delta_x + \frac{y_r+\Delta_y}{y_r-\Delta_y}\sqrt{\gamma^2-(y_r-\Delta_y)^2}. \tag{65}$$

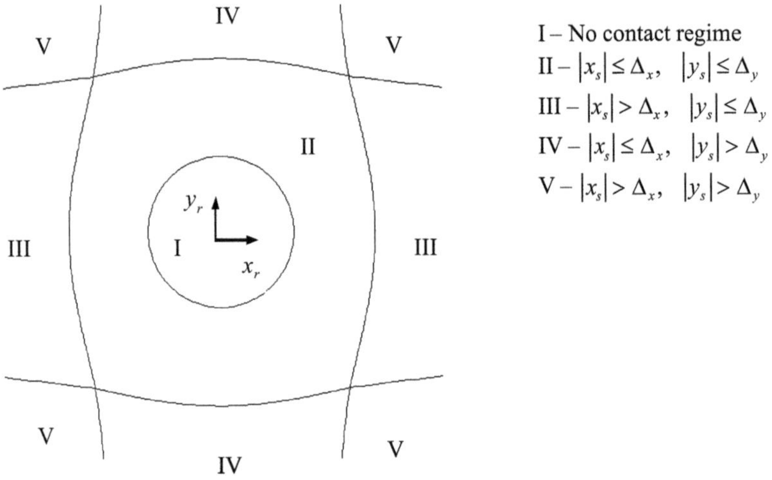

Fig. 34 Regions of operation for a rotor system with a symmetrically preloaded snubber ring in (x_r, y_r) plane [45]

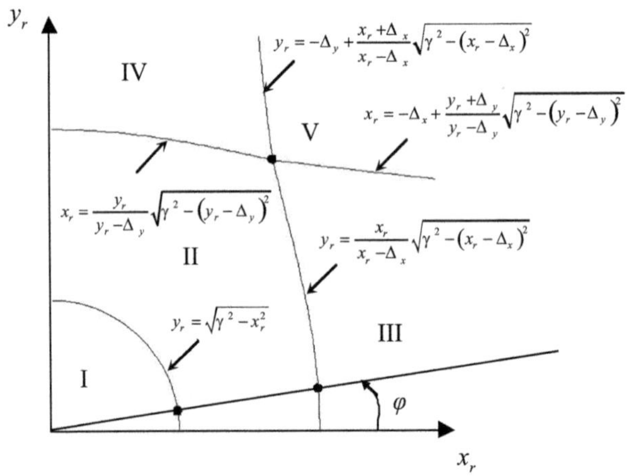

Fig. 35 Regions of operation and their boundaries for the first quadrant of (x_r, y_r) plane [45]

In the same way for the rotor moving in the region IV, one can obtain the inner (lower) border for the region IV as described by Eq. (64) and the outer (right) border as

$$y_r = -\Delta_y + \frac{x_r + \Delta_x}{x_r - \Delta_x}\sqrt{\gamma^2 - (x_r - \Delta_x)^2}. \tag{66}$$

Finally, the inner borders of the region V are described by Eq. (65) and (66).

4.3 Numerical Simulations

Numerical results presented in this section are to illustrate the use of the developed analytical formulas and to show the influence of the preloading on the dynamics of the rotor crossing different regions of operation. The calculations were performed in the dimensionless domain by defining the following dimensionless variables

$$\tau = \omega_n t, \quad \mathbf{f}_s = \frac{\mathbf{F}_s}{k_r \gamma}, \quad \hat{x}_r = \frac{x_r}{\gamma}, \quad \hat{y}_r = \frac{y_r}{\gamma}, \quad \hat{x}_s = \frac{x_s}{\gamma}, \quad \hat{y}_s = \frac{y_s}{\gamma}, \quad v_{x_s} = \hat{x}_s',$$

and parameters

$$\eta = \frac{\Omega}{\omega_n}, \quad \nu_1 = \frac{c_r}{2\sqrt{k_r M}}, \quad \nu_2 = \frac{c_s}{2\sqrt{k_r M}}, \quad \eta_m = \frac{m}{M}, \quad \hat{\rho} = \frac{\rho}{\gamma},$$
$$\hat{K} = \frac{k_s}{k_r}, \quad \hat{\varepsilon}_x = \frac{\varepsilon_x}{\gamma}, \quad \hat{\varepsilon}_y = \frac{\varepsilon_y}{\gamma}, \quad \hat{\Delta}_x = \frac{\Delta_x}{\gamma}, \quad \hat{\Delta}_y = \frac{\Delta_y}{\gamma}.$$

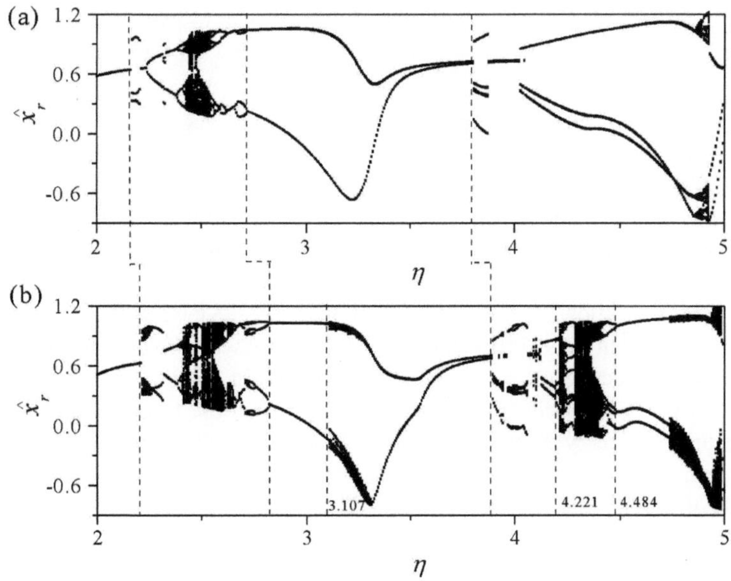

Fig. 36 Bifurcation diagrams showing the displacement of the rotor as function of frequency $\hat{x}_r(\eta)$ calculated for **a** $\hat{\Delta}_x = \hat{\Delta}_y = 0$; **b** $\hat{\Delta}_x = \hat{\Delta}_y = 0.1$; and $\nu_1 = 0.125$, $\nu_2 = 0.002$, $\hat{K} = 30$, $\eta_m = 0.0017$, $\hat{\rho} = 70$, $\hat{\varepsilon}_x = 0.9$ and $\hat{\varepsilon}_y = 0$. Adopted from [45]

In this study numerous bifurcation diagrams were constructed including the two shown in Fig. 36 for the displacement of the rotor $\hat{x}_r$ under varying the frequency ratio η for the unpreloaded (Fig. 36(a)) and the preloaded (Fig. 36(b)) cases. The control parameter η was set to the leftmost value 2. Starting with zero initial conditions first 300 cycles were disregarded to ensure that steady state solutions had been reached. The displacement $\hat{x}_r$ for the next 150 cycles was plotted. Then a small increment was added to the control parameter and the procedure was repeated until the control parameter reached the rightmost value $\eta = 5$. The parameters used in numerical computations were as follows: $\nu_1 = 0.125$, $\nu_2 = 0.002$, $\hat{K} = 30$, $\eta_m = 0.0017$, $\hat{\rho} = 70$, $\hat{\varepsilon}_x = 0.9$ and $\hat{\varepsilon}_y = 0$. The preloading was set to zero in both directions for Fig. 36(a), and $\hat{\Delta}_x = \hat{\Delta}_y = 0.1$ for Fig. 36(b). As can be clearly seen from Fig. 36 the preloading changes the bifurcation structure. Firstly, it shifts the bifurcation points towards higher frequencies; dash lines in Fig. 36 point out such behaviour. For instance, the period one observed in the beginning of the diagram bifurcates at $\eta = 2.165$ for unpreloaded and at $\eta = 2.213$ for preloaded case. The bifurcation of period four motion into period two motion moves from $\eta = 2.717$ to 2.824, and the period two bifurcates into period four at $\eta = 3.803$ and 3.893 for unpreloaded and preloaded cases respectively. Secondly, the introduction of the preloading changes the character of bifurcations. For example, the period one motion marked by the leftmost dash line, bifurcates into period three motion (see Poincaré map in Fig. 37(a)) for the unpreloaded case and into quasi-periodic motion (Fig. 37(b)) for the preloaded case. Also the preloading changes the structure of the chaotic attractor which be seen from Poincaré maps shown in Fig. 37(c) and 37(d) calculated at $\eta = 2.442$ for the unpreloaded and the preloaded cases. Finally and most importantly the preloading introduces new bifurcations and new regimes. For example, an additional bifurcation of the period two motion into quasi-periodic motion appears at $\eta \approx 3.107$ for the preloaded case.

The changes in dynamical behaviour are even more visible in $(\hat{x}_s, \hat{v}_{x_s})$ plane. The comparisons between trajectories of the snubber ring on the phase plane $(\hat{x}_s, \hat{v}_{x_s})$ for the system with and without preloading are presented in Fig. 38. The dynamics of the snubber ring is shown in Figs. 38(a) and 38(c) for the case without preloading ($\hat{\Delta}_x = \hat{\Delta}_y = 0$), and in Figs. 38(b) and 38(d) for the case with preloading ($\hat{\Delta}_x = \hat{\Delta}_y =$ 0.05 and 0.03 respectively). As can be seen from the Fig. 38 in both cases, velocity of snubber ring $\hat{v}_{x_s}$ experiences a jump at $\hat{x}_s = 0$, when the rotor hits the snubber ring. For the systems with preloading there is an additional jump of velocity $\hat{v}_{x_s}$, which appears at $\hat{x}_s = \hat{\Delta}_x$. It can be also observed that the preloading reduces the amplitude of the snubber ring vibrations.

4.4 Experimental Verification

The results of the mathematical modelling were verified on a purpose designed experimental rig. Figure 39(a) shows the rotor rig which comprises of essentially two main parts, a rigid rotor (1), visco-elastically supported by four flexural rods (2) and excited by the out-of-balance mass (3), and a snubber ring (4) also elastically supported using four compression springs. The rotor assembly consists of a steel rotor, running in two angular contact bearings. Holes (5) were drilled and tapped in both

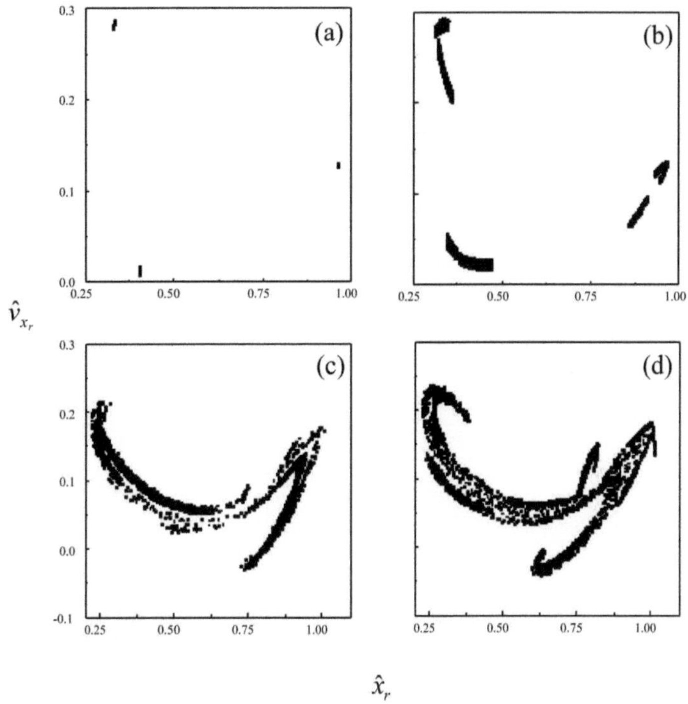

Fig. 37 Poincaré maps $\hat{v}_{x_r}(\hat{x}_r)$ [45] calculated for **a, c** $\hat{\Delta}_x = \hat{\Delta}_y = 0$; **b,d** $\hat{\Delta}_x = \hat{\Delta}_y = 0.1$; and $v_1 = 0.125$, $v_2 = 0.002$, $\hat{K} = 30$, $\eta_m = 0.0017$, $\hat{\rho} = 70$, $\hat{\varepsilon}_x = 0.9$ and $\hat{\varepsilon}_y = 0$; and **a** $\eta = 2.223$, **b** $\eta = 2.181$ and **c, d** $\eta = 2.442$

inner sleeves for the addition of imbalance weights. A pair of viscous dampers (6) was attached to the rotor to provide the system with heavier damping.

Four flexural rods (2) are clamped symmetrically at one end to the outer bearing housing and at the other to a large support block. The support block (7) is in turn bolted to a large cast iron bed. The stiffness of the snubber ring is provided by four compression springs (8), of much greater stiffness than that of the flexural rods. The rotor ran inside the ring, with a radial clearance between the ring (4) and the outer bearing housing (1). Two different outer rings were used in the experiments one with a 0.5 mm and another with 0.75 mm gap.

The rotor is driven by a variable speed DC motor (9). The shaft speed monitoring disc has a notch cut into it, which is aligned with the imbalance mass. As the notch passes a light-emitting-diode optoswitch, a once-per-revolution phase signal is obtained. The displacements of the rotor system are monitored by non-contacting eddy current probes. The displacement and forcing frequency signals were collected by a Labview data acquisition system with a custom written program controlling the rate of sampling, the number of samples, calibration and computation of the rotational frequency. The relative velocities of the rotor and the snubber ring $\dot{x}_r$, $\dot{y}_r$, $\dot{x}_s$ and $\dot{y}_s$

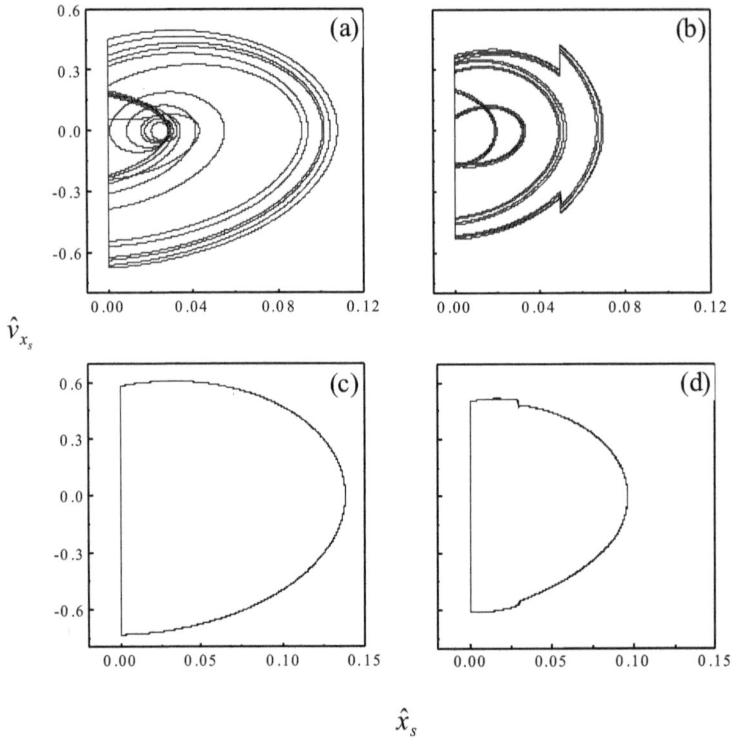

Fig. 38 Phase portraits $\hat{v}_{x_s}(\hat{x}_s)$ calculated for **a** $\hat{\Delta}_x = \hat{\Delta}_y = 0$; **b** $\hat{\Delta}_x = \hat{\Delta}_y = 0.05$; and $v_1 = 0.06$, $v_2 = 0.002$, $\eta = 2.5$, $\hat{K} = 30$, $\eta_m = 0.00289$, $\hat{\rho} = 70$, $\hat{\varepsilon}_x = 0.4$ and $\hat{\varepsilon}_y = 0.5$; and **c** $\hat{\Delta}_x = \hat{\Delta}_y = 0$; **d** $\hat{\Delta}_x = \hat{\Delta}_y = 0.03$ and $v_1 = 0.125$, $v_2 = 0.002$, $\eta = 3.0$, $\hat{K} = 30$, $\eta_m = 0.0017$, $\hat{\rho} = 70$, $\hat{\varepsilon}_x = 0.9$ and $\hat{\varepsilon}_y = 0$. Adopted from [45]

were calculated from the displacements measured by the eddy current probes. The data was collated on the computer, where it was scaled, plotted and analysed in the form of Poincaré maps and bifurcation diagrams.

Now a sample of extensive experimental studies [30] conducted to verify the mathematical model of Jeffcott rotor system with a preloaded snubber ring [45] is presented here. The following values of the system parameters were chosen: the rotor mass and mass of the out-of-balance were $M = 9.7$ kg and $m = 0.028$ kg respectively. The combined stiffness of the rods supporting the rotor was $k_{r_x} = k_{r_y} = 79$ kN/m, which yields a natural frequency of 14.4 Hz. The snubber ring stiffness was $k_s = 2354$ kN/m and the equivalent viscous damping from the rods and the dampers in the horizontal and vertical directions was the same and equal to $c_{r_x} = c_{r_y} = 1050$ N/s. The out-of-balance radius, was $\rho = 35$ mm.

When constructing the bifurcation diagrams, the forcing frequency (the shaft rotational speed) was varied between 7 and 30 Hz and for some tests up to 50 Hz to

Fig. 39 Photograph of the experimental Jeffcott rotor [34]

examine the global bifurcations. The system responses were investigated by collecting data with the forcing frequency steps of around 1 Hz. The continuation method was applied, so for each frequency the initial conditions were taken from the previous examined frequency discarding about 400 cycles in order to ensure the steady-state behaviour is reached.

The bifurcation diagrams presented in Fig. 40 give a comparison of the theoretical (Fig. 40(a)) and the experimental (Fig. 40(b)) responses showing a good degree of correspondence. In both figures a wide range of chaotic regimes is observed, separated by period one, two and three regimes. Here chaotic attractors shown as Poincaré maps were obtained for two different values of the frequency: $f = 30.1$ Hz, and $f = 37.1$ Hz keeping the remaining parameters constant: $k_s = 2354$ kN/m, $k_{r_x} = k_{r_y} = 79$ kN/m, $c_{r_x} = c_{r_y} = 105$ kg/s, $c_s = 3.5$ kg/s, $M = 9.7$ kg, $m = 0.028$ kg, $\rho = 35$ mm, $\gamma = 0.5$ mm, $\varepsilon_x = 0.5$ mm, $\varepsilon_y = 0$ mm and $\Delta_x = \Delta_y = 0.1$ mm. It is apparent that the theoretical and experimental attractors are topologically similar.

In the next presented experiment the eccentricity ratios were set up as $\varepsilon_x = 0.45$ mm and $\varepsilon_y = 0.05$ mm. The bifurcation diagrams constructed theoretically and experimentally for this case are presented in Fig. 41. Because only period one motion regime exists in the interval $f \in (30,50)$ Hz, the maximum forcing frequency for these diagrams was reduced to 30 Hz. Comparing with the previously shown diagrams of Fig. 40, here the eccentricity change leads the transition from period one motion to chaos through period doubling bifurcations. The experimental result of Fig. 41(b) follows all the basic bifurcations observed theoretically such as the period doubling bifurcation at $f \approx 14.74$ Hz and the boundary crisis at $f \approx 26.1$ Hz. Phase portraits for the periodic and chaotic cross-sections were plotted for $f = 17.6$ Hz and

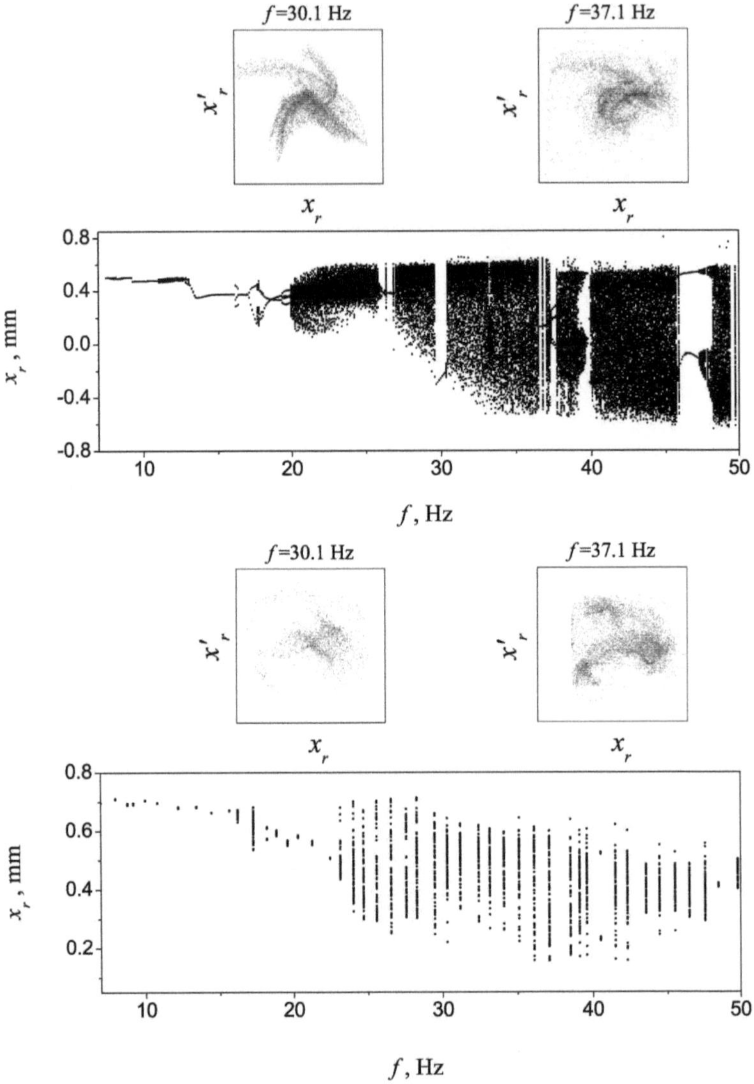

Fig. 40 Bifurcation diagrams for the forcing frequency [30] **a** theoretical and **b** experimental; $k_s = 2354$ kN/m, $k_{r_x} = k_{r_y} = 79$ kN/m, $c_{r_x} = c_{r_y} = 105$ kg/s, $c_s = 3.5$ kg/s, $M = 9.7$ kg, $m = 0.028$ kg, $\rho = 35$ mm, $\gamma = 0.5$ mm, $\varepsilon_x = 0.5$ mm, $\varepsilon_y = 0$ mm and $\Delta_x = \Delta_y = 0.1$ mm

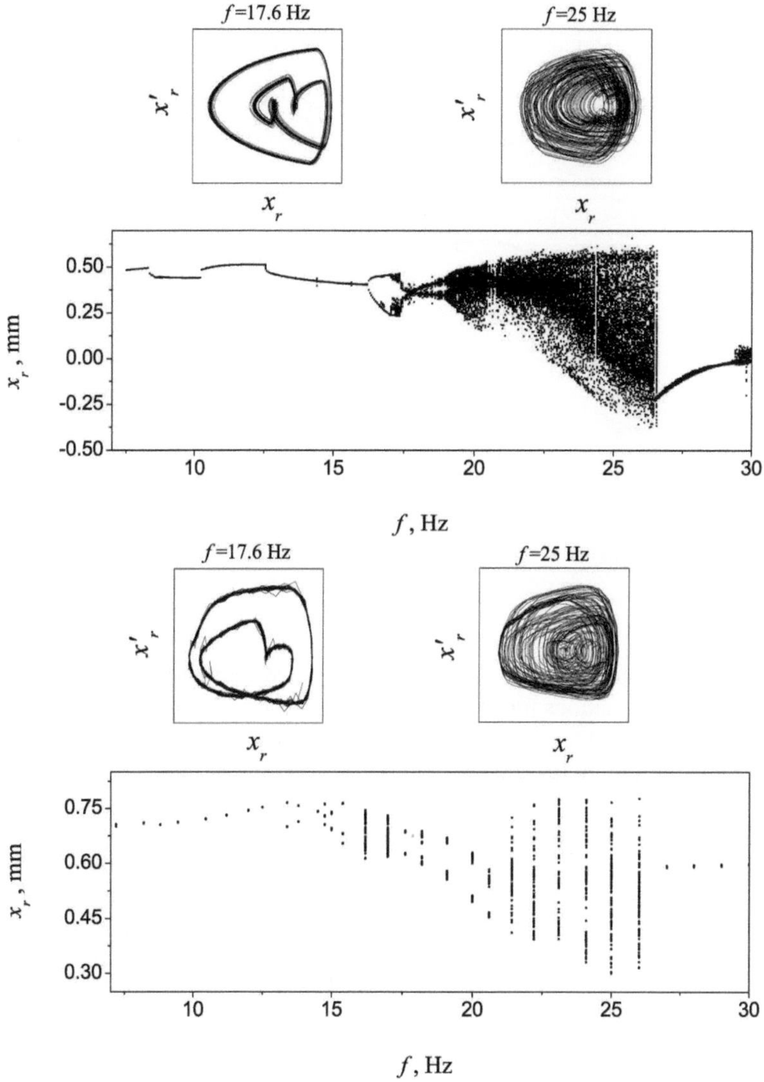

Fig. 41 **a** Theoretical and **b** experimental bifurcation diagrams [34] for the forcing frequency where $k_s = 2354$ kN/m, $k_{r_x} = k_{r_y} = 79$ kN/m, $c_{r_x} = c_{r_y} = 105$ kg/s, $c_s = 3.5$ kg/s, $M = 9.7$ kg, $m = 0.028$ kg, $\rho = 35$ mm, $\gamma = 0.5$ mm, $\varepsilon_x = 0.45$ mm, $\varepsilon_y = 0.05$ mm and $\Delta_x = \Delta_y = 0.04$ mm

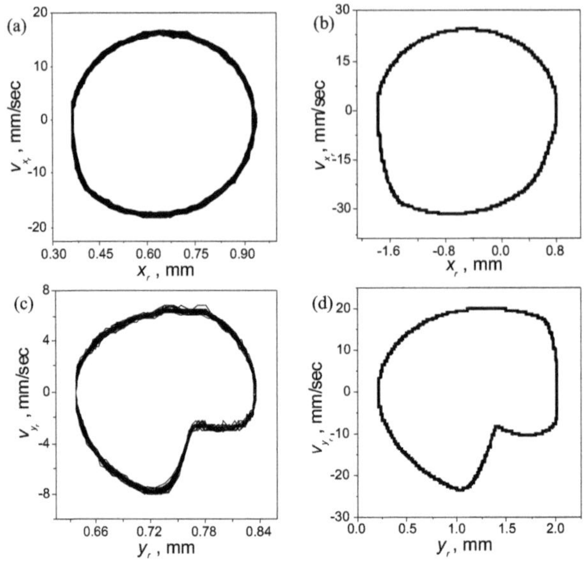

Fig. 42 **a**, **c** Experimental and **b**, **d** theoretical phase portraits for periodic behaviour of the Jeffcott rotor system in $x-$ and $y-$directions for the cross-section $f = 13.1$ Hz. Adopted from [34]

for $f = 25$ Hz showing also a good correspondence. Furthermore, the phase portraits in the x and y directions for periodic trajectories were examined at $f = 13.1$ Hz and are shown in Fig. 42. The experimental phase portraits in the $x-$ and $y-$directions are plotted in Fig. 42(a) and (c), and the corresponding to them theoretical graphs are depicted in Fig. 42(b) and (d). As can be seen, again the theoretical predictions correspond well to the experimental results.

Examining the system responses for different values of the forcing frequency shown in Figs. 40 and 41, it is clear from the bifurcation diagrams and phase planes that periodic regimes dominate at low and at high frequencies. The periodic regimes for the low frequency are caused by insufficient excitation of the rotor and as a result either there is no contact between the rotor and the snubber ring or just one contact per period. As the forcing frequency is increased, and the amplitude of oscillations rises the impacts between the rotor and the ring become stronger and the system generates chaotic vibration. The periodic regimes observed for the high frequencies have a wide range and lower amplitude of vibration than chaotic ones. In summary, the experimental results correspond well with the theoretical predictions.

4.5 Conclusions

In this section a rotor system with bearing clearance was investigated in order to gain understanding of its complex dynamic responses. First we discussed the mathematical model developed at the Centre for Applied Dynamics Research. The model which is a Jeffcott rotor with an elastic preloaded snubber ring was formulated as a piecewise nonlinear system operating in one of five distinct regimes of operation: one of no contact and four of different types of contact. The boundaries between those regions were determined analytically and the equations of motion were formulated.

The undertaken numerical simulation shows significant differences in the system behaviour for the cases without and with preloading of the snubber ring. Specifically, the constructed bifurcation diagrams show the tendency to shift the bifurcation points towards higher frequencies for the preloaded cases. It was also found that the character of the bifurcations changes for the cases with preloading, and new bifurcations and regimes were observed in these cases. In particular, chaotic vibrations within a wider range of excitation frequency were obtained. In addition, we found that the periodic response tends to occur at the lower and higher frequencies.

The theoretical predictions were verified experimentally on the rig developed in Aberdeen. A good degree of correlation was found for the wide range of system parameters.

5 Conclusions

In this chapter we studied an important phenomenon of non-smoothness occurring in dynamical systems, and very common in engineering applications. Mathematically, such systems can be described as piecewise smooth as suggested in Fig. 1. Hence, their global solutions can be obtained by stitching local solutions, which are easy to determine by standard methods. For example, a global solution for a piecewise linear smooth dynamical system often leads to a set of nonlinear algebraic equations.

Three mechanical engineering systems were modelled and analysed in order to illustrate an approach which has been developed by the Centre for Applied Dynamics Research at Aberdeen to study non-smooth dynamical systems. Firstly, the vibro-impact moling device was investigated in order to understand how to maximise its progression rates. Applying the developed methodology, in this case periodic trajectories were reconstructed as they go through three linear subspaces (*No contact*, *Contact with progression* and *Contact without progression*), and using combination of analytical and numerical methods the optimal range of the system parameters was identified. The conducted analysis revealed that the best progression rates are achieved for low frequencies and the ratio between the dynamic and static forces around 2.

In the second considered application the influence of opening and closing of a fatigue crack on the entire system dynamics was modelled and analysed. Specifically, we were interested in the aperiodic behaviour and therefore a novel apparatus to induce aperiodic loading to a beam specimen with a fatigue crack was developed. It was shown experimentally that fatigue life can be reduced few times if the sample is loaded aperiodically. The experimental rig was modelled as two and one mass system depending whether the contact between the loadings and the sample is lost or not. The analysis of the developed mathematical model shown that as a crack grows linearly before reaching its critical value, the response of the system remains periodic. When its size exceeds the critical value, the system behaviour becomes chaotic and then the crack growth increases exponentially. This phenomenon can be used in structural health monitoring.

Finally, we investigated a problem from rotordynamics, where nonlinear interactions between the rotor and the snubber ring were studied. We discussed the most up-to-date model and the experimental rig developed to understand the complex behaviour of this system. The rotor system can operate in one of five regimes, which were determined analytically. The influence of the preloading of the snubber ring on the system behaviour was investigated and the range of the system parameters where chaotic vibrations occur was identified. The results obtained from the developed mathematical model confronted with the experiments shown a good degree of correlation.

Acknowledgements. The financial support from EPSRC, Rolls-Royce, Scottish Enterprise and the European Commission is gratefully acknowledged. Also we would like to thank our former PhD students and collaborators including E.V. Karpenko (Aberdeen), W.F. Deans (Aberdeen), N.Jaksic (Ljubljana), M.Boltežar (Ljubljana), M.P. Cartmell (Glasgow), R.D. Neilson (Aberdeen), A.M. Krivtsov (St.Petersburg), C.Grebogi (Sao Paulo), K.C. Woo (Aberdeen), A.A. Rodger (Aberdeen), B. de Kraker (Eindhoven), V.M.T. Sin (Aberdeen), K. Li (Michigan).

References

1. Abraham, O.N.L., Brandon, J.: A piecewise linear approach for the modelling of a breathing crack. In: 17th International Seminar on Modal Analysis, Leuven, Belgium, pp. 417–431 (1992)
2. Actis, R.L., Dimarogonas, A.D.: Non-linear effects due to closing cracks in vibration beams. In: 12th ASME Conference on Mechanical Engineering, Vibration and Noise, Montreal, Canada, pp. 99–104 (1989)
3. Anderson, T.L.: Fracture mechanics - Fundamental and applications. CRC Press, Boca Raton (1994)
4. Childs, D.W.: Fractional-frequency rotor motion due to nonsymmetric clearance effects. Trans. ASME, Journal of Engineering for Power 104(3), 533–541 (1982)
5. Choy, F.K., Padovan, J.: Non-linear transient analysis of rotor-casing rub events. Journal of Sound and Vibration 113(3), 529–545 (1987)
6. Chu, Y.C., Shen, M.-M.H.: Analysis of forced bilinear oscillators and the application to cracked beam dynamics. AIAA Journal 30, 2512–2519 (1992)
7. Chu, F., Zhang, Z.: Periodic, quasi-periodic and chaotic vibrations of a rub-impact rotor system supported on oil film bearing. International Journal of Engineering Sciences 35, 963–973 (1997)
8. Chu, F., Zhang, Z.: Bifurcation and chaos in a rub-impact jeffcott rotor system. J. Sound Vibr. 210, 1–18 (1998)
9. Collins, K.R., Plaut, R.H., Wauer, J.: Detection of cracks in rotating Timoshenko shafts using axial impulses. Journal of Vibration, Acoustics, Stress, and Reliability in Design, Trans. ASME 113, 74–78 (1991)

10. Collins, K.R., Plaut, R.H., Wauer, J.: Free and forced longitudinal vibrations of a cantilvered bar with a crack. Journal of Vibration, Acoustics, Stress, and Reliability in Design, Trans. ASME 114, 171–177 (1992)
11. Ehrich, F.F.: Spontaneous sidebanding in high-speed rotordynamics. Trans. ASME, J. Vibr. Acoust. 114, 498–505 (1992)
12. Ebrahimi, S., Eberhard, P.: Rigid-elastic modeling of meshing gear wheels in multibody systems. Multibody System Dynamics 16(1), 55–71 (2006)
13. Feeny, B.: A non-smooth Coulomb friction oscillator. Physica D 59, 25–38 (1992)
14. Filippov, A.F.: Differential equations with discontinuous right-hand side. American Mathematical Society Translations 42(2), 199–231 (1978)
15. Foong, C.H.: Influence of fatigue crack growth on the dynamics of engineering components and structures, PhD thesis, University of Aberdeen (2004)
16. Foong, C.H., Wiercigroch, M., Deans, W.F.: Novel dynamic fatigue-testing device: Design and measurements. Measurement Science and Technology 17, 2218–2226 (2006)
17. Foong, C.H., Jakšić, N., Wiercigroch, M., Boltežar, M.: Parameter identification of the fatigue-testing rig. International Journal of Mechanical Sciences 50, 1142–1152 (2008)
18. Foong, C.H., Pavlovskaia, E., Wiercigroch, M., Deans, W.F.: Chaos caused by fatigue crack growth. Chaos, Solitons and Fractals 16, 651–659 (2003)
19. Foong, C.H., Wiercigroch, M., Pavlovskaia, E., Deans, W.F.: Nonlinear vibration caused by fatigue. Journal of Sound and Vibration 303, 58–77 (2007)
20. Friswell, M.I., Penny, J.E.T.: A Simple Nonlinear Model of a Cracked Beam. In: 10th International Modal Analysis Conference, San Diego, California, USA, pp. 516–521 (1992)
21. Ganesan, R.: Dynamic response and stability of a rotor-support system with non-symmetric bearing clearances. Mechanism Machine Theory 31, 781–798 (1996)
22. Gonsalves, H.D., Neilson, R.D., Barr, A.D.S.: A study of the response of a discontinuously nonlinear rotor system. Nonlinear Dynamics 7, 451–470 (1995)
23. Grabec, I.: Chaotic dynamics of the cutting process. International Journal of Machine Tools and Manufacture 28, 19–32 (1988)
24. Gudmundson, P.: The dynamic behaviour of slender structures with cross-sectional cracks. Journal of the Mechanics and Physics of Solids 31, 329–345 (1983)
25. Guinea, G.V., Pastor, J.Y., Elices, M.: Stress intensity factor, compliance and CMOD for a general three-point-bend beam. International Journal of Fracture 89, 103–116 (1998)
26. Hsu, C.S.: Cell-To-Cell Mapping: A Method of Global Analysis for Nonlinear Systems. Springer, Heidelberg (1987)
27. Ibrahim, A., Ismail, F., Martin, H.R.: Modelling of the dynamics of a continuous beam including nonlinear fatigue crack. International Journal of Analytical and Experimental Modal Analysis 2, 76–82 (1987)
28. Ismail, F., Ibrahim, A., Martin, H.R.: Identification of fatigue cracks from vibration testing. Journal of Sound and Vibration 140, 305–317 (1990)
29. Kahraman, A., Singh, R.: Non-linear dynamics of a spur gear pair. Journal of Sound and Vibration 142(1), 49–75 (1990)
30. Karpenko, E.V.: Nonlinear dynamics of a Jeffcott rotor with imperfections, PhD Thesis, University of Aberdeen (2003)
31. Karpenko, E.V., Pavlovskaia, E.E., Wiercigroch, M.: Bifurcation analysis of a preloaded Jeffcott rotor. Chaos, Solitons and Fractals 15, 407–416 (2003)
32. Karpenko, E.V., Wiercigroch, M., Cartmell, M.P.: Regular and chaotic dynamics of a discontinuously nonlinear rotor system. Chaos, Solitons and Fractals 13, 1231–1242 (2002)
33. Karpenko, E., Wiercigroch, M., Pavlovskaia, E.E., Cartmell, M.P.: Piecewise approximate solutions for a Jeffcott rotor with a snubber ring. International Journal of Mechanical Sciences 44, 475–488 (2002)

34. Karpenko, E.V., Wiercigroch, M., Pavlovskaia, E.E., Neilson, R.D.: Experimental verification of Jeffcott rotor model with preloaded snubber ring. Journal of Sound and Vibration 298, 907–917 (2006)
35. Krivtsov, A.M., Wiercigroch, M.: Dry friction model of percussive drilling. Meccanica 34, 425–435 (1999)
36. Krivtsov, A.M., Wiercigroch, M.: Penetration rate prediction for percussive drilling via dry friction model. Chaos, Solitons and Fractals 11, 2479–2485 (2000)
37. Kunze, M.: Non-Smooth Dynamical Systems. Springer, New York (2000)
38. Litak, G., Friswell, M.I.: Dynamics of a gear system with faults in meshing stiffness. Nonlinear Dynamics 41, 415–421 (2005)
39. Lok, H.P., Neilson, R.D., Rodger, A.A.: Computer-based model of vibro-impact driving. In: Proceedings of ASME DETC: Symposium on Nonlinear Dynamics in Engineering Systems, Las Vegas (1999)
40. Muszynska, A., Goldman, P.: Chaotic responses of unbalanced rotor/bearing/stator systems with looseness or rubs. Chaos, Solitons and Fractals 5(9), 1683–1704 (1995)
41. Natsiavas, S.: Periodic response and stability of oscillators with symmetric trilinear restoring force. Journal of Sound and Vibration 134(2), 313–331 (1989)
42. Pavlovskaia, E., Wiercigroch, M.: Periodic solutions finder for vibro-impact oscillator with a drift. Journal of Sound and Vibration 267, 893–911 (2003)
43. Pavlovskaia, E., Wiercigroch, M.: Analytical drift reconstruction for impact oscillator with drift. Chaos, Solitons and Fractals 19(1), 151–161 (2004)
44. Pavlovskaia, E.E., Wiercigroch, M.: Low dimensional maps for piecewise smooth oscillators. Journal of Sound and Vibration 305(4-5), 750–771 (2007)
45. Pavlovskaia, E., Karpenko, E.V., Wiercigroch, M.: Nonlinear dynamic interactions of a Jeffcott rotor with a preloaded snubber ring. Journal of Sound and Vibration 276, 361–379 (2004)
46. Pavlovskaia, E., Wiercigroch, M., Grebogi, C.: Modeling of an impact system with a drift. Physical Review E 64, 56224 (2001)
47. Pavlovskaia, E., Wiercigroch, M., Grebogi, C.: Two dimensional map for impact oscillator with drift. Physical Review E 70, 36201 (2004)
48. Pavlovskaia, E.E., Wiercigroch, M., Woo, K.-C., Rodger, A.A.: Modelling of ground moling dynamics by an impact oscillator with a frictional slider. Meccanica 38, 85–97 (2003)
49. Peterka, F., Vacik, J.: Transition to chaotic motion in mechanical systems with impacts. Journal of Sound and Vibration 154(1), 95–115 (1992)
50. Rodger, A.A., Littlejohn, G.S.: A study of vibratory driving in granular soils. Geotechnique 30, 269–293 (1980)
51. Shaw, S.W., Holmes, P.J.: A periodically forced piecewise linear oscillator. Journal of Sound and Vibration 90(1), 129–155 (1983)
52. Shen, M.-H.H., Chu, Y.C.: Vibrations of beams with a fatigue crack. Computers & Structures 45, 79–93 (1992)
53. Sin, V.M.T., Wiercigroch, M.: A symmetrically piecewise linear oscillator: design and measurement. Proceeding of the Insitute of Mechanical Engineers Part C 213, 241–249 (1999)
54. Spektor, M.: Principles of Soil Tool Interaction. Journal of Terramechanics 18, 51–65 (1981)
55. Wiercigroch, M.: Comments on the study of a harmonically excited linear oscillator with a Coulomb damper. Journal of Sound and Vibration 167, 560–563 (1993)
56. Wiercigroch, M.: Dynamics of discrete mechanical system with discontinuities, p. 127. Silesian Universtity Press, Gliwice (in Polish)

57. Wiercigroch, M.: Chaotic vibrations of a simple model of the machine tool-cutting process system. Journal of Vibration and Acoustics Trans. ASME 119(3), 468–475 (1997)
58. Wiercigroch, M.: Modelling of dynamical systems with motion dependent discontinuities. Chaos, Solitons and Fractals 11(15), 2429–2442 (2000)
59. Wiercigroch, M.: Applied nonlinear dynamics of non-smooth mechanical systems. Journal of the Brazilian Society of Mechanical Sciences and Engineering XXVIII (4), 519–526 (2006)
60. Wiercigroch, M., de Kraker, B. (eds.): Applied nonlinear dynamics and chaos of mechanical systems with discontinuities. Nonlinear Science Series A. World Scientific, Singapore (2000)
61. Wiercigroch, M., Pavlovskaia, E.E.: Nonlinear dynamics of vibro-impact systems: theory and experiments. In: Glasgow, Cartmell, M. (eds.) Proceedings of the 5th International Conference on Modern Practice in Stress and Vibration Analysis, September 9-11, pp. 513–520 (2003)
62. Wiercigroch, M., Sin, V.M.T.: Experimental study of base excited symmetrically piecewise linear oscillator. Journal of Applied Mechanics, Trans. ASME 65, 657–663 (1998)
63. Wiercigroch, M., Sin, W.T.V., Li, K.: Measurement of chaotic vibration in symmetrically piecewise linear oscillator. Chaos, Solitons and Fractals 9(1-2), 209–220 (1998)
64. Wiesenfeld, K., Tufillaro, N.B.: Suppression of period doubling in the dynamics of a bouncing ball. Physica D 26, 321–335 (1987)
65. Yorke, J.A., Nusse, H.E.: Dynamics. Springer, New York (1998)
66. Zastrau, B.: Vibration of cracked structures. Archive of Mechanics 37, 736–743 (1985)

Author Index

MIX
Papier aus verantwortungsvollen Quellen
Paper from responsible sources
FSC® C105338

If you have any concerns about our products,
you can contact us on
ProductSafety@springernature.com

In case Publisher is established outside the EU,
the EU authorized representative is:
Springer Nature Customer Service Center GmbH
Europaplatz 3, 69115 Heidelberg, Germany

Printed by Libri Plureos GmbH
in Hamburg, Germany